D1156403

123 Advances in Polymer Science

Synthesis and Photosynthesis

With contributions by
L. Angiolini, C. Carlini, I. R. Dunkin, M. Kaneko,
A. Matsumoto, K. Müllen, R. Ramaraj, U. Scherf,
D. Sherrington, J. Steinke

With 106 Figures, 37 Tables and 81 Schemes

Springer

ISBN 3-540-58908-2 Springer-Verlag Berlin Heidelberg NewYork
ISBN 0-387-58908-2 Springer-Verlag NewYork Berlin Heidelberg

Library of Congress Catalog Card Number 61-642
Printed in Germany

Typesetting: Macmillan India Ltd., Bangalore-25
SPIN: 10477704 02/3020 - 5 4 3 2 1 0 - Printed on acid-free paper

Editors

Table of Contents

The Synthesis of Ladder Polymers

U. Scherf and K. Müllen
Max-Planck-Institut für Polymerforschung, Ackermannweg 10,
D-55021 Mainz, FRG

This article provides an overview and a critical evaluation on the synthesis and characterization of ladder polymers, with regard to such aspects as structural regularity and processibility of the materials. Hereby, the review focuses on synthetic principles defining further progress in the field of ladder polymer synthesis (Diels-Alder-polymerizations as a concerted process of ladder formation, the polymer-analogous bridging of single-stranded precursors as an example for the stepwise generation of ribbon structures). Scope and limitations of the novel synthetic routes are discussed in detail.

Furthermore, particular consideration is given to the tuning of attractive electronic properties for potential applications (nonlinear optics, electroluminescence).

Advances in polymer Science, Vol. 123

1 Introduction

A comprehensive definition of the term "ladder polymer" has been proposed by Winslow [1] as: "an uninterrupted series of rings connected by sterically restrictive links around which rotation cannot occur without bond rupture". This definition includes the so-called "classical" ladder polymers as well as polymers with spiro- or allene-type structural units forming a ladder-type framework. The "classical" ladder polymers consist of two independent strands connected via rigid bridges without formation of crossing-points, e.g. double bonds or spiro centers. The early attempts at synthesizing ladder polymers were made in the late 1950s and the early 1960s. At that time, and to the present day, two general routes for preparing ribbon-type molecules have been used [2]. They first employs the polymerization of multifunctional monomers, in which both strands of the ladder structure are generated in a single reaction. The second utilizes the cyclization of suitably functionalized open-chain (single-stranded) precursor polymers via a polymer-analogous process.

Both strategies pre-suppose certain essentials to arrive at structurally defined ladder polymers, especially the exclusion of side-reactions and an almost quantitative conversion of the starting materials. One important limitation of the first route is the possibility of branching and/or crosslinking during the stepwise condensation process of the multifunctional monomers: this results primarily in insoluble products with structural defects [3]. Another problem is connected with the necessity of reaching a nearly complete turnover of the monomers reacted– a demand which is not often realized using the common types of condensation. Nevertheless, the success of such "multifunctional" condensation methods is the result of the pronounced tendency to form five- and six-membered rings in an intramolecular reaction forcing the generation of the double-stranded ladder structure.

A more recent, and particularly successful, variant of the multifunctional coupling process to ladder polymers is the use of repetitive cycloaddition reactions that start from bifunctional monomers, for instance dienes and dienophiles (Diels-Alder-type cycloadditions) [2–4]. The fact, that both chains are generated simultaneously in a concerted process constitutes the important progress associated with such a route.

The difficulties of the second, stepwise, synthesis of ladder polymers involve, in particular, the final step, namely the polymer-analogous formation of the double-stranded structure. Here, an optimal design of the open-chain precursor polymer is mandatory to arrive at a quantitative conversion with high chemo- and regioselectivity in order to minimize structural defects. These problems have raised serious doubts in the literature [3] about the feasibility of this concept.

Nevertheless, the route is attractive, because many interesting ladder structures are not accessible using the polycondensation or polyaddition of multifunctional monomers [2, 5]. Besides the specific problems of ladder polymer synthesis discussed above, another problem is almost invariably associated with ladder polymer chemistry: the rigid structure of the polymeric backbone, with its strong intermolecular interactions, causes poor solubility of the products [2, 3, 6, 7]. Therefore, to make progress in the synthesis and characterization of double-stranded ribbon-type polymers, it is mandatory to increase the solubility of the products. One possible way to reach this goal is by the introduction of solubilizing side chains [2, 3, 7, 8].

An alternative approach would generate precursor macromolecules with some disorder of the rigid geometry. A possible way of achieving this is the generation of polymers that have structural units which cause an angular shape of the double-stranded chain. Such precursor polymers then have to be converted into the final ladder macromolecules by means of an elimination or rearrangement step [2, 5].

The review presented here provides a classification and a critical evaluation of ladder polymer synthesis, with special regard to such aspects as structural regularity and processibility, and the tuning of attractive electronic properties for potential applications. Special consideration will be given to synthetic strategies (repetitive Diels-Alder-reactions, polymer-analogous ring-closure via electrophilic substitution) defining further progress in this field. The review focuses on "classical" ladder-type structures, spiro- and allene-type ribbons are considered only in an incidental manner.

2 Polycondensation Methods of Preparing Ladder Polymers (Multifunctional Condensation)

Besides the 'synthesis-specific' drawbacks discussed above, one-step polycondensations to generate ladder polymers conceptually represent the most simple way of preparing such macromolecules. The particular challenge of following this route is to find monomers and conditions that allow synthesis of structurally defined products. To undertake such a synthetic sequence successfully it is necessary that the intramolecular formation of the second chain is greatly favored over intermolecular condensation that would result in branched or crosslinked products. The driving force toward the predominant generation of double-stranded molecules is (1) the high tendency to form five- or six-membered rings and/or (2) the gain of resonance energy when generating aromatic sub-structures. Careful design of the monomers and a control of the reaction conditions are thereby of crucial importance.

The first attempts to control a synthetic process using multifunctional monomers were published in 1967/68 by Stille and co-workers [9–12]. They condensed 2,5-disubstituted 1,4-benzoquinones **2** with various tetra-functional aromatic compounds **1** containing amino-, hydroxy- and chloro- functionalities. When carrying out the reaction at high temperature in polar, aprotic solvents such as HMPT (hexamethyl phosphoric acid triamide) they immediately ob-

1a-c + **2a-c** → **3**

X: $-NH_2$, -OH, -SH, -Cl

Y: -OH, -Cl, $-OCOCH_3$

Z: -NH-, -O-, -S-

solvent: polyphosphoric acid; 250°C

tained black, insoluble materials, for which a ladder structure **3** was predicted [9, 10]. A detailed characterization, however, was very difficult; indeed, the assumption of the double-stranded structure was only based on elemental analysis and an IR-spectroscopic comparison with low molecular weight model compounds. Alternatively, it is possible to generate defined and soluble single-stranded intermediates when the condensation reaction is carried out at low temperatures (room temperature). This process is then followed by thermal treatment of these precursors to form the predicted ladder structures (see the synthesis of **9** and **10**).

The authors feel, that the classification of the synthetic principles applied here is somewhat arbitrary. Multifunctional polycondensations which are conducted in a two-step manner (generation of single-stranded intermediates, followed by cyclization), could be classified with the same justification as stepwise processes. On the other hand some of the stepwise syntheses of ladder structures constitute condensations of multifunctional monomers (e.g. the use of butadiynes as starting compounds, see Sect. 4.1.).

In order to ensue a clear presentation of the results the authors decided to segregate both synthetic principles: All synthetic strategies developed from the multifunctional condensations of Stille and Marvel were assigned to this general type of reaction. At the same time the first multistep sequences (polymer-analogous cyclization of poly(methyl vinyl ketone) and polyacrylonitrile) are used as point of reference for the classification of the other type of synthesis (stepwise procedures).

Various polyheteroaromatic ladder structures were obtained, for example polymers that consist of phenoxazine- (Z: –O–), phenothiazine (Z: –S–) or quinoxaline- (Z: –NH–) subunits with different patterns of the heteroatoms incorporated [6].

An indication that an imperfect ladder is formed with this route is the fact that additional water molecules are eliminated when the products are heated to 600°C, as shown by Marvel and co-workers [13–15]. They have applied the principles developed by Stille and co-workers to the generation of related aromatic ladder polymers. The thermal stability of the materials such as **3** (Z = –NH–) is poor since decomposition starts at about 275°C under nitrogen, indicating incomplete formation of the second strand and/or a low molecular weight of the polymers. Stille's group has also studied the polycondensation of tetraketo compounds **4** and **5** with aromatic tetramines such as **1a** (X = $-NH_2$), to form various ladder structures (**6**, **7**) with quinoxaline-subunits [10, 11]. The ladder poly(quinoxaline)s formed were not significantly more thermally stable than comparable single-stranded polymers. This behavior may be caused by structural defects – remaining functional groups like amino functionalities – in the condensation products that produce favored sites for decomposition to start.

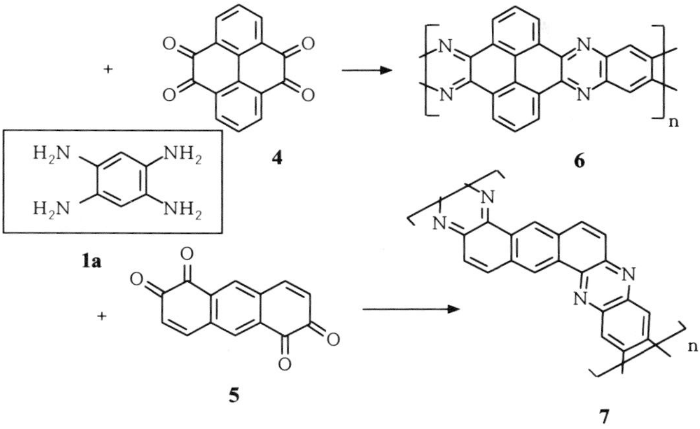

solvent: HMPT, 180°C

Electrical conductivity measurements give values of σ between 10^{-8} and 10^{-5} S/cm. These uncommonly high conductivities for undoped conjugated polymers probably arise from structural defects of an ionic nature and/or contamination with impurities [16, 17]. The ladder polymer with phenothiazine subunits can be doped oxidatively, whereby the conductivity is increased by only 2–5 orders of magnitude. Further progress in this field is handicapped by the very poor solubility of the materials which prevents the synthesis and characterization of structurally defined new materials as well as excluding good processibility.

Generally, two ways have been pursued to overcome these drawbacks. The first is to suppress the strong interchain interactions which cause the poor solubility and processibility [16]. These interchain interactions are drastically reduced when the distance between the polymeric backbones is increased, for instance via the introduction of bulky side-groups. The second possibility is to separate the formation of the first and second chain of the double-stranded polymer, and to process the materials at the stage of the soluble single-stranded intermediates. These precursors can be converted into the (insoluble) ladder polymers during the final step that is a favored thermal process carried out in the solid state.

A synthetic sequence to poly(quinoxaline)s with (dialkylamino)vinyl substituents as solubilizing side-groups was developed by Yu and Dalton by combining the two approaches discussed [18, 19]. The condensation of a bis-(dialkylamino)vinyl-2,5-dichloro-1,4-benzoquinone (**8**) with tetraaminobenzene

1a at room temperature, initially leads to soluble open-chain precursor polymers **9**. This 'prepolymer' **9** was then used to cast films. Thermal treatment (300°C) of these films in the solid state provides the final ladder polymer **10**. IR- and NMR-spectroscopic characterizations, as well as elemental analysis, indicate that the final product contains ladder-type segments, but that polymer-analogous ribbon formation is incomplete. The products show a very high thermal stability (thermogravimetric analysis) up to 690°C (no weight loss occurs under nitrogen).

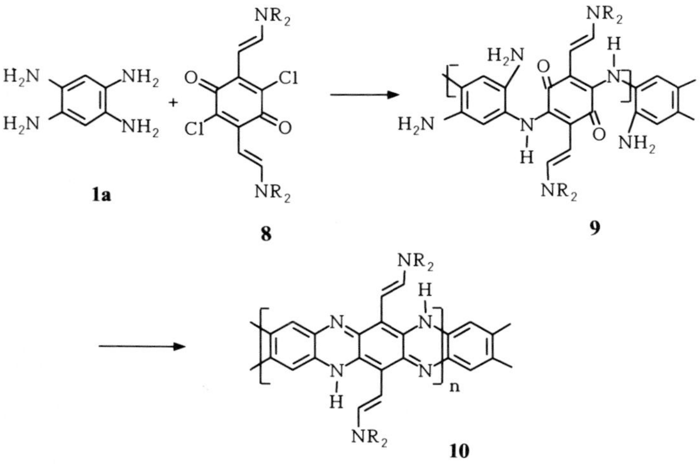

R: $-C_2H_5$, $-CH(CH_3)_2$, $-C_6H_{11}$

A comparable synthetic sequence was accomplished by Müllen and coworkers [20] to prepare ladder-type oligomers with phenothiazine subunits. The generation of the structurally defined soluble oligomers succeeded by the introduction of solubilizing alkyl chains and separated the formation of the first and second chain of the double-stranded molecules. These molecules represent novel electron-donating molecules that forming stable donor-acceptor complexes, for instance with TCNQ (7,7,8,8-tetracyanoquinodimethane). If the nitrogen atoms in the molecules are non-conjugatively linked (*meta*-position), the generation of stable oligo-radicals should be possible and these are attractive for the formation of organic high-spin systems.

A new quality in ladder polymer synthesis via multifunctional polycondensation was reached in the late 1960s when poly(benzimidazobenzophenanthroline) (BBL) **12** was prepared by Arnold and van Deusen [21–24]. The polymer was synthesized from naphthalene-1,4,5,8-tetracarboxylic acid dianhydride (**11**) and 1,2,4,5-tetraaminobenzene (**1a**) in strong acidic media (polyphosphoric acid, sulfuric acid). BBL is completely soluble in concentrated sulfuric acid or methanesulfonic acid and is processible into durable films and

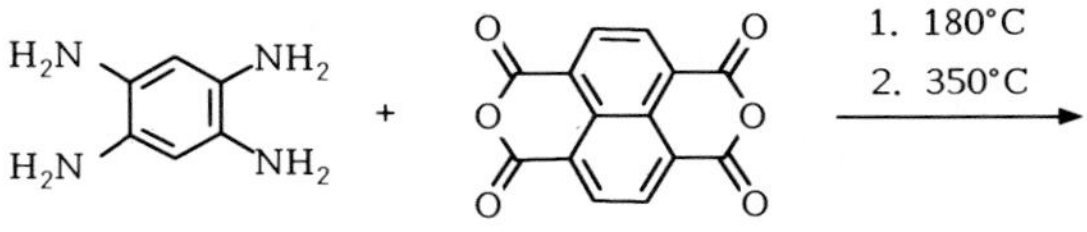

1. 180°C
2. 350°C

1a + 11

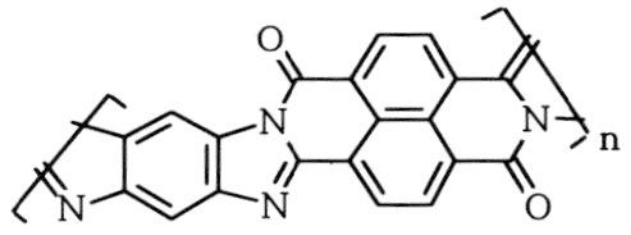

12 BBL

layers, of dark color with a golden luster. A comparison of infrared spectra and elemental analyses with those of model compounds indicate a high degree of ladder formation. BBL is thermally stable (to 550 °C in air, 700°C under nitrogen), and shows semi-crystallinity with a compact interchain packing of the polymeric molecules. Aggregates are formed in solution as well as in the solid state [25].

Doping of the insulator BBL (conductivity of pristine material: ca. 10^{-12} S/cm) is possible oxidatively or reductively. Oxidative doping with SO_3 or H_2SO_4, AsF_5 or BF_3 leads to electrically conductive materials which retain their flexibility [17, 26]. Doping with alkali metals (sodium, potassium) resulted in products with conductivities of ca. 1 S/cm. BBL can also be doped using the ion-implantation technique (implantation of boron or rare gas atoms– Ar, Kr) [27]. The black or silver-gray products exhibit conductivities of up to 200 S/cm while the mechanical behaviour (flexibility) remains nearly unchanged. With selective ion-implantation it is possible to create conducting regions and structures in an insulating, polymeric BBL-matrix. Several other variations of ladder polymers of the BBL-type were synthesized using different monomers, for example pyromellitic anhydride [28–30] or naphthalene-1,4,5,8-tetracarboxylic acid dianhydride [31] (as the carbonyl component) and 1,2,4,5-tetraaminobenzene [28, 29, 30] or 1,4,5,8-tetraaminonaphthalene [31] (as the amino component) as set of starting compounds. In addition, Arnold and van Deusen [31, 32] reported the polycondensation of an AB-type monomer **13**, incorporating both types of reactive functionalities into one and the same molecule. The polymers obtained **14** are thermally stable up to 600°C under nitrogen. A further achievement in this field is marked by the first synthesis of an *soluble* BBL-type polymer starting from 1,6,7,12-tetrakis(4-tert-butylphenoxy)perylene-3,4,9,10-tetracarboxylic acid dianhydride and 1,2,4,5-tetraaminobenzene as monomers [33]. The deeply colored, soluble material obtained possesses a molecular weight M_n of 10000 and is characterized by a longest wavelength absorption λ_{max} of ca. 689 nm. The solubility in common organic solvents (chloroform) is good and allows processibility into thin films and layers (spincoating).

The majority of ladder polymers, however, synthesized by multifunctional polycondensation do not have a well-defined structure, since side reactions such as branching, crosslinking or incomplete cyclization cannot be avoided [3, 7]. The poor solubility of the products often prevents the generation of high molecular weight products since the materials precipate from the reaction media before reaching desirable chain lengths. In addition, impurities are often co-precipitated during the condensation process. Despite the problems associated with the synthetic processes, the materials often show attractive thermal, optical, and electronic properties [7].

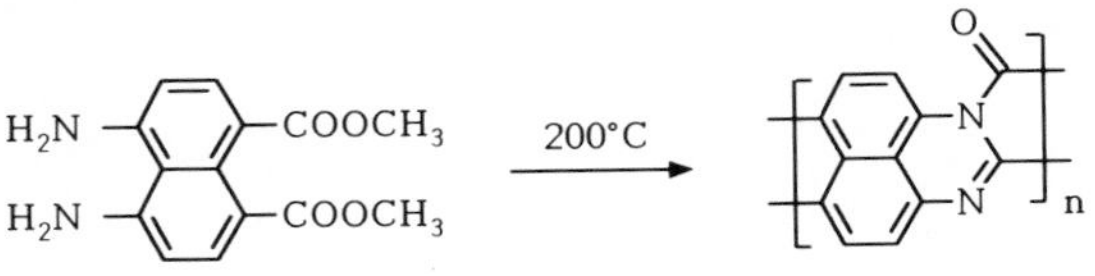

An additional interesting approach for further investigations may be the incorporation of short rigid ladder segments into polymers with a flexible main chain – the generation of so-called 'step-ladder' structures. In these, the flexible segments of the copolymers guarantee a sufficient solubility of the materials. The aromatic, ladder-type segments must be long enough to reach the desired optical and electronic properties. In this way, such a strategy circumvents the problem of poor polymer solubility discussed above. Preliminary results following such a concept were published by Yu and Dalton [34] for polyethers, polyamides/imides, and polyurethanes containing rigid phenoxazine- and pyrazine-type ladder segments. The copolymers show large NLO effects. Müllen and co-workers [35] have synthesized copolyamides/imides and poly(phenylene-arylenebisthiazole)s with ribbon segments of the phenothiazine-type. This synthetic approach provides one possible way to utilize unusual optical properties and yet realize a sufficient solubility necessary to prepare thin films of high optical quality.

3 Repetitive Cycloaddition – The Concerted Formation of Both Chains

3.1 [4 + 2] Cycloaddition Reactions

One special drawback of the multifunctional polycondensation route to ladder polymers is the possibility of crosslinking or branching during the polymer building process. This disadvantage results from the independent formation of both chains of the double-stranded macromolecules [3]. Besides the dominant intramolecular coupling, which forms the second chain, a significant number of functionalities in the single stranded intermediates undergo intermolecular condensation (crosslinking). Side-reactions of this type cause incomplete ladder formation and drastically decrease the solubility of the products generated.

A novel synthetic concept to overcome these shortcomings is the simultaneous and concerted formation of both chains (linkages) of the double-stranded ladder polymers, which excludes undesirable branching or crosslinking. Repetitive cycloadditions, e.g. the Diels-Alder-reaction, represent a reaction type particularly suitable for this concept. The important advantage of using cycloaddition reactions is the concerted formation of the ladder-type framework, which in most cases will have a well-defined stereochemistry [2, 3, 47]. Following this general concept, several approaches to develop a repetitive cycloadditon sequence of various bifunctional monomeric starting compounds to create ladder-type polymers were undertaken. The first attempts were made by Bailey and co-workers [4] in the early 1960s. They reacted 1,4-benzoquinone (**15**) as a bifunctional dienophile with 1,2,4,5-tetramethylenecyclohexane (**17**) or 2-vinylbuta-1,3-diene (**16**) as diene components (or potential diene component, in the latter case forming an AB-type monomer after the first cycloaddition step). The polymeric coupling product **18** of benzoquinone with **16** [36, 37] exhibits a molecular weight of about 4 000 and is soluble in dioxane or hexafluoroisopropanol (higher molecular weight products). It is possible to reduce about 95% of the keto functions with zinc/zinc chloride/sodium chloride to obtain a saturated hydrocarbon ladder polymer. This material shows a high thermal stability and only 6% weight loss when heated up to 900°C [4].

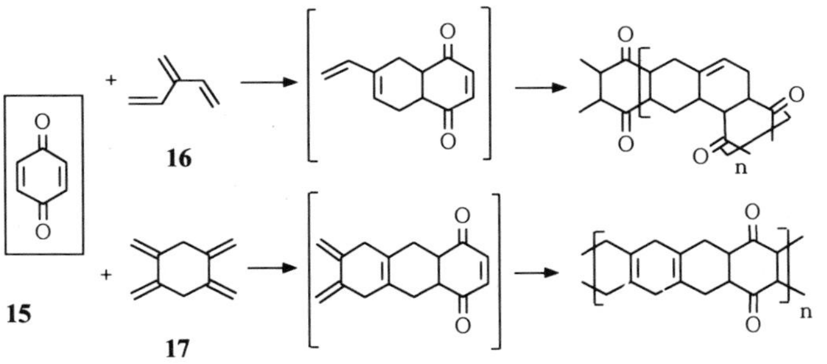

15 16 17 18 19

The poly-Diels-Alder-adduct **19** of 1,4-benzoquinone and 1,2,4,5-tetramethylenecyclohexane has only very limited solubility, for example in hexafluroisopropanol, that results from high crystallinity of the linear cycloaddition product which consists exclusively of double-stranded linked six-membered rings [37].

The general synthetic strategy outlined above was continued by several groups. The progress of these extensive investigations allows for the following conclusions to be drawn: The synthesis of novel monomers of the AA/BB-(bisdienophile/bisdiene) or AB-(dienedienophile) type [4, 38–44] and the subsequent polyaddition paves the way to achieve more soluble products with an increased molecular weight. At the same time, detailed characterization of structure and properties becomes possible. Some of the monomers investigated are summarized in Table 1. The following, however, should be noted: (1) the use of monomers **20** and **21** (as bisdienes) and monomer **23** (as bisdienophile) brings about the incorporation of flexible or angular structural units (ethene-units of the eight-membered rings [4, 44], oxo-bridges [38, 39, 41, 42]) in the products, which increase the solubility of the ladder-type products, (2) the introduction of flexible side-groups (alkyl, alkoxy chains) in monomers **22**, **23**, **24** and **26** is a very successful way of producing polymers with a significantly increased solubility.

AA	BB	AB
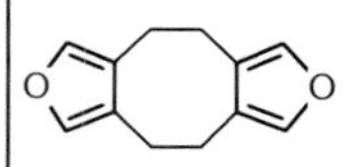 **20**	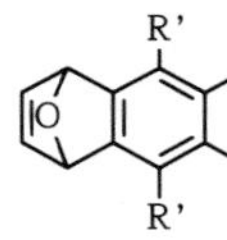 **23** (R', R')	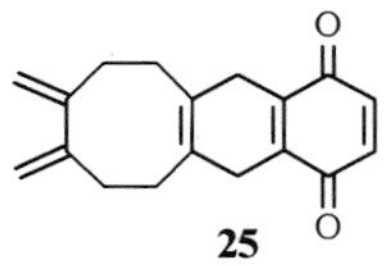 **25**
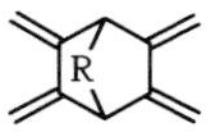 **21** (R)	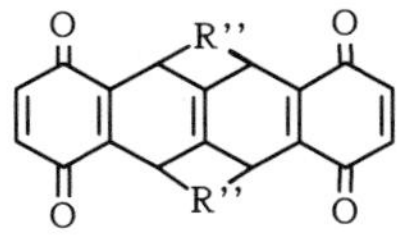 **24** (R'', R'')	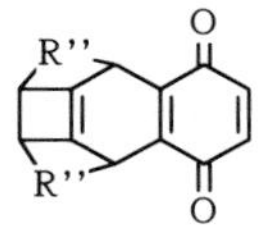 **26** (R'', R'')
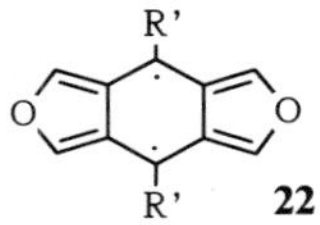 **22** (R', R')		

Many polymeric structures were synthesized starting from the monomers listed in Table 1 and some of the most attractive targets will be outlined and discussed. The main criterion for initial selection is the possibility of the cycloaddition products first formed being used as precursor structures to create unsaturated, aromatic ladder polymers with attractive optical and electronical properties (like extended Π-systems) or to have redox active polymers. However, a key problem exists when the Diels-Alder-route is used to prepare ladder-type polymers [2]. The repetitive formation of the double-stranded structure with fused six- or eight-membered rings can be achieved in high structural regularity (regioselective, stereoselective), but the subsequent conversion into fully conjugated structures is very problematical in most cases. As a result, all attempts to synthesize polymeric, fully unsaturated structures following the Diels-Alder-route until now have been unsuccessful or only partially successful [2, 4].

Stoddard and co-workers [45–47] describe the synthesis of cyclic and linear ribbon-type oligomers starting from the monomers **21** [39] and **23** [42]. The double stranded macrocycles, e.g. **27**, generated are intermediates in the preparation of cyclic oligoarenes (cycloacenes) – attractive compounds with two dimensional cyclic Π-systems of the Hückel-type. Synthetic approaches to remove the oxo-bridges reductively and to generate the final fully unsaturated hydrocarbons, lead only to intermediates, e.g. **28**, which are partially hydrogenated [46].

21 (R:–O–)

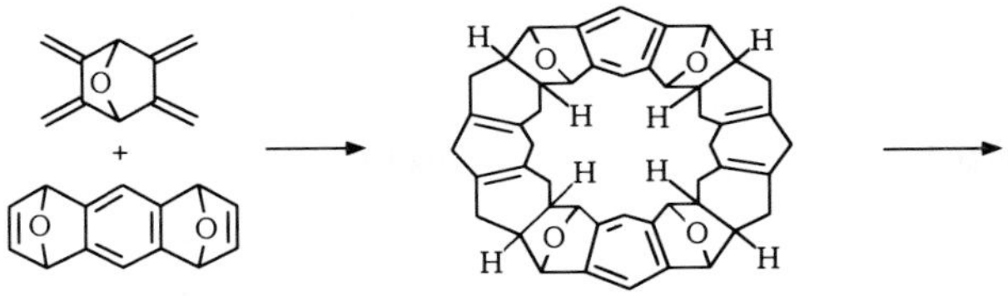

23 (R′:–H)

27

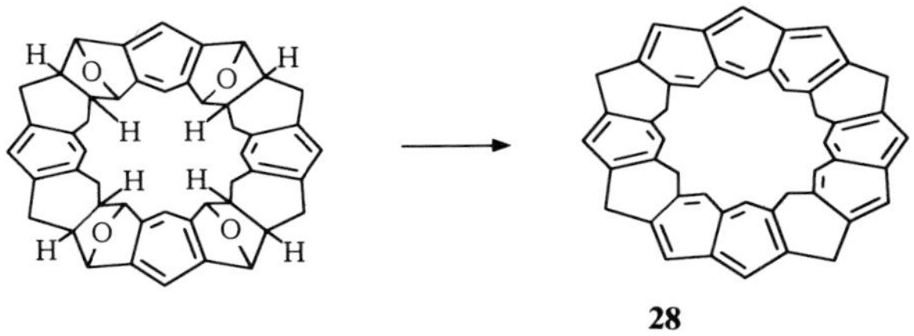

28

Schlüter and co-workers [3, 44, 48] have reacted **26**, an AB-type monomer, that contains a benzoquinone-substructure (dienophile) and a substituted tricyclic cyclobutene system, as it represents a potential diene (pseudodiene). Under the conditions of the Diels-Alder-cycloaddition, the cyclobutene ring-opens (in a disrotatory fashion) with the formation of butadienylene-subunits. The polymer formed (**29**) was found to contain repeating units in both *endo*- or *exo*-stereoisomeric forms as shown by comparison of ^{13}C-NMR spectra of polymeric and oligomeric model compounds of uniform stereochemistry; the *exo*/*endo* ratio is approximately 1:1. With the help of computer simulations some assumptions about the secondary structure of the non-planar ladder molecules are available. An adequate solubility of the polycycloaddition products is the result of the cyclic oligoethene chains in the periphery of the ribbon-type molecules that form flexible cycloalkane rings. The number average molecular weight was determined as ca. 20 000 [3], representing about 50 repeating units of the double-stranded ladder polymer (determined via vapor pressure osmometry).

Müllen and co-workers [2] have investigated the use of tetramethylenebicyclo [2.2.2]octene (**21**; R: –CH =CH–) [40] and a bisethylene-bridged difuran **20** [38] as difunctional dienes in repetitive cycloaddition reactions. Reacting the difuran with different acetylene dicarboxylic esters provides ribbon-type structures (**31**), whose solubility can be controlled by choice of the alcohol component of the ester monomer [8]. A detailed investigation of this reaction revealed a characteristic side-reaction of repetitive cycloadditions: intramolecular cage formation occurs after generation of the primary 1:1 adduct and this is enhanced at increased temperatures [8]. Two types of cages are generated and they result from a condensation of one or two molecules of the 1:1 condensed intermediate. The oligomeric products formed (**31**) – besides the cage molecules – consist of 25–30 repeating units (number average molecular weight).

26 **29**

R'': $-C_6H_{12}-$

20 **30** **31**

R: $-COOC_nH_{2n+1}$, $-CN$

The major advantage of using the bicyclo [2.2.2] octene derivative **21** [40] as bisdiene in Diels-Alder polycycloadditions is that a two-fold cycloaddition leads to the incorporation of barrelene (bicyclo [2.2.2]octatriene) substructures in the ladder macromolecules. The concave angular shape of the barrelene-substructure brings about a significantly improved solubility of the Diels-Alder-adducts. On the other hand, when reacting this bisdiene with less reactive bisdienophiles, high pressure conditions are necessary to synthesize higher molecular weight material [2, 8]. Thus, the reaction with 1,4,5,8-tetrahydro-1,4,5,8-diepoxyanthracene: (**23**; R′: –H) [42] results in the formation of soluble ladder-type molecules as a series of oligomers (**32**), consisting of up to 33 six-membered rings [2, 49]. By chromatographic separation of the polydisperse into monodisperse products, defined oligomers are available. Investigations of model compounds provide insight in the stereochemistry of the cycloaddition. The bisdiene **21** (R: –CH = CH–) gives rise exclusively to an *exo*- attack relative to the oxo-bridges of the bisdienophile **23**, as is expected under high-pressure conditions. The stereochemistry in relation to the vinylene-bridge is not uniform, *exo*- and *endo*-configured substructures are generated. Cage formation is not observed. In contrast, when using the oxo-bridged bisdiene **21** (R: –O–), the generation of cage molecules is favored, as described by Stoddard and coworkers [45, 46, 47].

An improved solubility of the Diels-Alder-adduct **32** can be achieved by the introduction of solubilizing alkyl sub-units into the 9- and 10-positions of the anthracene-bisdienophile used (**23**, R′: -alkyl). High molecular weight products (**32**), consisting of up to 120 six-membered rings, have been observed [2].

A spiro-type ladder polymer is produced when 1,4,5,8-tetrahydro-1,4,5,8-diepoxyanthracene (**23**), as the bisdienophile component, is reacted with the bicyclic silaspirodiene **33** [111]. The products formed (**34**) in the repetitive cycloaddition (under high pressure conditions) possess number average molecular weights of up to 11 000 (D_p = 13). Polymers **34** are promising candidates for

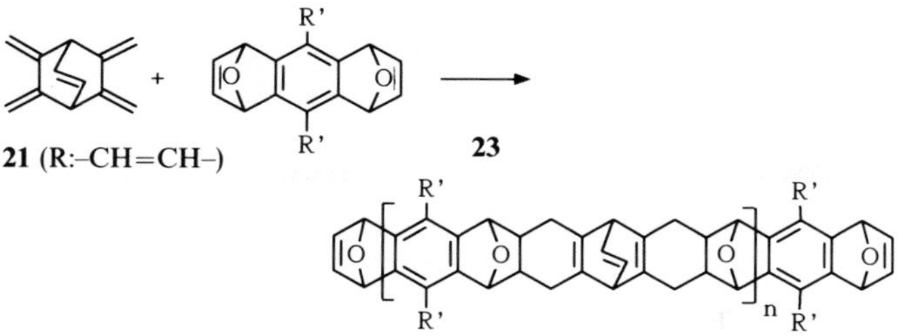

21 (R: –CH=CH–) **23**

R': $-C_nH_{2n+1}$ **32**

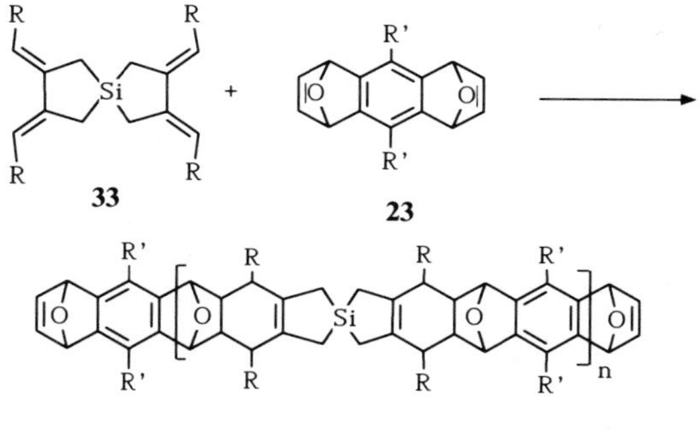

33 **23**

R: $-Si(CH_3)_3$ R': $-C_6H_{13}$ **34**

the development of structures with orthogonally arranged Π-subunits. Therefore, the possibility of a further polymer-analogous conversion (aromatization) is of primary concern.

The key steps towards generation of fully unsaturated (acenic) structures require the removal of the heteroatomic oxo-bridges followed by subsequent aromatization (dehydrogenation). In this respect, Müllen and co-workers [2] described the first successful experiments in removing the –O– bridges using HCl gas as the dehydrating agent in dioxan solution. In contrast Schlüter and co-workers [48] investigated the use of trimethylsilyl iodide as a reagent to eliminate water. The second step in the aromatization sequence, e.g. of **32**, would be the dehydrogenation of cyclohexadiene rings adjacent to the barralene moieties, to give acene (anthracene)-fused barrelene substructures in the case of precursor **32**. Unfortunately, the execution of this reaction has met with problems, especially in the second (dehydrogenation) step. However, the synthetic approach via the formation of barrelene substructures is promising for obtaining oligo- and polyacenes and spiro-type ladder polymers. Once generated, oligo- and polyacenes are expected to be very reactive species [50]. The reactivity increases with the number of fused aromatic rings. Because of the instability of the target systems, the route via barrelene-bridged intermediates is worth an additional comment. The last step, the formation of the fully unsaturated system, can be performed without adding reagents, in the immobilized solid state (film, layer, crystal) as a retro-Diels-Alder-elimination and, in principle, there are two possible ways of doing this. The first involves the direct elimination of acetylene while the second reduces the ethene- to ethano-bridges before elimination of ethene. The second approach is favored because of the milder reaction conditions necessary for the retro-Diels-Alder-reaction.

A fruitful idea for further progress in this field is to replace the bisdienophile monomer **23**, containing the heteroatomic oxo-bridges, by other starting materials. Thus, the aim is to generate ladder-type intermediates without the oxo-bridges. Based on this idea, is the use of potential (or psuedo-) bisarynes as reactive bisdienophiles. Using the bisdiene **21** (R: –CH = CH–) as one component and the "pseudo-bisaryne" **35** (R′: -hexyl) as the second, and generating the aryne sub-structure *in situ* in the reaction mixture with butyl lithium, one can successfully prepare ladder-type polymers **36** consisting only of fused six-membered rings with bicyclic vinylene-bridges in every fourth ring [51]. The subsequent steps necessary to generate the polyacenic structure are (1) dehydrogenation of the rings adjacent to the barralene moiety and (2) the retro-Diels-Alder-reaction as outlined previously. The potential of this novel route was shown for pentacene (**37**) as a low molecular weight model compound whose preparation followed the general synthetic sequence outlined above.

The Diels-Alder-route to ladder polymers is clearly a powerful method for the formation of the primary cycloaddition products. The key problem, however, is the polymer-analogous transformation of the primary macromolecules, containing heteroatomic bridges and saturated sp^3-centers in the double-stranded main chain, into the desired optically and/or electronically attractive target structures [2, 3, 48]. Thus, the optimal design of the starting compounds as well as the optimization of reaction conditions will play an important role. It must be concluded, therefore, that the key problem in the Diels-Alder-approach to ladder polymers, the polymer-analogous aromatization of the primary ribbon-type polyadducts, is still unsolved. To make further progress, it is necessary to target in detail the polymer-analogous derivatization reactions.

3.2 [2 + 2 + 2] Cycloaddition Reactions

The use of Diels-Alder-type cycloaddition reactions is the most intensively investigated cycloaddition approach to the design of ladder polymers in a concerted process. Another methodology was published by Tsuda and co-workers [52, 53, 54]. They developed a nickel (O)-catalyzed [2 + 2 + 2] cycloaddition copolymerization of cyclic diynes **38** with heterocumulenes (like carbon dioxide or isocyanates **39**). The soluble ladder-type products – poly(2-pyrone)s and poly(2-pyridone)s **40** – possess molecular weights M_n of up to 60 000, corresponding to a $D_p > 200$. Unfortunately, the products formed were contaminated by nickel salts originating from the catalyst used $Ni(COD)_2$.

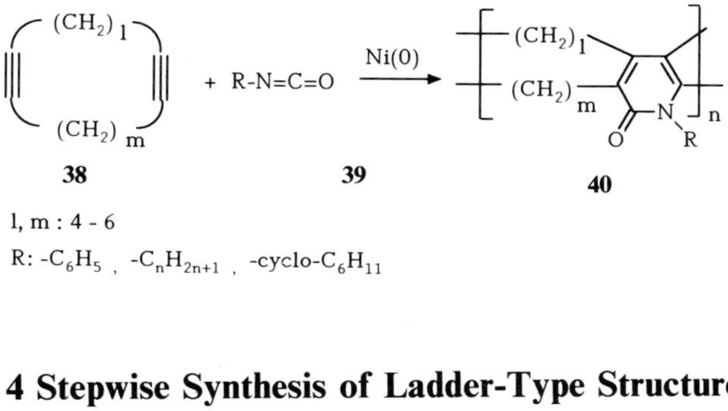

$(CH_2)_l$, $(CH_2)_m$ + R-N=C=O $\xrightarrow{Ni(0)}$ [$(CH_2)_l$, $(CH_2)_m$, N, R, O]$_n$

38 **39** **40**

l, m : 4 - 6

R: $-C_6H_5$, $-C_nH_{2n+1}$, -cyclo-C_6H_{11}

4 Stepwise Synthesis of Ladder-Type Structures

To overcome the problems in carrying out the polymer-analogous derivatization sequences when following the Diels-Alder-route to ladder polymers, the use of "classical", stepwise reaction procedures have been favored recently in the case of several target structures [2, 55]. A step-wise synthetic process to ladder polymers proceeds in two main steps; (1) chain formation via polymerization or polycondensation of suitable monomers, and (2) polymer-analogous cyclization with generation of the double-stranded structure. The demands on the second step to synthesize structurally defined ladder structures are high degrees of conversion (cyclization) coupled with chemo- and regioselectivity to minimize structural defects. In the literature, these problems have cast serious doubts on the possibility of bringing such a concept to fruition [3, 48].

The first attempts to carry out a stepwise synthesis of ribbon-type macromolecules followed one general concept. A functionalized polymer was synthesized using a well-known polymerization (or polycondensation) reaction which, for the majority of the examples known, is a polymerization of vinylic monomers. Every second atom of the single-stranded main chain is substituted in such a way that a reaction of neighboring side-groups is possible. This forms the second part of the ladder structure [6, 7]. This linking or "zipping-up" process then proceeds in a way that allows for the formation of a ladder structure consisting of fused (exclusively six-membered) rings. The "zipping-up" process is possible in several ways.

In this paper, we shall discuss, first by a polymerization of unsaturated side-groups (side-chains), second by the polymer-analogous condensation of suitable functional groups, third by ring-closures (cyclization) via electrocyclic reactions and, fourth by cyclization via electrophilic substitution reactions.

The first route (a two-fold polymerization) seems to be optimal on paper because of the great variety of monomers and target structures potentially possible. The major drawback, however, concerns the "zipping-up" reaction. The formation of the second strand begins statistically at one of the functionalities present in one of two possible directions. Because of this the formation of structural defects (like unreacted functional groups), or ruptures within the ladder is inherently coupled with the process. An additional problem is the strict exclusion of crosslinking reactions, these may be possible to some extent by carrying out the cyclization under high dilution.

4.1 Formation of the Second Strand via Polymerization

The scope and limitation of a two-fold, stepwise, polymerization process toward ladder structures is discussed for three representative examples.

1. Polymer-analogous Cyclization of Poly-1,2-butadiene and Poly-3,4-isoprene
Poly(1,2-butadiene) (**41**) and poly(3,4-isoprene) can be prepared in high regioselectivity using the butyllithium/tetramethylenediamine (TMEDA)-complex or transition metal complexes [56] as initiators. A "zipping-up" polymerization of the remaining vinylic side-chains is possible under cationic initiation (sulfuric acid [6] or $POCl_3$ [57]) at high dilution (0.25 mol %). IR- and NMR-spectra exhibit typical signals of the ladder-type segments formed. The soluble products **42** have a slightly increased molecular weight that indicates the occurence of some crosslinking as a side-reaction. Nevertheless, a comparison of thermal (T_g) and mechanical parameters show drastic changes (increase of T_g and loss of the rubber elasticity) during the transition from the rubber-like single-chain starting materials to the rigid (partially cyclized) products. This behavior illustrates the enormous structural change from a flexible precursor polymer to the geometrically fixed double-stranded target molecule.

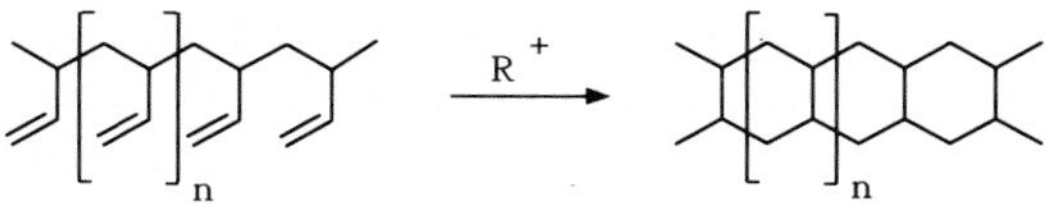

41 $\xrightarrow{R^+}$ **42**

Aromatization experiments show some evidence for relatively small aromatic substructures (naphthalene, anthracene, tetracene) and this independently demonstrates the success of the polymer-analogous cyclization process. The occurence of up to 20% non-cyclized vinylic side-groups is an indication of the statistical nature of the ring-closure reaction [58].

2. Cyclization Products of Polyacrylonitrile

A series of investigations published since 1959 concern the thermal cyclization of polyacrylonitrile (PAN) precursors [59–64]. When PAN is heated, either under an inert atmosphere or in air, to temperatures of 220–300°C, deeply colored products are formed. While the material **43** obtained under inert atmosphere does not possess aromatized (dehydrogenated) ladder segments, the thermal process in air produces **44** which contains fully aromatized structural units [62, 65]. Therefore, the oxidative cyclization of PAN generates (thermally) stable products. Oxygen plays a very important role during the thermal cyclization. An IR-spectroscopic analysis shows, that hydroxy- and keto-groups are formed via oxidation of the saturated main-chain [62]; possible chemical structures resulting from thermal treatment of PAN under oxidative or non-oxidative conditions are given in the corresponding scheme. However, the cyclization process to ladder-type segments is accomplished by extensive crosslinking reactions.

An important property of the oxidatively formed, partially cyclized, material **44** starting from PAN is the behaviour observed upon further thermal treatment

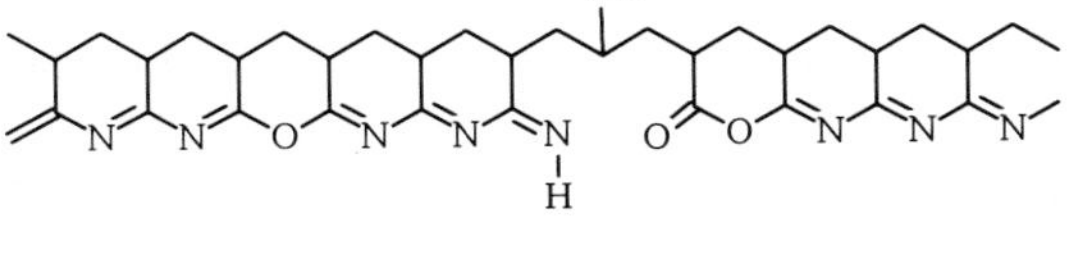

220°C in nitrogen **43**

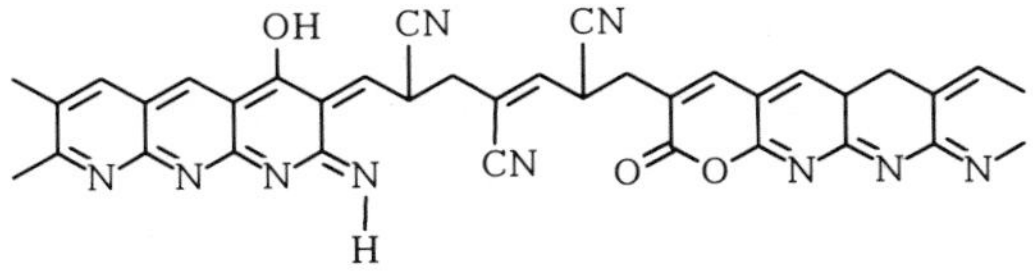

220°C in air **44**

at temperatures to 900°C. With evolution of volatile compounds, chain scission and partial decomposition, a complete reorganization of the structure occurs. The final products are fully aromatic, very thermally stable materials that posseses a two-dimensional network structure and contain small amounts of nitrogen. These products (in fiber form) are widely used as starting materials for the fabrication of carbon fibers and films at temperatures up to 2500°C [66, 67]. While PAN is an insulator, the pyrolyzed samples show an increased electrical conductivity with increased temperature of pyrolytic treatment. Products oxidatively generated at temperatures up to 900°C exhibit conductivities of 5 S/cm [68] to 500 S/cm [66]; the increase in electrical conductivity can be correlated with the formation of aromatic (conjugated) C = C and C = N– structural units [68].

3. Butadiynes as Precursors for Polyacene Synthesis

The preparation of polyacenes starting from butadiyne-type monomers via a twofold polymerization seems to be an extraordinary elegant approach but unfortunately, up to now, only on paper. A series of attempts to develop such a two-step synthetic sequence involve, 1,2-polymerization to prepare single-chain precursors of the poly(acetylene)-type followed by generation of the second chain via a "zipping-up" polymerization of the pendant acetylenic side-groups. Seher [69], Bohlmann [70] and Snow [71] describe experiments concerned with the generation of unsubstituted polyacene starting from butadiyne (**45**). Upon contact with an inert surface (for instance: polytetrafluoroethylene), butadiyne gas provides colored coatings. The product (**46**) lacks a well-defined structure and contains unpaired electrons (EPR-spectroscopy). Pendant acetylenic side-groups have been detected by IR-Spectroscopy [71]. Teysie and Korn-Girard [72] have published investigations using 1,4-diphenylbutadiyne (**47**; R = Ph) as starting material. A polymerization with Ziegler/Natta-catalysts was followed by a thermal treatment or by contact of the intermediates **48** (R = Ph) formed with Lewis-acids. Here again, insoluble, black products (**49**, R = Ph) are generated and no reliable structural information has been obtained. One must conclude, that crosslinked networks are the result of the synthetic sequence investigated.

A more recent paper [73] describes a slightly improved procedure starting from (triethylsilyl)butadiyne (**50**). The polymerization leads to poly(triethylsilylethinylacetylene) (**51**) which was desilylated (tetrabutylammonium fluoride) to poly(ethinylacetylene) (**52**). A thermal treatment of this precursor should produce the polyacene **53**. As in the previous examples, however, no indications for the generaton of a defined polyacene-type ladder polymer are available. Nevertheless, this method gives a highly conducting material (σ: ca. 1 S/cm). Obviously, in order to undertake a synthesis of structurally defined polyacenes successfully, a more sophisticated process is required in light of the high chemical reactivity expected of the unsaturated target products (see Sect. 3.).

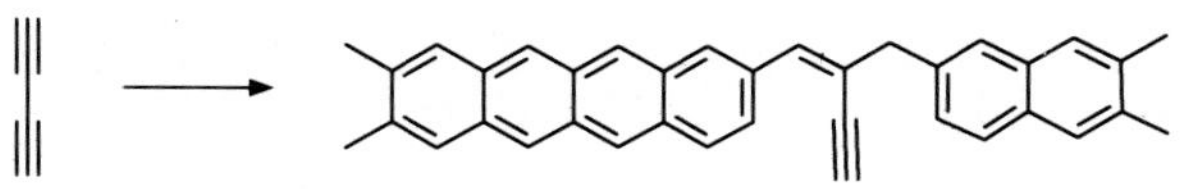

45 **46**

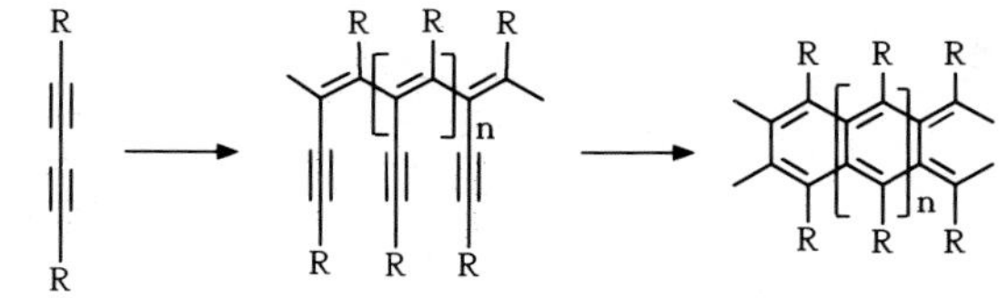

R R R R R R R R R R R R R R

R: $-C_6H_5$

47 **48** **49**

R_4N^+ Δ n n n

$Si(C_2H_5)_2$ $Si(C_2H_5)_2$ H

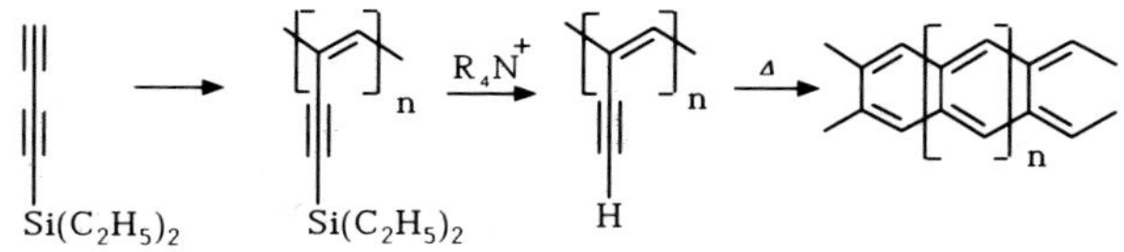

50 **51** **52** **53**

4.2 Condensation-Type Cyclization Reactions

An alternative to the generation of the second chain via "zipping-up" polymerization is a polymer-analogous condensation reaction of suitably functionalized single-stranded precursors. Following such a procedure, Marvel and co-workers [74, 75] investigated the formation of a ribbon-type polymer via a polymer-analogous condensation reaction. They studied the cyclization of poly(methyl vinyl ketone) (**54**) via polymer-analogous Aldol-condensation. Heating poly(methyl vinyl ketone) (**54**) to 300°C produces a red, amorphous solid **55**, during the condensation water is eliminated. A chemical analysis of **55** shows that ladder-type segments with dihydroacene structural units are formed. The polymer-analogous condensation, however, proceeds randomly and not in a concerted fashion. In this way the degree of cyclization is limited statistically – a maximum of 86% of the functional groups can participate in the Aldol-type condensation [76]. Of the keto substituents 14% or more remain unreacted in the polymer causing incomplete ladder formation.

54 55

Besides the thermal condensation, many publications describe the use of acid catalysts (CF_3COOH [77], $POCl_3$, polyphosphoric acid [78]) to carry out the polymer analogous cyclization. The occurrence of soluble products shows that the intramolecular cyclization is greatly favored over an intermolecular condensation step (crosslinking).

4.3 Electrocyclic Ring Closure of Single-Stranded Aromatic Precursor Polymers

The above examples of a stepwise synthesis of a ladder polymer involve the formation of single-stranded polymers via polymerization of suitable monomers to functionalized precursors. These consist of substituted poly(ethylene)- or (polyacetelyne)-type macromolecules, from which attempts are made to carry out a defined polymer-analogous cyclization reaction.

The next two approaches for synthesizing structurally defined ladder-type oligomers and polymers, discussed in sects. 4.3. and 4.4., rest on the formation of the single-stranded intermediates via polycondensation of (aromatic) monomers, again followed by a polymer-analogous ring-closure sequence.

Aromatic hydrocarbons that consist of double-stranded, *peri*-fused naphthalene subunits (so called "rylenes") have been synthesized as unsubstituted compounds up to quaterrylene ($n = 2$) [79, 80]. However, the insolubility of larger members of the homologous series prevents their synthesis and characterization [80]. On the other hand, a few papers have been published concerning a thermochemically induced formation of poly(*peri*-naphthalene) **57** (polyrylene) via pyrolysis of perylene-3,4,9,10-tetracarboxylic acid dianhydride **56** in vacuo or under an inert atmosphere [81–85]. A high conversion temperature of 550–900°C provides black and insoluble products from which a detailed structural analysis indicates the generation of two-dimensional, carbonaceous networks (comparable to pyrolytic carbon). This assumption is in good agreement with X-ray or electron diffraction analysis and Raman spectroscopy of the films or clusters formed [83, 84]. There is no significant difference in the spectroscopic behavior when comparing highly oriented pyrolytic graphite (HOPG) with the products of perylene tetracarboxylic acid dianhydride pyrolysis. Despite the fact that the formation of structurally defined poly(*peri*-naphthalene) is not probable by this process, highly conductive materials do result (conductivities of up to 15 S/cm [85]). Such conductivities, however, are typical for pyrolytic carbon coupled with a metal-like temperature dependence of conductivity.

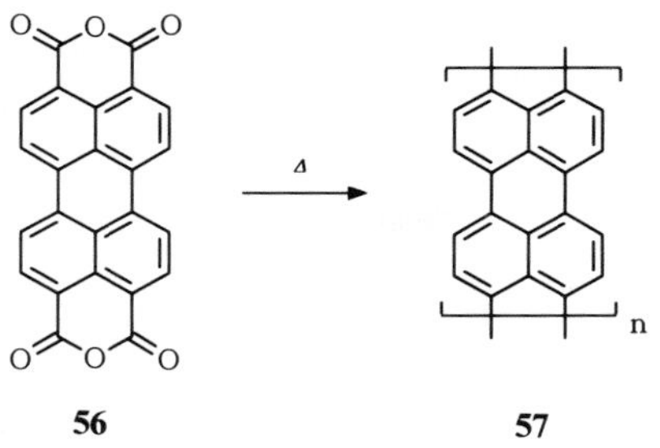

56 57

To obtain structurally defined, soluble, higher homologues of the rylene series (oligomers or polymers) the development of a more subtle synthetic approach seems to be necessary. Therefore, Müllen and co-workers [86, 87] developed an approach to generate alkylated (*tert*-butylated) oligorylenes **58** using a stepwise generation process. The method involves the electrocyclic ring-closure of open-chain oligo(naphthylene)s that lead directly to the aromatic ribbon-type compounds without involving difficult polymer-analogous dehydration and dehydrogenation steps. The *tert*-butyl substituents at the terminal positions of the oligomers cause sufficient solubility to undertake a detailed characterization and to guarantee the processibility of the materials. The synthesis proceeds in three steps. This first step is the generation of open-chain oligo(naphthylene)s via palladium(O) catalyzed cross-coupling (so-called Suzuki-coupling) of naphthalenes containing bromo and boronic acid functionalities (**59** and **60**). In this way tetra-*tert*-butylated ter-, quater- and quinquenaphthyls **61** (n: 1,2,3) are obtainable as soluble precursors for rylene-type molecules. The synthetic approach to cyclize these precursors is a two-step procedure. Firstly, a reductive (electrocylic) step using alkali metals (potassium is preferred) is followed by re-oxidation of the anionic species thus formed to partially cyclized products. There are some indications that the mechanism is an electron-transfer-induced electrocyclic condensation reaction [88]. Unfortunately, this method only allows the generation of terrylene structural units as the most extended *peri*-fused naphthalene segments in the oligomers. To design the higher oligorylenes a subsequent and second cyclization step is necessary. This can be done under rather mild oxidative (Kovacic-) conditions (aluminium trichloride, copper (II) chloride). Using the three-step synthetic process just outlined [generation of single-stranded oligo(naphthalene)s followed by a twofold cyclisation sequence (first reductive, then oxidative)], alkylated, soluble oligorylenes **58** up to pentarylene (**58**, n = 3) have been generated and characterized.

Finally, removal of the solubilizing groups (*tert*-butyl) is possible by retro-Friedel-Crafts alkylation or by heating. These lead to unsubstituted oligorylenes. The oligomers possess outstanding optical and electronic properties. The UV/VIS-absorption spectra exhibit distinct bathochromic shifts in the longest wavelength absorption band as the number of naphthalene rings incorporated increases. Pentarylene **58** (n = 3) shows a λ_{max} of 745 nm. Convergence of the electronic properties in the rylene series is, however, not reached with pentarylene. That is why the next members of the homologous series (hexa-, heptarylene) are very attractive compounds, since the absorption maxima should be shifted into the near-infrared region. An extrapolation of the HOMO/LUMO energy difference (band gap) towards the polymer poly(*peri*-naphthalene) or polyrylene predict a value of about 1.0 eV [87]. Considering the high chemical and thermal stability of the oligorylenes **58**, this class of compounds represents a nearly optimal combination of (intrinsic) low band gap character and high chemical stability. This behavior is promising, especially when considering the low stability of other classes of real and potential low band gap-molecules (polyacetylene, poly(acene)s, poly(arylenemethide)s [89]). These (single- or double-stranded) conjugated polymers which possess comparable band gap energies exhibit drastically lowered chemical and thermal stabilities. NLO-measurements (third harmonic generation) of the oligorylenes **58** reveal a dependence of the χ^3-coefficient upon linear optical properties such as the longest wavelength absorption band [90]. The measurements were carried out under non-resonant conditions, but the low energy absorption maxima of the higher homologs suggest that investigations of NLO-properties should be performed under resonance conditions, too. The oligorylenes **58** are also very attractive electrophors. A number of fully reversible electron-transfer steps occur under cyclovoltametric conditions depending on the size of the molecule investigated. For quaterrylene **58** (n = 2), for instance, seven reversible steps are observed (four reduction, three oxidation steps) [86, 87]. An essential feature in investigating oligomeric systems is the occurrence of fully separated, and quantifiable, reduction and oxidation waves; in contrast to measurements using polymeric systems. This is very helpful in postulating and understanding structure-property relationships and it demonstrates the importance of model studies on oligomeric compounds. However, in spite of the great success in preparing the oligorylenes **58**, the generation of polyrylene [**57**, poly(*peri*-naphthalene)] adapting the stepwise synthetic procedure of Müllen and co-workers is still to be achieved. In such a synthetic process, the synthesis of substituted poly(naphthylene)s **62** is followed by a polymer-analogous cyclization (carried out reductively and/or oxidatively). The introduction of alkyl substituents into the main-chain naphthalene-repeating units (monomer **63**) is necessary to guarantee a sufficient solubility of the product. The introduction of the alkyl substituents into the so-called "bay" positions, however, brings about a steric hindrance to the planned intramolecular cyclization and also decreases the chemical stability of the cyclized (rylene) subunits.

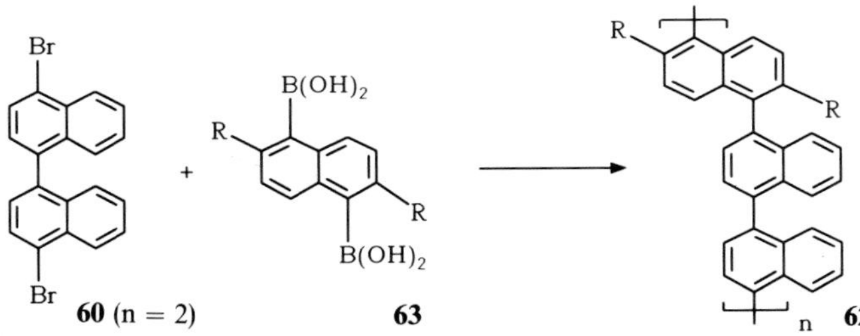

Br Br **60** (n = 2) + B(OH)$_2$ R R B(OH)$_2$ **63** → R R n **62**

Cyclization experiments using *n*-alkyl substituted, soluble poly(naphthylene)s **62** gives rise to macromolecules **64** with ladder-type segments up to the quaterrylene subunit; unfortunately it is impossible to conduct the ring-closure to give a complete cyclization [91]. Some progress is made if *n*-butyl substituted poly(perylene) **65**, containing perylene units as "preformed" dimeric ladder segments (representing a so-called "step-ladder" structure), is used as the open-chain precursor instead of the *n*-alkylated poly(naphthylene) **62**. In this case there is no steric hindrance to ring-closure and the cyclization products ($SbCl_5$ as oxidating agent) contain more extended ladder segments; *peri*-fused

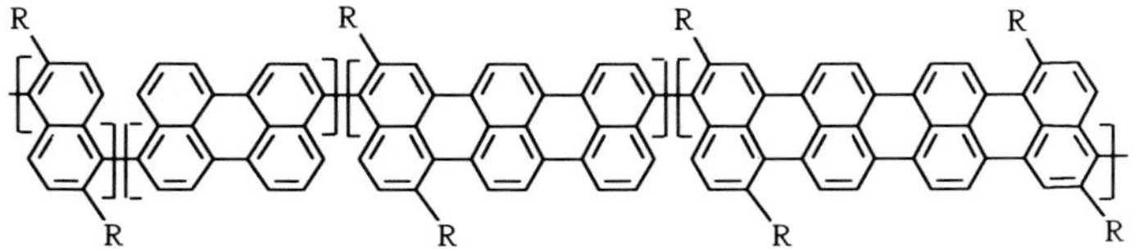

R R R R R R R R

64

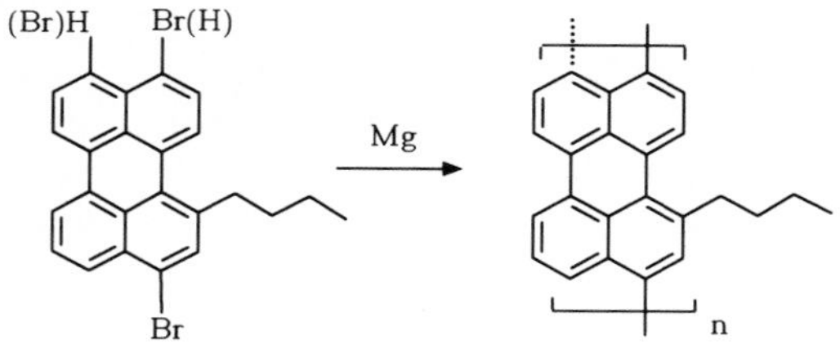

naphthalenes up to the hexarylene subunit are detectable [92]. The longest wavelength absorption maximum of this (partially) cyclized ladder-type polymer is shifted into the NIR-region to a value of ca. 900 nm, corresponding to the absorption of the hexarylene subunit.

4.4 Ladder Polymers via Polymer-Analogous Electrophilic Substitution Reactions

The following example describes the first successful synthesis of a soluble and structurally defined ladder polymer by the stepwise route. It illustrates the synthetic potential of the "classical" route to ribbon-type macromolecules.

Besides the still unsolved problems concerning the polymer-analogous derivatization steps, a series of ladder-type topologies does not seem to be accessible by following the concerted synthetic procedure (polyaddition route) outlined in Sect. 3. This aspect is of current importance for rylene-type molecules (*peri*-fused naphthalenes). Bridged, planar poly(*para*-phenylene)s (PPP) consisting of alternatively linked five- and six-membered all-carbon rings represent another important synthetic goal in this context. For such planar, conjugated ladder-type PPP's (LPPP) a synthesis via a concerted route (repetitive cycloaddition) is not conceivable. Therefore, one has to concentrate on the development of a "classical", stepwise synthetic procedure.

The target structure, a polymeric fluoreneacene **71**, is of primary importance as a planarized poly(phenylene) as well as a starting material to synthesize fully unsaturated ladder-type poly(arylenemethide)s. Some general remarks, concerning the electronic structure and chemical stability of (arylenemethide)-type polymers as potential low band-gap polymers will be given. A stepwise synthesis of the desired ladder-type structures was developed and published by Scherf and Müllen [93]. By means of an aryl-aryl cross-coupling according to Suzuki [94] the bifunctional aromatic dibromodiketone **67** was reacted with an aromatic diboronic acid **68**. To guarantee sufficient solubility of the primary condensation products **69**, poly(phenylene)s with two benzoyl side-groups on alternating aromatic rings, solubilizing alkyl chains were introduced in both monomers. The use of solubilizing side-chains is a well-established method to increase the solubility of rigid polymers [95, 96]. The Suzuki-type condensation is a very

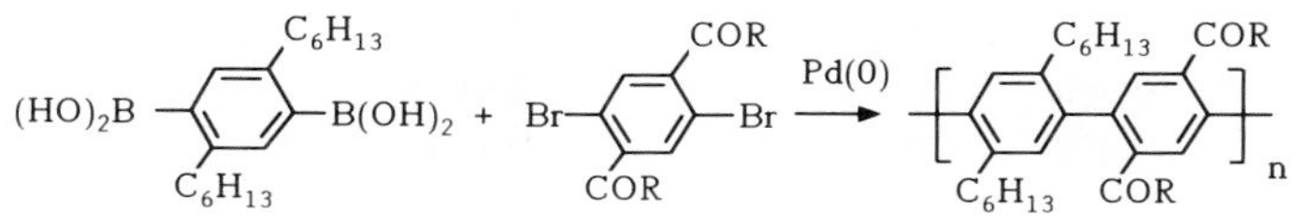

68 **67** **69**

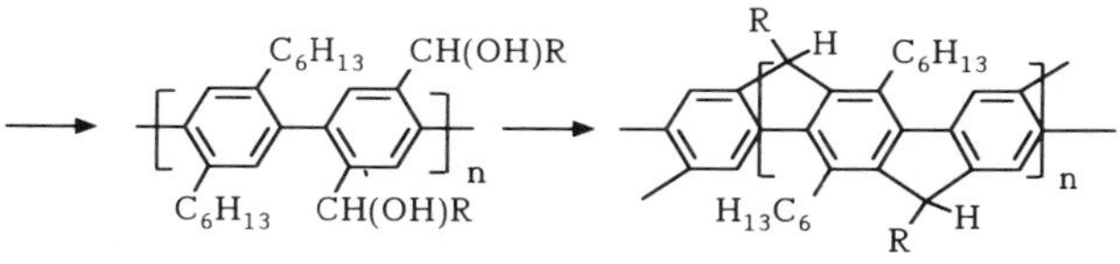

70 **71**

R: -1.4-C_6H_4-R'

R': -C_nH_{2n+1} , -OC_nH_{2n+1}

efficient method for coupling starting materials with functional groups. For example, monomers containing nitro-, keto- or ether-functions can be condensed without any complication. In addition, relatively high degrees of polymerization can be reached as shown by Schlüter and Wegner and their co-workers [96, 97]. In the case presented here, open chain poly(phenylene) precursors **69** with number average molecular weights of up to 20 000 (about 50 phenylene-subunits!) and weight average molecular weights up to 38 000 (Mw/Mn about 2.0) are obtained. The electron-accepting keto substituents appear to increase the rate of the aryl-aryl-coupling considerably. These linear single-stranded intermediates can then be converted into soluble ladder-type poly(phenylene)s, via formation of methylene bridges, by carrying out a simple sequence of two polymer-analogous reaction steps. In the following synthetic sequence, the optimal design of the precursor structure guarantees a regioselective and quantitative conversion of the functional groups.

The first step involves the quantitative reduction of the ketone to an alcohol function by means of lithium aluminium hydride. In the second, the corresponding polyalcohol **70** undergoes ring-closure to the desired target structure **71** under very mild conditions. In a few seconds the double-stranded, fully soluble poly(fluoreneacene) **71** is generated using boron trifluoride as catalyst. The cyclization to give five-membered rings proceeds completely as shown by ^{1}H and ^{13}C NMR-spectroscopy; no indication of structural irregularities such as incomplete cyclization and/or intermolecular crosslinking was found.

The key problem of such intramolecular cyclization reactions is the strict exclusion of side-reactions like intermolecular crosslinking or β-elimination while effecting a nearly quantitative conversion of the functional groups. However, as shown, this problem can be solved by the choice of carefully tailored starting compounds. The lack of α-H-atoms (as for phenyl substituents) prevents β-eliminations, and bulky *n*-alkylated phenyl substituents do not allow for intermolecular crosslinking. The quantitative conversion (ring-closure) is in agreement with earlier results obtained for low-molecular weight compounds. Thus, the completeness of such Friedel-Crafts-type cyclizations was reported for spiro-type compounds derived from fluorene as illustrated by the reaction of fluorenone with 2-phenylphenyl magnesium iodide and subsequent ring-closure to 9,9′-spirobifluorene [98].

With respect to the synthetic sequence outlined for LPPP, it is noteworthy that a polymer-analogous ring-closure fails when starting from polyalcoholic precursors that do not contain substituents at the benzylic bridge position. In such cases, cross-linked insoluble products are formed via intermolecular condensation upon treatment with BF_3, $AlCl_3$ or H_2SO_4. The high number of alkyl groups attached to the macromolecules **71** is necessary in order to guarantee a sufficient solubility during the polymer-analogous reaction steps. The less soluble species is the primary, open-chain, polyketone **69** formed during the Suzuki-type cross-coupling reaction.

To reduce the number of alkyl side-chains incorporated and to increase the chromophorically active part of the macromolecules **71** (the double stranded poly(phenylene) chromophor) an alternative synthetic sequence to the ribbon-type poly(phenylene)s **71** has been designed. When using 2,5-dibromoterephthalic acid dialdehyde (**72**) instead of the aromatic dibromodiketone **67**, the aryl-aryl coupling reaction with the dialkylated phenylenediboronic acid leads to open-chain poly(phenylene)s with main chain aldehyde functions (**73**; M_n up to 7 000; degree of polymerization ca. 30–40 phenylene groups) [99]. Surprisingly, side reactions, especially of the Cannizaro-type, do not take place under the strong alkaline conditions of the coupling reaction. Starting from this single-stranded precursor polymer, a two-step polymer-analogous reaction sequence (addition of a metallo-organic species followed by electrophilic ring-closure) also leads to poly(fluoreneacene)s **71** (LPPP) that possess an identical basic structure to that outlined above. The main advantage of this modified reaction sequence lies in the great variety of side-groups which can then be attached to the macromolecules. In this way, a drastic reduction of the number of solubilizing alkyl chains is possible by reacting the polyaldehyde with, for example, phenyllithium. The products formed are characterized by an increased molar value of chromophorically active substructures incorporated in the macromolecules.

C_6H_{13} — $(HO)_2B$ — $B(OH)_2$ — C_6H_{13} + CHO — Br — Br — CHO → C_6H_{13}, CHO, $H_{13}C_6$, CHO, n

68 **72** **73**

On the other hand, when reacting the polyaldehydes with alkyl-magnesium or -lithium species, side-reactions occur during the final Friedel-Crafts-type electrophilic cyclization reaction, thereby competing with the ring-closure step. About 10–15% of the secondary alcohol groups undergo β-elimination upon treatment of the precursor polymer with the Lewis-acid catalyst (BF_3 of $AlCl_3$) resulting in olefinic side-groups and causing incomplete ladder formation. This behavior is detectable by means of NMR spectroscopy. UV/VIS spectra of this incompletely cyclized material (containing shorter, planarized ladder segments) are characterized by a drastic decrease of the intensity of the two longest wavelength absorptions and the occurrence of intense, shorter wavelength absorptions that correspond to ladder segments with reduced dimension. It appears, that aromatic substituents such as phenyl at the methylene bridge positions are structurally and geometrically optimal in respect to a quantitative and uniform ring-closure reaction to the desired ribbon-type polymers.

The main physical changes during the conversion of the single-stranded precursors to the ladder-type molecules concern the optical and electronic properties. Figure 1 shows the distinct bathochronic shift of the longest wavelength absorption maximum of the double-stranded, planarized molecules **71** compared to the open-chain, strongly distorted precursors **70**. The 0–0 transition lies in the range of 438–450 nm about (2.7–2.8 eV) and is largely independent of the kind of methylene bridge substituents. These drastic changes in the

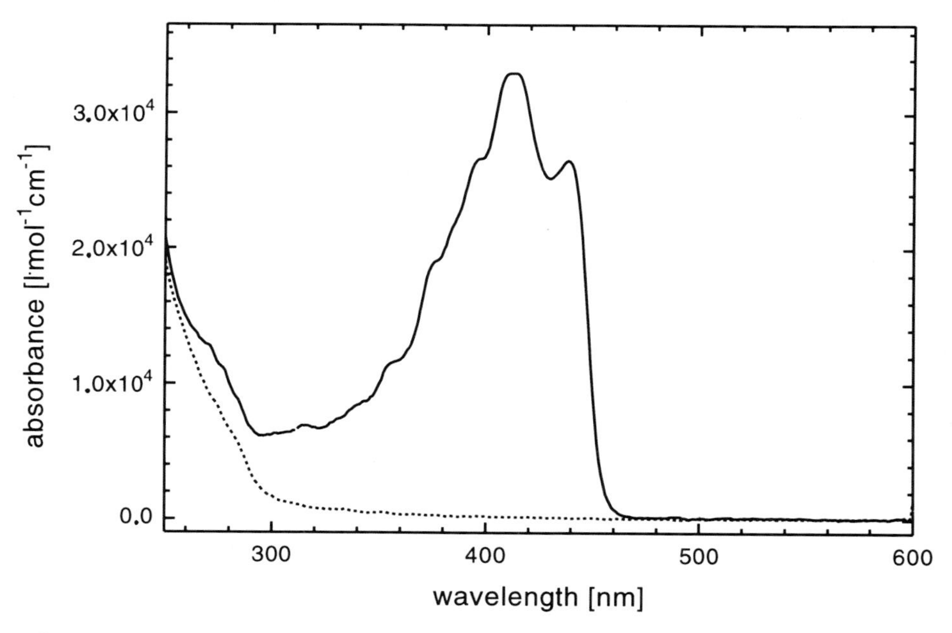

Fig. 1. UV-VIS absorption spectrum of the poly(fluoreneacene) ladder polymer **71** and the open-chain polyalcoholic precursor **70** (*dashed line*; R: -1.4-C_6H_4–$C_{10}H_{21}$; solvent: dichloromethane; ε in $l \cdot mol^{-1} cm^{-1}$)

electronic properties during the cyclization (bridging) to the ladder-type polymers illustrate the novel nature of conjugated polymers defined by the planar poly(phenylene) ribbons.

Another outstanding feature of the ladder poly(phenylene)s **71** concern their fluorescense characteristics. Corresponding to the fixed, rigid geometry of the macromolecules, an extraordinarily small Stokes-loss of the photoluminescence is detectable in solution (dichloromethane), with a mirror-symmetrical shape of the emission and absorption peaks. A new situation occurs when the photoluminescence behavior of the materials is examined in the solid state. There, an additional low energy emission (about 2.2 eV) is detectable, and this becomes the most intense in the spectrum when the films or layers are annealed at a temperature of 150°C. The origin of this emission is not fully understood but time-resolved measurements should bring more insight to the solid state fluorescence behavior, especially after an investigation of the supramolecular structure [100]. Electroluminescence measurements using films of the methylene bridged poly(phenylene)s have demonstrated the occurrence of a highly efficient yellow electroluminescence (efficiency ca. 1%; 600 nm = 2.2 eV) which can be assigned to the low energy photoluminescence emission observed in the solid state [100]. The outstanding optical properties of the novel polymers **71** argue strongly for an investigation of photoelectrical and non-linear optical parameters. The photoconductivity of the poly(*para*-phenylene) ladder polymers is nearly as high as that of poly(*para*-phenylenevinylene), one of the best polymeric photoconductors yet known. The main advantage of these ladder-type polymers are their high chemical purity and their solubility in common organic solvents that allow good processibility. The non-linear optical behavior will be discussed later and will include other structures derived from those described above.

A further attractive property of the ladder-type poly(phenylene)s **71** is the possibility of achieving some supramolecular ordering of the macromolecules. A parallel arrangement of the rigid chromophors is detectable when using X-ray diffractometry via side chain ordering (crystallization) of the alkyl groups attached to the methylene bridges. A small-angle diffraction peak occurs whose intensity can be increased by annealing the samples to 120–150°C. Thereby, the value of the X-ray peak, representing the spacing of the rigid chromophors, is a function of the length of the alkyl side-chains. Modelling the supramolecular arrangement of main- and side-chains now becomes possible with the help of these results. In full agreement with these findings, DSC experiments display sharp endothermic peaks of the so-called side-chain molding in the temperature range of 70–100°C. These results explain why an annealing temperature of 120–150°C is necessary to increase the supramolecular ordering [101]. A disadvantage that prevents a further supramolecular ordering is the statistical arrangement of the optically active methylene-bridge centers along the double-stranded chain. When achieving a defined stereochemical arrangement (tacticity) a drastically increased supramolecular ordering of the macromolecules should occur. To this end it is necessary to carry out the synthetic procedure with stereochemically pure starting compounds.

The main advantage of the novel synthetic sequence is its great versatility. One attractive possibility concerns the coupling of *meta*-(1,3-)linked monomers instead of the *para*-(1,4-)substituted ones. Following this idea, ladder-type poly(phenylene)s **74** consisting of alternatively *para*- and *meta*-linked phenylene subunits or all-*meta*-linked subunits can be synthesized [2, 102]. The reduction of the initially formed polyketones into polyalcohols, and their subsequent cyclization as described above, leads to ladder-type poly(fluoreneacene)s with a different geometric arrangement of the all-carbon five- and six-membered rings. An interesting feature from a structural viewpoint is the expected helical structure of the all-*meta*-linked poly(phenylene) ribbons **75** that contrasts to the linear shape of the all-*para*- and the alternating *para*-*meta*-linked ladder-type poly(phenylene)s.

The planarized, ladder-type poly(phenylene)s containing sp^3-methylene bridges represent attractive precursors to generate the fully unsaturated double-stranded hydrocarbons **76** and **77** via dehydrogenation. Therefore, intensive investigations were carried out to dehydrogenate the methylene groups of the all-*para*-linked, fully conjugated ladder-type poly(phenylene) precursor, and the copolymer with alternating *para*-*meta*-linked subunits; fully unsaturated ladder polymers with quinoid substructures, so called poly(arylenemethide)s, should result. In addition, the all-*meta*-linked ladder polymer constitutes an attractive precursor for polyradicalic materials because its formation of quinonoid substructures is structurally impossible. Such compounds with high-spin character are of interest as potential organic ferromagnets [103].

The fully unsaturated poly(arylenemethide)s –PPM– consist of alternating aromatic (benzoid) and quinonoid subunits. In the case of the ladder-type molecules described here the poly(arylenemethide) substructure is constrained in a two-dimensional framework. The attempts to generate these poly(arylenemethide)s via dehydrogenation of the benzylic bridge positions were successful for the case of the polymer **74** containing alternating *para*- and *meta*-subunits. In a surprisingly simple fashion, the polymer-analogous oxidation is achieved using 2,3-dichloro-5,6-dicyanobenzoquinone (DDQ) as dehydrogenating agent. Thereby, according to ^1H NMR-spectroscopy, more than 90% of the bridge positions are converted, into a deep blue, fully soluble poly (arylenemethide) **76**.

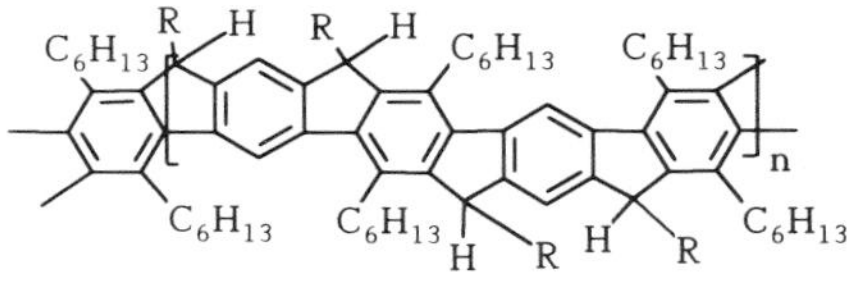

74

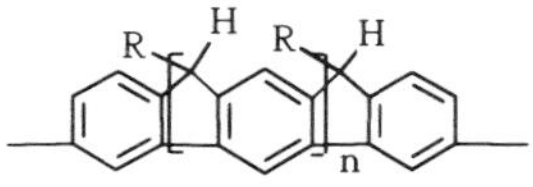

75

R: -1.4-C_6H_4-C_nH_{2n+1}

The polymeric quinonedimethide is ESR-inactive reflecting the absence of unpaired spins (radicals). The electron absorption spectrum of **76** displays a new broad absorption in the visible range (λ_{max}: 605 nm) due to the benzoquinonedimethide chromophors generated in the dehydrogenation step. In the near infrared region (NIR) no additional absorption features are detectable, which supports the existence of undoped, neutral species. In comparison with a low molecular weight model compound, 9,12-di(4-methylphenyl) indeno(1,2-b) fluorene (**78**) (λ_{max}: 543 nm), the longest wavelength absorption of the polymer **76** has a slight bathochromic shift and the absorption is broadened (tailing into the NIR region). This is a consequence of the conjugative interaction between the quinonoid sub-structures along the double-stranded polymeric backbone. The quinonoid ladder polymer is stable in solution or the solid state (film) in the absence of oxygen but decomposes slowly over several weeks in air [102].

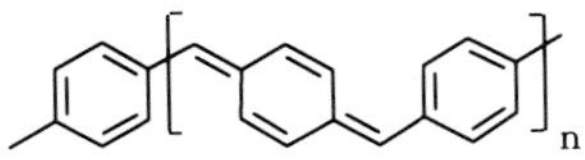

PPM

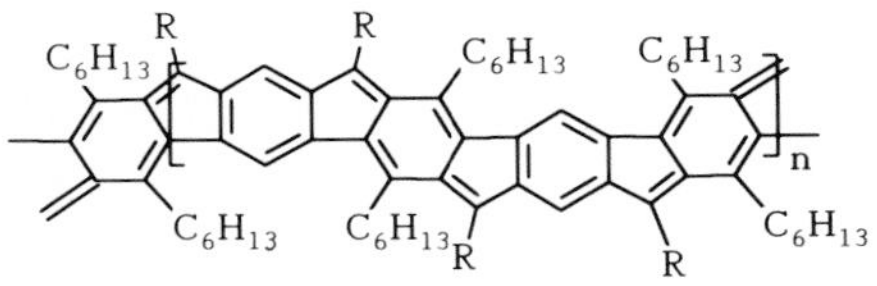

76

R: $-1.4-C_6H_4-C_nH_{2n+1}$

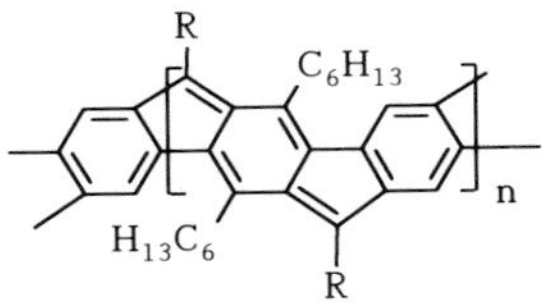

77

R: $-1.4-C_6H_4-R'$

R': $-C_nH_{2n+1}$, $-OC_nH_{2n+1}$

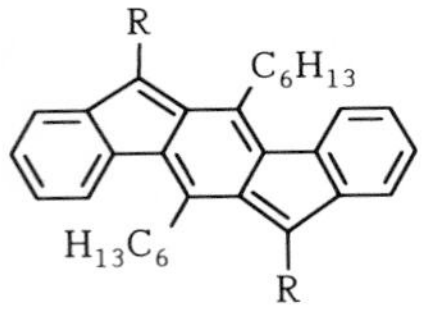

R: $-1.4-C_6H_4-CH_3$ **78**

While the poly(arylenemethide) **76** represents a polymer with localized quinonoid substructures, the poly(arylenemethide) **77** that is derived from the all-*para*-linked, fully conjugated poly(phenylene) precursor possesses a degenerate electronic ground state with a delocalized (degenerate) quinonoid character. This electronic structure causes a drastically decreased band gap energy and is coupled with a decreased chemical stability of the desired target macromolecules [89, 101]. All reported attempts to synthesize single-stranded poly(arylenemethide)s possessing a degenerate electronic ground state have led to structurally non-defined products [104, 105].

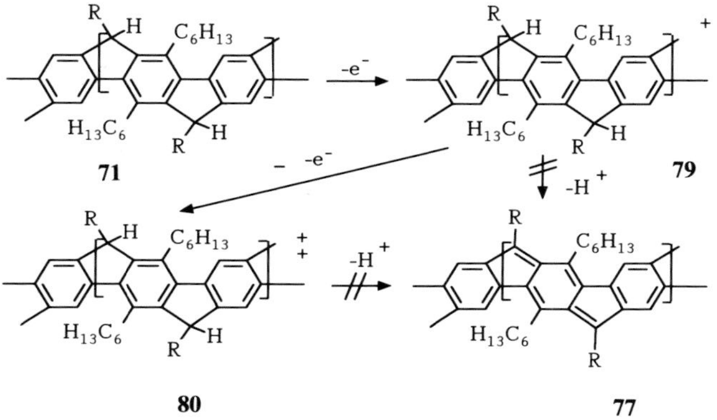

A direct dehydrogenation of **71** to the corresponding ladder-type poly(arylenemethide) **77** using DDQ fails even though it is successful for alternating *para-meta*-linked polymer as described above. The materials formed cannot be fully characterized as the presence of unpaired electron spins prevents a detailed (NMR spectroscopic) analysis. Therefore, we have tested two other ways for generating the target ladder-type poly(phenylenemethide)s with a degenerate ground state. Firstly, doped (cationic) species were generated by oxidation in the hope, that a subsequent deprotonation at the benzylic bridge positions would be favored. Secondly, the oxidation of species metallated at the benzylic bridge position, for instance with iodine, cadmium chloride or silver trifluoroacetate, was attempted. When following the route via oxidation of the poly(phenylene) precursors, the phenylene-type ladder-polymer is treated with oxidizing agents such as antimony pentachloride or iron trichloride. Deeply colored solutions are immediately formed and the characteristic absorption of the neutral polyconjugated material in the UV/VIS region disappears. The new absorptions are in the VIS and NIR range of the optical absorption spectrum. The presence of a visible range absorption (λ_{max} ca. 540 nm) coupled with a broad, intense NIR peak (maximum ca. 2000 nm) [106] is in accordance with the electronic theory of doping, and is characteristic of radical cationic species **79** that possess a singly occupied mid-gap state. Further treatment with the oxidant leads to the formation of a new level of doping – the dicationic (bipolaronic) stage **80** – characterized by the complete disappearance of the visible absorption and by a hypsochromic shift of the NIR absorption maximum to a value of about 1700 nm. An outstanding feature of the doped species, and especially the radical cationic ones, is their high chemical stability, the radical cations are stable in contact with water and dilute acids. Unfortunately, however, the oxidative doping process is not accompanied by subsequent deprotonation to **77** as demonstrated by model studies on the corresponding polymeric species lacking the methylene-bridge hydrogens.

Electrochemical (cyclovoltametric) investigations of the ladder-type poly-(*para*-phenylene) species **71** support the results of the chemical oxidation (doping) experiments both in solution and in the solid state (film). A reversible oxidation takes place and it is well-separated into two waves especially in the solid-state experiment. These are assigned to the formation of radical cationic (**79**) and dicationic species (**80**), respectively. The halfwave potential ($E_{1/2}$) for the first oxidation wave lies between 0.75 V (solution experiment) and 0.95 V (solid state – film) – versus a standard calomel electrode SCE) [106]. Consequently, one has to search for an alternative synthetic process to generate the ladder-type poly(phenylenemethide)s **77** or polymers containing extended segments of the fully unsaturated structure desired. The oxidation of polymeric carbanions appeared suitable, but it proved necessary to work under conditions which completely exclude water and air.

The formation of the polymeric carbanions **81** of the fluorenyl-type is successful starting from the poly(*para*-phenylene) ladder polymer **71** with butyllithium as metallating agent. The degree of lithiation lies in the range of 90–95% (NMR). The UV/VIS absorption spectrum of these polymeric anions (**81**) is comparable with that of the 9-phenylfluorenyl anion and indicates the presence of mostly localized (anionic) sub-structures [101].

The most successful oxidizing agent for the generation of neutral fully unsaturated (arylenemethide) species starting from the polymeric anions, is $CdCl_2$. The heterogeneous reaction allows for a simple separation of products **82** and prevents over-oxidation (doping). A detailed NMR spectroscopic investigation of the products, however, is not possible due to the generation of some unpaired spins that are localized at benzylic bridge positions. Nevertheless,

71 —LiC_4H_9→ **81** —$CdCl_2$→ **82**

R: $-1.4\text{-}C_6H_4\text{-}C_nH_{2n+1}$

the UV/VIS absorption spectra displays a series of sharp absorptions of decreasing intensity at 544, 687, 828 and 950 (s) nm. Compared with the UV spectrum of the model compound 9,12-di(4-methylphenyl)indeno(1,2-b)fluorene **78** (λ_{max}: 543 nm), the first absorption of the ladder-type polymer in the blue range (λ_{max}: 544 nm) clearly can be ascribed to the occurrence of a "dimeric" (arylenemethide) sub-structure. The other absorption maxima of **82** represent more extended oligo-(arylenemethide) subunits (trimeric, tetrameric etc.) [101].

Products **82** are of low chemical stability and rapidly decompose (several minutes) in air; decomposition is slower (several days) in the absence of water and oxygen. Thus, these investigations have defined general limitations for poly(arylenemethide) synthesis, set by their chemical instability. A structural analysis of a decomposed ladder-type poly(arylenemethide) **76** has clearly shown that the bridge (methide position) is the site of attack with formation of hydroxyl- or hydroperoxyl-groups [102].

Measurements of the NLO properties have been performed on the three topologically different ladder-type poly(phenylene)s synthesized with the double-stranded backbone – all-*para* (**71**), alternating *para-meta* (**74**), and all-*meta* (**75**) – and also on the poly(phenylenemethide) ladder polymer **76** formed via dehydrogenation of **74**. They reveal a dependence of the hyperpolarizability upon linear optical properties, particularly on the longest wavelength absorption maximum. The absolute values of the χ^3-coefficients, determined via third harmonic generation, increase with decreasing optical gap energy [102]. The parameters are comparable with those of other single-stranded conjugated polymers (like PPV) that possess an optical band-gap in the same region. However, the quinonoid polymer **76** shows a χ^3-value one order of magnitude smaller than those of linear conjugated polymers without double-stranded, two-dimensional structure that possess comparable absorption characteristics, e.g. polyacetylene or polydiacetylene. The reason may be the predominate existence of electronically localized chromophors without large amounts of conjugative interaction in the direction of the chain [102].

The synthetic sequence to methylene-bridged poly(phenylene)s **71** represents the first successful employment of the stepwise process to ladder-type macromolecules involving backbone formation and subsequent polymer-analogous cyclization. As shown, however, such a procedure needs carefully tailored monomers and reaction conditions in order to obtain structurally defined materials. The following examples demonstrate that the synthesis of structurally defined double-stranded poly(phenylene)s **71** (LPPP) via a non-concerted process is not just a single achievement, but a versatile new synthetic route to ladder polymers. By replacing the dialkyl-phenylenediboronic acid monomer **68** by an *N*-protected diamino-phenylenediboronic acid **83**, the open-chain intermediates **84** formed after the initial aryl-aryl cross-coupling can be cyclized to an almost planar ladder-type polymer of structure **85**, as shown recently by Tour and co-workers [107].

Another fruitful synthetic variation has been the replacement of the aryl-aryl single bonds of the poly(phenylene) sub-structure by heteroaromatic bridges

83 **67** **84**

R: $-C_nH_{2n+1}$

85

(–O– or –S–) to obtain ladder polymers with a conjugatively decoupled backbone (Scherf and Müllen). The nucleophilic aromatic substitution reaction of the substituted dichlorodibenzoylbenzene **86** (R = 1.4-C_6H_4–$C_{10}H_{21}$) with 1,3-benzenedithiole (**87**) leads to polythioetherketones **88** (R = 1.4-C_6H_4–$C_{10}H_{21}$) with the keto function localized in the side chain. The materials are characterized by very high molecular weights, $M_n > 100\,000$, one order of magnitude higher than those of the aryl-aryl cross-coupling products [108]. Addition of an alkyl magnesium compound or reduction with LAH to **89** followed by an Friedel-Crafts-type intramolecular ring-closure gives ladder polymers of structures **90**. This high molecular weight ladder polymers **90** should be used to investigate the often claimed, but hitherto not proved specific rheologic and mechanical properties of macromolecules with a ribbon-type topology. In addition, the ladder structures with R′: –H are attractive intermediates for preparing fully unsaturated ribbons (dehydrogenation of the methylene bridges).

The target structures in the final example are fully aromatic polymeric hydrocarbons, consisting of all-carbon six-membered rings – so-called angularly annulated polyacenes **91** [55]. The structural difference between those and the methylene-bridged poly(phenylene)s is the replacement of the benzylic methylene bridges by vinylene moieties.

Surprisingly, such structures (**91**) are also available via a simple two-step synthetic procedure, as shown by Scherf [55]. While the generation of the methylene-bridged poly(phenylene)s **71** described above involves an AA/BB-type cross-coupling process of an aromatic diboronic acid **68** and an aromatic dibromodiketone **67** in the polymer-forming step [2], this synthesis is a very simple AA-type reductive condensation of the dibromodiketone component **67** as monomer (terminally alkoxy substituted 2,5-dibromo-1,4-dibenzoylbenzenes). In this case, the coupling reaction employs a procedure first described by Yamamoto for several aromatic and heteroaromatic dihalides. It uses the nickel (O)/cyclooctadiene complex as the dehalogenating agent [109]. The

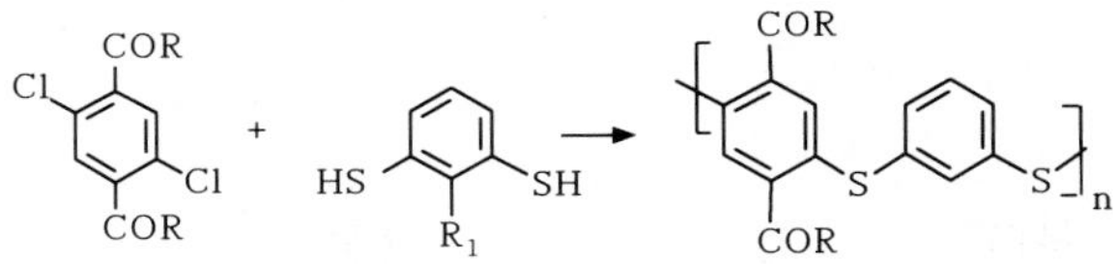

COR, Cl, Cl, COR + HS, SH, R_1 → COR, S, S, COR, n

86 **87** **88**

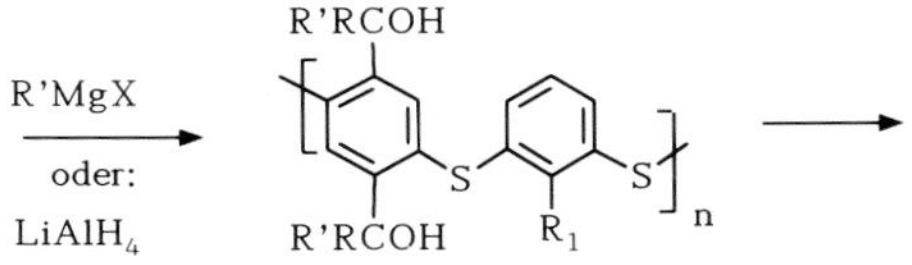

$R'MgX$ → oder: $LiAlH_4$ R'RCOH, S, S, R'RCOH, R_1, n →

89

R_1, S, S, R' R, R' R, S, S, R' R, R' R, R_1, n

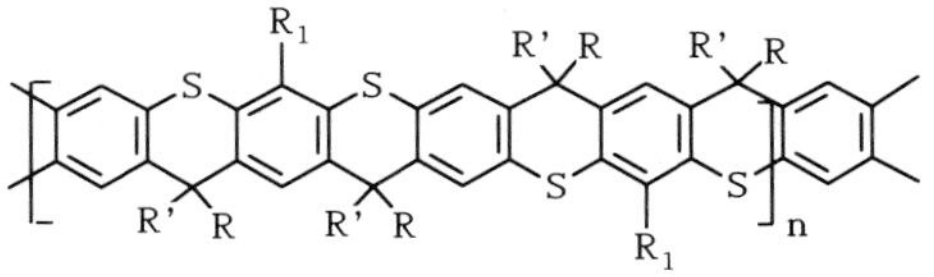

90

R: $-1.4-C_6H_4-C_nH_{2n+1}$

R': $-CH_3$, -H

R_1: $-CH_3$

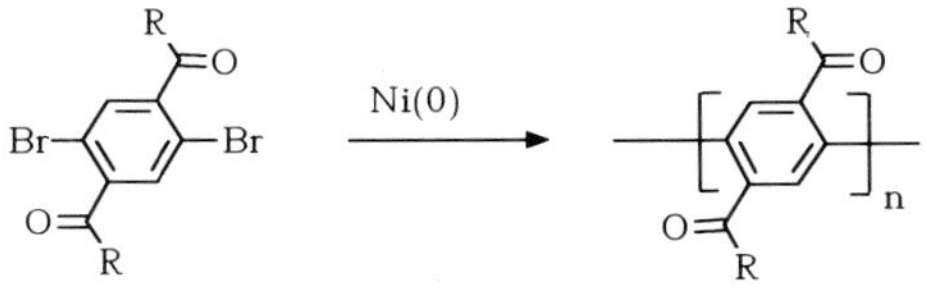

R, O, Br, Br, O, R → Ni(0) → R, O, O, R, n

67 **92**

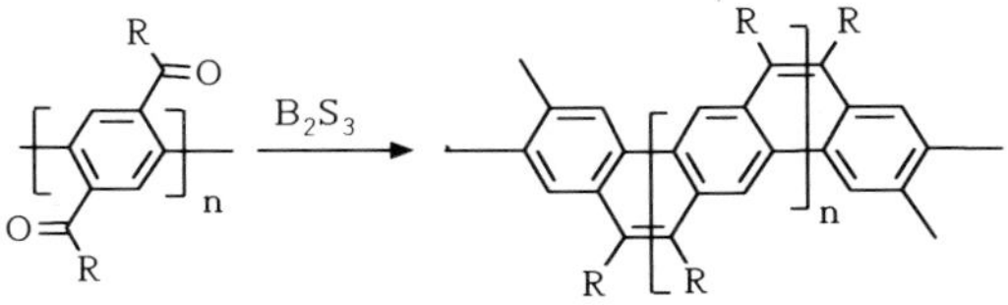

R, O, O, R, n → B_2S_3 → R, R, R, R, n

92 **91**

R: R', R', R'

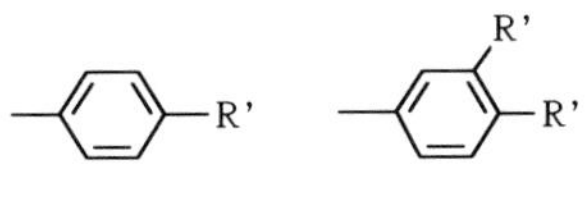

R': $-OC_nH_{2n+1}$

polyketones **92** formed in the coupling reaction are soluble, structurally defined polymers, with number average molecular weights of up to 12000 (corresponding to ca. 20 repeating units).

Starting from the single-stranded precursor **92** the generation of the desired ladder-type framework **91** is possible by adapting a carbonyl-olefination procedure using B_2S_3 as coupling agent. This is formed in situ by reacting boron trichloride with bis(tricyclohexyl) stannyl sulfide [110]. In this coupling reaction, the immediate formation of thioketones is followed by subsequent dimerization and elimination of S_2. The soluble product formed, **91**, was characterized as a structurally well-defined polymer using ^1H and ^{13}C NMR spectroscopy. The planarization (ladder formation) is accompanied by a drastic bathochromic shift of the longest wavelength absorption maximum to 431 nm, a position comparable to that of the methylene-bridged poly(phenylene)s **71** (λ_{max}: 438–450 nm). This sharp, structured absorption feature of **91** illustrates that convergence of the electronic properties is reached (D_p: ca. 20) [55]. The fully aromatic ladder polymers **91** are characterized by a high thermal stability up to 300°C with no weight loss detectable.

An extension of the synthetic concept to the corresponding *meta*-substituted starting materials is possible and commences with the corresponding alkoxy substituted 1,3-dibenzoyl-4,6-dibromobenzenes as monomer. In contrast to the linear structure of the ladder-type molecules **91**, the ladder polymers derived from the *meta*-substituted monomers should possess a helical shape and contain the solubilizing alkoxy groups completely outside the helix. Besides the monomers discussed several other types of starting compounds seem to be very promising in the design of novel ladder topologies.

5 Conclusions

The chemical and physical properties of ladder polymers are expected to differ greatly from those of their linear counterparts, and this aspect has attracted considerable attention in the literature:

- one may intuitively anticipate that ladder structures exhibit a higher thermal and chemical stability than their linear analogs. While bond cleavage within a single stranded chain brings about an immediate decrease of the molecular weight, such a decrease would require cleavage of two neighbouring bonds of the same building block in a ladder polymer. Such a twofold cleavage process is much less likely to occur;
- the presence of two parallel, but interconnected strands in a ladder polymer is expected to produce a high degree of rigidity and should therefore be reflected in the physical and chemical properties of the ribbon-type macromolecules (light scattering behavior, persistence length).

– the limited conformational freedom of ladder polymers is particularly relevant in the case of conjugated ribbon structures since the steric inhibition of electron delocalization is drastically reduced. Furthermore, the unique π-topology of conjugated ladder polymers (planar structure) affects the band structure and thus the resulting electronic properties (absorption, photo- and electroluminescence and nonlinear optical properties, for example).

It is clear, however, that such physical aspects are intimately related to the quality of ladder polymer synthesis, which is evidenced by such facts as structural homogeneity and solubility of the resulting products.

A single deviation from the perfect ladder structure, and thus the formation of single standed subunits, produces the point of attack for chemical decomposition as, induced by thermolysis or hydrolysis. An additional complication arises because one often compares macroscopic properties, deduced from TGA or DSC measurements, for example, with the spectroscopic description of the molecular structure. Even if the spectroscopic characterization of the ladder polymere, say by ^{13}C-NMR spectroscopy, is supported by the inclusion of well-defined low molecular weight model components, the limit of detection for structural defects will not exceed about 1%. Such a degree of inhomogeneity, however, may be disastrous for many material properties.

In describing the state of the art of research devoted to two-dimensional polymer structures, one may well conclude that the lack of structurally perfect ribbon polymers with sufficiently high and, possibly controllable molecular weights, has so far excluded a reliable correlation of structure and properties such as thermal stability or light scattering, and rheologic properties. Attempts toward a systematic comparison of single and double stranded polymers would greatly benefit from the availability of different fractions possessing a narrow distribution of the molecular weight for both open-chain and ladder-type polymers. In the two-step approach toward ladder polymers, the attainable molecular weight is limited by the molecular weight of the open-chain precursors. The most commonly used condensation methods allow number-average molecular weights of only up to 50 000–100 000 to be attained. A similar problem may arise for the "concerted" ladder polymer synthesis via repetitive cycloadditions. As a result of the increasing viscosity of the solution (i.e. decreasing molecular mobility) and of increasing steric hindrance of the reaction partners, the rate constant of the cycloaddition may well decrease with on-going degrees of polymerization. Thus, even if the formation of belt-type (cyclic) structures can be excluded, a further coupling of more extended ribbon structures would be inhibited and would result in a limited degree of polymerization.

The decoration of the ribbon structure with various substituents has appeared as a crucial step for obtaining sufficient polymer solubility. On the other hand, the presence of the long alkyl or alkoxy substituents typically used may have unwanted consequences: (1) the presence of lateral substituents drastically changes the molecular dimensions of the macromolecules (length/width ratio); the mechanical and rheological properties, as well as the chemical and thermal stability, are not only a result of the specific structure of the ladder-type backbone, because the side groups have been shown to dramatically affect the macroscopic behaviour; (2) side-chain ordering (crystallization) can give rise to the formation of characteristic supramolecular aggregates which influence specific physical properties such as photo- or electroluminescence (generation of aggregates); (3) in the case of a conjugated ladder polymer the electronic characteristics of the π-system is "diluted" by the chromophorically inactive side chains.

The unique structure of ribbon polymers poses a challenge to classical methods of polymer synthesis and characterization. Additionally, there exists an urgent need for polymers with unconventional structure in order to achieve macromolecular materials with tailor-made chemical and physical properties.

6 References

1. Winslow FH (1955) J Polym Sci 16: 101
2. Scherf U, Müllen K (1992) Synthesis 1992: 23
3. Schlüter A-D (1991) Adv Mat 3: 282
4. Bailey WJ (1972) Diels-Alder polymerization. In: Solomon DH (ed) Step growth polymerization Marcel Dekker. New York, p 279
5. Packe R, Enkelmann V, Schlüter A-D (1992) Makromol Chem 193: 2829
6. Overberger CG, Moore JA (1970) Adv Polym Sci, Vol 7: 113
7. Yu L, Chen M, Dalton LR (1990) Chem Mater 2: 649
8. Fahnenstich U, Koch K-H, Pollmann M, Scherf U, Wagner M, Wegener S, Müllen K (1992) Makromol Chem, Macromol Symp 54/55: 465
9. Stille JK, Mainen EL, Freeburger ME, Harris FW (1967) Polymer Prep 8: 244
10. Stille JK, Freeburger ME (1968) J Polym Sci Al 6: 161
11. Stille JK, Mainen EL (1968) Macromolecules 1: 36
12. Stille JK, Mainen EL (1966) J Polym Sci B 5: 39, 665
13. Okada M, Marvel CS (1968) J Polym Sci Al 6: 1259
14. Wolf R, Okada M, Marvel CS (1968) J Polym Sci Al 6: 1503
15. Jadamus H, DeSchryver F, DeWinter W, Marvel CS (1966) J Polym Sci Al 4: 2831
16. Kim O-K, J Polym Sci (1985) Polym Lett Ed 23: 137
17. Kim O-K (1984) Mol Cryst Liq Cryst 105: 161
18. Yu L, Dalton LR (1989) Synth Met 29: E463
19. Yu L, Dalton LR (1990) Macromolecules 23: 3439
20. Kistenmacher A, Adam M, Baumgarten M, Pawlik J, Räder H-J, Müllen K (1992) Chem Ber 125: 1495
21. VanDeussen RL (1966) J Polym Sci B4: 211
22. VanDeussen RL, Goins OK, Sicree AJ (1968) J Polym Sci Al 6: 1777
23. Arnold FE, Van Deussen RL (1969) Macromolecules 2: 497
24. Arnold FE, Van Deussen RL (1971) J Appl Polym Sci 15: 2035
25. Berry GC (1978) J Polym Sci, Polym Symp 65: 143
26. Liepins R, Aldissi M (1984) Mol Cryst Liq Cryst 105: 151
27. Jenekhe SA, Tibbetts SJ (1988) J Polym Sci, Polym Phys 26: 201
28. Dawans F, Marvel CS (1965) J Polym Sci Al 3: 3549
29. Bell VL, Jewell RA (1967) Polym Prep 8: 235
30. Colson JG, Michel RH, Paufler RM (1966) J Polym Sci Al 4: 59
31. Arnold FE, VanDeussen RL (1968) Polym Lett 6: 815
32. Arnold FE, VanDeussen RL (1968) AFML-Tr-68-1
33. Quante H, Müllen K, to be published
34. Yu L, Dalton LR (1989) J Am Chem Soc 11: 8699

35. Müllen K, Kistenmacher A, to be published
36. Bailey WJ, Economy J, Hermes ME (1962) J Org Chem 27: 3295
37. Bailey WJ, Fetter EJ, Economy J (1962) J Org Chem 27: 3497
38. Garatt PJ, Neoh SB (1979) J Org Chem 44: 2667
39. Gassmann PJ, Gennick I (1980) J Am Chem Soc 102: 6864
40. Gaboiud R, Vogel P (1980) Tetrahedron 36: 149
41. Luo J, Hart H (1988) J Org Chem 53: 1341; ibid (1989) 54: 1762
42. Hart H, Lai C, Nwokogu GC, Shamouilian S (1987) Tetrahedron 43: 377
43. Wagner M, Wohlfarth W, Müllen K (1988) Chimia 42: 377
44. Godt A, Schlüter A-D (1991) Chem Ber 124: 149
45. Kohnke FH, Slawin AMZ, Stoddart JF Williams DJ (1987) Angew Chem 99: 941; Angew Chem Int Ed Engl 26: 892
46. Ashton PR, Isaacs NS, Kohnke FH, Slawin AMZ, Spencer CM, Stoddart JF, Williams DJ (1988) Angew Chem 100: 981; Angew Chem Int Ed Engl 27: 966
47. Ashton PR, Isaacs NS, Kohnke FH, Mathias JP, Stoddart JF (1989) Angew Chem 101: 1266; Angew Chem Int Ed Engl 28: 1258
48. Löffler M, Schlüter A-D (1992) GIT Fachz Lab 1992: 1101
49. Wegener S, Müllen K (1991) Chem Ber 124: 2101
50. Fang T (1986) PhD Thesis, University of California, Los Angeles
51. Horn T, Scherf U, Wegener S, Müllen K (1992) Polym Prep 33: 190
52. Tsuda T, Maruta K, Kitaike Y (1992) J Am Chem Soc 114: 1498
53. Tsuda T, Maruta K, (1992) Macromolecules 25: 6102
54. Tsuda T, Hokazono H, (1993) Macromolecules 26: 5528
55. Chmil K, Scherf U (1993) Makromol Chem Rapid Commun, 14: 217
56. Kiji J, Iwamoto M (1968) Polym Lett 6: 53
57. Angelo RJ, Wallach ML, Ikeda RM (1967) Polym Prep 8: 221
58. Angelo RJ, Wallach ML, Ikeda RM (1963) Polym Prep 4: 32
59. Houtz RC (1950) Text Res 20: 786
60. Grassie N, McNeil IC (1956) J Chem Soc 1956: 3929; ibid (1958) J Polym Sci 27: 207; 33: 171; (1959) 39: 211
61. Grassie N, McGuchan R (1971) Eur Polym J 7: 1357, 1503; ibid (1972) 8: 243, 257
62. Goodhew PJ, Clarke AJ, Bailey WJ (1975) J Mater Sci Eng 17: 3
63. Arey G, Chadda SK, Poller R (1983) Eur Polym J 19: 313; ibid (1982) J Polym Sci Polym Chem Ed 20: 2249
64. Brokman A, Wager M, Maeson G (1980) Polymer 21: 1114
65. Bailey JE, Clarke AJ (1971) Nature 234: 529
66. Renschler CL, Sylwester AP (1987) Appl Phys Lett 50: 1420
67. Topchiev AV, Geiderich MA, Kargin A, Krenzel BA, Polak LS, Kustanovitch IM (1959) Doklady Akad Nauk SSR 128: 312
68. Teoh H, Metz PD, Wilhelm WG (1982) Mol Cryst Liq Cryst 83: 297
69. Seher A (1952) Fette Seifen Anstrichm 54: 544
70. Bohlmann F, Inhoffen E (1956) Chem Ber 89: 1276
71. Snow AW (1981) Nature 292: 40
72. Teysie P, Korn-Girard AC (1964) J Polym Sci A 2: 2849
73. Kobayashi N, Mikitoshi M, Ohno H, Tsuchida E, Matsuda H, Nakanishi H, Kato M (1987) New Polym Mater 1:3
74. Marvel CS, Levesque CL (1938) J Am Chem Soc 60: 280
75. Marvel CS, Cormer JO, Riddle EH (1942) J Am Chem Soc 64: 92
76. Flory PJ (1939) J Am Chem Soc 61: 1518
77. Kitamura T, Hasumi K, Fujisawa N (1987) JPN Kokai Tokkyo Koho, Japanese Patent 52, 179, 509 (87, 178, 509)
78. Schulz RC, Vielhaber H, Kern W (1960) Kunststoffe 50: 500
79. Clar E, Kelly W, Laird RM (1956) Monatsh Chemie 87: 391
80. Kerr KA, Ashmore JP, Speakman JC (1975) Proc R Soc 311: 199
81. Kaplan ML, Schmidt PH, Chen CH, Walsh Jr WM (1980) Appl Phys Lett 26: 867
82. Forrest SR, Kaplan ML, Schmidt PH, Venkatesan T, Lovinger AJ (1982) Appl Phys Lett 41: 708
83. Murakami M, Yoshimura S (1984) J Chem Soc, Chem Commun 1984,: 1649
84. Murakami M, Yoshimura S (1985) Mol Cryst Liq Cryst 118: 95
85. Iqbal Z, Ivory DM, Marti J, Bredas JL, Baughman RH (1985) Mol Cryst Liq Cryst 118: 103

86. Bohnen A, Koch K-H, Lüttke W, Müllen K (1990) Angew Chem 102: 548; Angew Chem Int Ed Engl 29: 525
87. Koch K-H, Müllen K (1991) Chem Ber 124: 2091
88. Solodovnikov SP, Ioffe ST, Zaks YB; Kabachnik MI (1968) Bull Acad Sci USSR Div Chem Sci 1968: 442
89. Scherf U, Müllen K (1993) Makromol Chem Macromol Symp, 69: 23
90. Schrader S, Koch K-H, Mathy A, Bubeck C, Müllen K, Wegner G (1991) Synth Met 41–43: 3223
91. Anton U, Müllen K (1993) Macromolecules, 26: 1248
92. Anton U, Müllen K (1993) Makromol Chem Rapid Commun, 14: 223
93. Scherf U, Müllen K (1991) Makromol Chem Rapid Commun 12: 489
94. Miyaura N, Yanagi T, Suzuki A (1981) Synth Commun 11: 513
95. Heitz W (1986) Chem-Ztg 110: 385
96. Rehahn M, Schlüter A-D, Wegner G, Feast WJ (1989) Polymer 30: 1060
97. Rehahn M, Schlüter A-D, Wegner G (1990) Makromol Chem 191: 1991
98. Clarkson RG, Gomberg M (1930) J Am Chem Soc 52: 2881
99. Scherf U, Müllen K (1992) Macromolecules 25: 3546
100. Grem G, Leising G (1993) Synth Met, 55–57: 4105
101. Scherf U (1993) Synth Met, 55–57: 767
102. Scherf U, Müllen K (1992) Polymer (Commun) 33: 2443
103. Baumgarten M, Müller U, Bohnen A, Müllen K (1992) Angew Chem 104: 482; Angew Chem Int Ed Engl 31: 448
104. Jenekhe SA (1986) Nature 322: 345
105. Jira R, Bräunling H (1987) Synth Met 17: 691
106. Scherf U, Bohnen A, Müllen K (1992) Makromol Chem 193: 1127
107. Tour JM. Lamba JJS (1993) J Am Chem Soc 115: 4935
108. Scherf U, Müllen K (1993) Synth Met 55–57: 739
109. Yamamoto T, Morita A, Miyazaki Y, Marayama T, Wakayama H, Zhou Z, Nakumura Y, Kanbara T, Sasaki S, Kubota K (1992) Macromolecules 25: 1214
110. Steliou K, Salama P, Yu X (1992) J Am Chem Soc 114: 1456
111. Horn T, Müllen K (1993) Macromolecules 26: 3472

Editor: Prof. T. Saegusa
Received February 1994

Free-Radical Crosslinking Polymerization and Copolymerization of Multivinyl Compounds

Akira Matsumoto
Department of Applied Chemistry, Faculty of Engineering, Kansai University, Suita, Osaka 564, Japan

The network formation in the free-radical crosslinking polymerization and copolymerization of multivinyl compounds, especially including diallyl esters and dimethacrylates, was dealt with by focusing our attention on the mechanistic discussion on deviation from Flory-Stockmayer's (F-S) theory. First, the reasons for the greatly delayed gelation in diallyl polymerizations were discussed mechanistically in detail and the following conclusion was reached: The primary factor is the significance of the thermodynamic excluded volume effect on the intermolecular crosslinking reaction between the growing polymer radical and prepolymer, especially at high molecular weight. Beyond the theoretical gel point, a secondary factor is related to the intramolecular crosslinking which becomes progressively important with conversion. The latter leads to the restriction of segmental motion of the prepolymer and, moreover, imposes steric hindrance, inducing the significance of the reduced reactivity of prepolymer as a tertiary factor. This discussion was satisfactorily extended to the network formation in common multivinyl polymerizations and some problems involved are described in detail such as the validity of F-S theory, intramolecular cyclization, intramolecular crosslinking, microgelation, and the solvent effect on gelation. Finally, our preliminary attempts to control the network formation are introduced on the basis of the above mechanistic elucidation.

Advances in Polymer Science, Vol. 123

List of Abbreviations and Symbols

ABz	allyl benzoate
APP	allyl propyl phthalate
BuAc	butyl acetate
BzMA	benzyl methacrylate
DAA	diallyl adipate
DAD	diallyl diphenate
DAI	diallyl isophthalate
DAP	diallyl phthalate
DAT	diallyl terephthalate
DEGBAC	diethylene glycol bis (allyl carbonate)
DMA	docosyl methacrylate
DVB	divinylbenzene
EDMA	ethylene dimethacrylate
EtAc	ethyl acetate
HEMA	2-hydroxyethyl methacrylate
LALLS	low angle laser light scattering
LMA	lauryl methacrylate
LS	light scattering
MALLS	multi-angle laser light scattering
MMA	methyl methacrylate
MPEGMA-23	methoxytriicosaethylene glycol methacrylate
MWD	molecular-weight-distribution
PEGDMA	polyethylene glycol dimethacrylate
RMA	alkyl methacrylate
SEC	size exclusion chromatography
SMA	stearyl methacrylate
St	styrene
TMPTMA	trimethylolpropane trimethacrylate

1 Introduction

For a long time, the network formation mechanism and the fine structure of cured resins have been controversial problems because of the complexity of the reactions involved and the insolubility of the products. In particular, the free-radical crosslinking polymerization and copolymerization of multivinyl compounds are the problem cases. Since the pioneering theoretical and experimental works of Flory and Stockmayer [1, 2] and Walling [3], many papers have been published, especially on ethylene dimethacrylate (EDMA) [3–18], divinylbenzene (DVB) [19–28], and diallyl dicarboxylates [18, 29–34] as typical divinyl compounds. Dusek [35] reviewed the network formation by chain crosslinking (co)polymerization, especially emphasizing the importance of cyclization leading to the formation of microgel-like particles. Quite recently, Dotson et al. [17, 36] tested the predictions of classical Flory–Stockmayer's theory for the establishment of a network by free-radical crosslinking polymerization and Hamielec et al. [37] summarized their studies on the elucidation of the crosslinking mechanism and kinetics, the characterization of network microstructures, and the development of kinetic gelation models; both research groups have dealt with the copolymerization of methyl methacrylate (MMA) with EDMA, a historical copolymerization system which was first investigated in 1945 by Walling [3]. Styrene (St)-DVB copolymers have also been under investigation for more than 50 years because of their commercial importance as the resins for SEC and ion exchangers. Apparently, the interpretation of studies of the free-radical crosslinking polymerization and copolymerization of multivinyl compounds have been complicated by the great number of research groups and gelling systems, frequently providing scattered data, although several factors, including cyclization, reduced reactivity of pendant vinyl groups of the prepolymer, intramolecular crosslinking, and microgelation have been proposed in order to interpret the observed discrepancy from theory for each polymerization system. Still, it seems that a more comprehensive study is required for a full understanding of network formation processes.

We have extensively investigated the free-radical polymerization of a variety of symmetric or asymmetric divinyl compounds including diallyl dicarboxylates [38], dimethacrylates [39], bis(alkyl fumarate)s [40], and allyl unsaturated carboxylates [41] in terms of cyclopolymerization and gelation, although our research goal is aimed at the elucidation of the crosslinking reaction mechanism and the control of network formation in order to "molecular-design" three-dimensional vinyl-type polymers with high performance and high functionality. Thus the present review is mainly concerned with the mechanistic discussion of the three-dimensional network formation in the radical polymerization of multivinyl compounds, based on the experimental results obtained mainly in our laboratory.

As the first step, we have taken the gelation of diallyl dicarboxylates for the following reasons: (a) Simpson et al. have unsuccessfully attempted to check Gordon's equation, including a formal allowance for cyclization [30], against diallyl dicarboxylates other than diallyl phthalate (DAP) [31]. Note that a satisfactory correlation was observed fortuitously in the case of DAP [30]. (b) Even in bulk homopolymerization, gelation was observed at a conversion of 20–25%, quite high compared with the polymerizations of common divinyl compounds, because of a short primary chain length and thus, these polymerization systems have some advantages for obtaining many useful experimental data for the elucidation of network formation processes up to the gel point. (c) The reactivity of pendant allyl groups of the prepolymer is quite low, and thus the prepolymer is stable, leading to almost no occurrence of spontaneous post-polymerization of the prepolymer providing reproducible data instead of the scattered ones. (d) Even beyond the gel point, we are able to follow the conversion dependency of network formation until a completely cured resin is obtained, since it is easy to obtain useful information even from a gel polymer by hydrolyzing ester crosslinkages. (e) No gel effect is observed, and thus the primary chain length is kept constant. It is noteworthy that each chain is formed by a single active center because monomer chain transfer is an essential termination reaction in allyl polymerization [42].

Secondly, we have extended the investigation of network formation in allyl resins further to common multivinyl systems typically exemplified by EDMA.

2 Network Formation in Allyl Resins

In general, the polymerization of divinyl compounds leads to the formation of a gel polymer, except for the rather special cases in which only cyclic, linear polymers are obtained [43]. Therefore, the intramolecular cyclization and intermolecular crosslinking reactions are characteristic of the polymerization of divinyl compounds as compared with monovinyl compounds. At an early stage of polymerization, the prepolymer consisting of cyclized and uncyclized units, the latter having pendant double bonds, is formed. The crosslinking reaction between the pendant double bond of the prepolymer and the growing polymer radical gradually becomes more important with the progress of polymerization and, eventually, gelation occurs and the viscosity of the system suddenly increases sharply. This was attributed by Carothers [44] to the formation of a three-dimensional network of indefinitely large size.

Flory [1] has outlined a general method for determining the extent of the reaction at which such a network becomes possible, and has derived a simple general equation (Eq. [1]) connecting the critical conversion at gelation (α_c) with the weight-average functionality ($\bar{f}_w$) of the initial system for the case of polycondensation reactions.

$$\alpha_c = 1/(\bar{f}_w - 1) \tag{1}$$

In the particular case of a vulcanization reaction, where the initial polyfunctional material consists of chains of units, each bearing a single functionality (crosslinking position), $\overline{f}_w$ merely signifies the weight-average degree of polymerization, $\overline{P}_w$, of the initial material. Thus,

$$\alpha_c = 1/(\overline{P}_w - 1) \qquad (2)$$

Although Eqs. (1) and (2) cannot generally be applied to chain reactions, the latter has been extended to the special case of chain polymerizations of symmetrical divinyl compounds by Stockmayer [2]. He applied Flory's procedure to a mixture of polyfunctional components with a generalized distribution of functionality and obtained an expression, Eq. (3), for predicting the gel point in addition copolymerizations of monovinyl and divinyl monomers, in which it was assumed that the structure of both monomers is so closely related that all double bonds present have the same reactivity and moreover, cyclization is ruled out.

$$\alpha_c = (1/\rho)(\overline{P}_w - 1)^{-1} \qquad (3)$$

where $\overline{P}_w$ is the weight-average degree of polymerization of the primary chains which would result if all crosslinks in the network at the gel point were cut, and ρ is the fraction of all double bonds residing on divinyl units in the initial system.

Walling [3] investigated systems of this type experimentally, i.e., MMA-EDMA and vinyl acetate-divinyl adipate, and obtained results in reasonable agreement with Eq. (3) only in the presence of 0.2 mol% or less of divinyl monomer. In all other cases, gelation occurred at higher conversion than that predicted by theory. The results obtained for higher fractions of the divinyl monomer were explained qualitatively on the basis of the idea that the reaction mixture consists of discrete swollen polymer molecules whose rate of diffusion is slower than that of polymer chain growth. Walling also assumed that the extent of cylization was small, and that its effect on quantitative work was of minor significance.

On the other hand, Simpson et al. [29] investigated the structure of DAP prepolymer in detail, demonstrating that cyclization is an important phenomenon in DAP polymerization and should have a major bearing on gelation theory. It was suggested that the existence of extensive formation of cyclic structures may be partially responsible for the poor correlation between the Stockmayer equation and actual degree of conversion at gelation.

Gordon [30] has pointed out that even Simpson's treatment considerably underestimates the importance of the effect and has extended the network theory to include a formal allowance for cyclization. He predicted that the conversion at gelation for a polymerizing monomer which undergoes cyclization is given by

$$(1 - b_c) = 1 - (\{r(2\overline{P}_w - 3) - 1\}/\{r(2\overline{P}_w - 3) + 1\})^2 \qquad (4)$$

where $(1 - b_c)$ is the conversion of the monomer at the gel point, r is the fraction of monomer units having pendant double bond in the polymer, and $\overline{P}_w$ is the weight-average number of divinyl monomer units per chain.

Gordon applied Eq. (4) to the polymerization of DAP and obtained a good correlation only by assuming uniform chains. On the basis of these results, and including those for the copolymerization of polyethylene maleate with MMA

[4], he also stated that a simple network theory suffices to predict the gel point in chain polymerization, and the substantial delays in gel point predicted by Walling's theory of diffusion control in such reactions are not encountered.

Subsequently, Simpson et al. [31] attempted to check Gordon's equation against diallyl dicarboxylates other than DAP. In all cases, gelation occurred at a conversion higher than the value predicted, the discrepancy being considerably greater than could be accounted for by experimental error.

On the other hand, Oiwa [32] has carried out theoretical studies of the reaction kinetics of polymerization involving chain cyclization and multiple crosslinking processes. In addition, the sterically reduced reactivity of pendant allyl groups in the polymer chain was considered.

Here it should be noted that in the theoretical calculation of gel point, the experimental value for the degree of polymerization has a predominant influence, as is evident from Eq. (3) or (4). In this connection, a comparison of the experimental data presented by different investigators reveals a very wide divergence of values, although Simpson and Oiwa made their measurements cryoscopically. Such a discrepancy in $\overline{P}_n$ for allyl polymers is also sometimes encountered in the case of monoallyl compounds. Moreover, Laible [45] has stated in his review of allyl polymerization that the available molecular weight information for allyl polymers is quite inadequate, and the development of a simple method for obtaining the molecular weight of polymers in the intermediate region (10^3 to 10^4) would considerably facilitate the progress of theoretical and applied research.

Later, we rechecked the molecular weights of the prepolymers obtained in the radical polymerization of a variety of diallyl dicarboxylates by using a vapor pressure osmometer [46] which has recently become rather widely used. Thus, we realized that the real value of $\overline{P}_n$ was quite high compared with those reported by Simpson and Oiwa. So we decided to recheck the gelation behavior of diallyl dicarboxylates, and then extended the discussion to the gelation of common multivinyl compounds. During our research, many papers have been published for the elucidation of three-dimensional network formation mechanism in free-radical multivinyl polymerizations as described in the Introduction. At the start, we intended to propose a new kinetic treatment as one of our research goals, but we abandoned the idea because the reactions involved seemed to be too complicated for establishing a comprehensive kinetic model. On the other hand, Eqs. (3) and (4) based on the simplified assumption may be favorable for the following detailed discussion on the deviation from the ideal network formation to clarify a real mechanistic nature in terms of reaction, although further theoretical developments have been carried out by several workers [35, 37, 47–53]; thus this article is not aimed at analysis of recent theoretical discussions, especially those supporting either classical or percolation behavior.

2.1 Determination of the Actual Gel Point

First, we tried to determine the gel point precisely. Figure 1 shows the graph for determination of the gel point in the bulk polymerization of DAP as a typical example [33]. The gel point was estimated as the conversion at the time when the gel starts to be formed. The gel point was determined as 22.3%, being somewhat lower than the 25% stated by Simpson [31]. This author estimated the gel point from the extrapolation of the conversion-time curve obtained in the early stage of polymerization to the time at which the fluidity of the reaction mixture was lost by gelation.

Here, it should be noted that the gel is defined as the crosslinked polymer which should be insoluble in all solvents. According to this definition, the choice of an extracting reagent used to separate a sol from a gel becomes very important. In this connection, we examined by light scattering and intrinsic viscosity the properties of a high-molecular-weight DAP prepolymer fraction obtained by fractionation, indicating that acetone, used as the extracting reagent in our previous work [33], is not always a good solvent [34]. So the gel point rechecked by using benzene or tetrahydrofuran as the extracting reagent was found to be 23.5%.

As can be seen in Fig. 1, no gel effect was observed even beyond the gel point in contrast to the polymerization of common multivinyl compounds such as EDMA [6]. This unusual polymerization behavior is characteristic of allyl polymerization, in which monomer chain transfer is essentially a termination reaction [42] in contrast to bimolecular termination of growing polymer radicals in common vinyl polymerization. In other words, primary chain length is quite short and kept nearly constant during polymerization as is required for the application of Eq. (4).

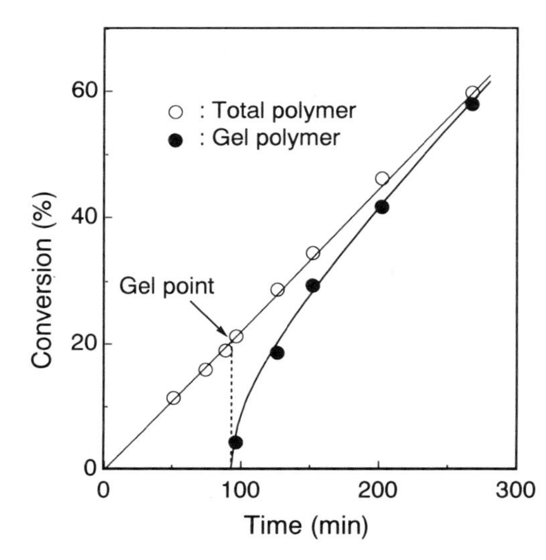

Fig. 1. Determination of gel point in the bulk polymerization of DAP using 0.1 mol l^{-1} of BPO at 80 °C

2.2 Estimation of the Theoretical Gel Point Using Gordon's Equation

We tried to estimate the theoretical gel point by using Gordon's equation [30]. Here, it should be noted that in the theoretical calculation of gel point, the experimental value of the weight-average degree of polymerization $\overline{P}_w$ has a predominant influence. However, a comparison of the experimental data presented by different investigators [29, 32, 46, 54] reveals a very wide divergence of values as described above, although a vapor pressure osmometer employed by us is widely used especially for the molecular weight measurement of oligomers such as the DAP prepolymer. Moreover, notice must be taken of the difficulty of isolating pure polymer by precipitation from the reaction mixture without loss of the low-molecular-weight polymer [55].

We re-examined the degree of polymerization and also attempted to check it by measuring the primary chain length [33] according to Simpson's method [29]. First, checks on measurement of the primary chain length were done by using three kinds of poly(allyl benzoate) having different degrees of polymerization; thus poly(allyl benzoate) was saponified to give poly(allyl alcohol) which was subsequently acetylated to poly(allyl acetate) and then the degree of polymerization of the derived acetate (confirmed by its spectral and/or elemental analyses) was reasonable compared with that of the original benzoate. Then, the acetates derived from the DAP and its isomers diallyl isophthalate (DAI) and diallyl terephthalate (DAT) homopolymers were also characterized similarly and the primary chain lengths were determined as the degree of polymerization of the derived acetates.

Table 1 shows the experimental results of DAP, including the residual unsaturation R_{us} and the number-average degree of polymerization $\overline{P}_n$ of DAP prepolymer and the primary chain length $\overline{P}_{n(ch)}$. In Table 2 the results of $R_{us,0}$, $\overline{P}_{n,0}$, and $\overline{P}_{n(ch),0}$ as extrapolated to zero conversion from the conversion-dependences based on the data of Table 1 are summarized along with those of DAI and DAT polymerizations.

In the early stage of polymerization the crosslinking reaction should be negligible. Thus, the structure of the initially produced prepolymer is represented as follows:

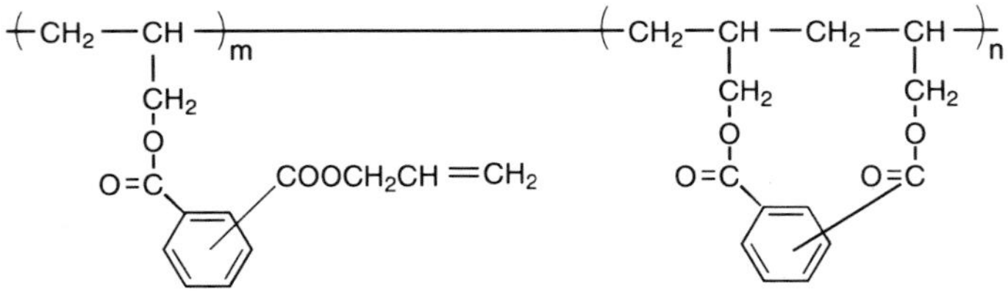

$-(CH_2-CH)_m-$ (side chain: $CH_2-O-C(=O)-C_6H_4-COOCH_2CH=CH_2$) —— $-(CH_2-CH-CH_2-CH)_n-$ (both side chains: $CH_2-O-C(=O)-$, joined through one C_6H_4 ring)

Then, $\overline{P}_{n,0} = m + n$, $R_{us,0} = m/2(m + n)$, $\overline{P}_{n(ch),0} = m + 2n$. So, we calculate the initial degree of polymerization of the primary chain as $\overline{P}_{n(ch),0} = 2\overline{P}_{n,0}(1 - R_{us,0})$; these calculated values are also shown in Table 2.

Table 1. Polymerization of DAP[a]

[M] mol l^{-1}	Conversion, %	R_{us}[b]	$\overline{P}_n$[c]	$\overline{P}_{n(ch)}$[d]
4.55 (Bulk)	4.2	0.278	32.8	28.1
	9.3	0.271	34.3	27.9
	12.6	0.266	35.8	26.5
	17.2	0.260	41.6	25.2
	22.8	0.255	48.3	22.7
2.27	5.9	0.210	24.1	19.5
	9.8	0.202	24.8	
	16.5	0.198	24.5	19.6
	18.0	0.197	25.6	20.3
	25.8	0.192	26.6	18.7
	29.1	0.188	27.1	17.1
1.52	6.4	0.171	20.2	18.7
	12.2	0.168	20.5	17.0
	17.1	0.163	20.3	
	19.3	0.164	20.7	16.7
	23.5	0.163	21.6	18.2

[a] Polymerization at 80 °C, using 0.1 mol l^{-1} benzoyl peroxide as the initiator and benzene as solvent.
[b] The residual unsaturation is the degree of unsaturation of the polymer expressed as a percentage of the corresponding pure DAP monomer.
[c] Degree of polymerization of DAP prepolymer.
[d] Primary chain length expressed as the number of reacted allyl units per chain, corresponding to the degree of polymerization of poly(allyl acetate) obtained after saponification and reesterification of DAP prepolymer.

Table 2. Values of $R_{us,0}$, $\overline{P}_{n,0}$, and $\overline{P}_{n(ch),0}$

Monomer	[M] mol l^{-1}	$R_{us,0}$	$\overline{P}_{n,0}$	$\overline{P}_{n(ch),0}$		
				obs	calc	obs/calc
DAP	4.55	0.284	32	29	46	0.63
	2.27	0.209	24	21	38	0.55
	1.52	0.170	20	18	33	0.54
DAI	4.56	0.410	31	34	37	0.93
	2.28	0.345	24	27	31	0.86
	1.52	0.310	22	23	30	0.76
DAT	4.56	0.455	32	36	35	1.03

It is noteworthy that the observed values are lower than the calculated ones except for DAT and the decreasing tendency is enhanced with the lower value of $R_{us,0}$; this may be ascribed to the intramolecular chain transfer reaction accompanying abstraction of an allylic hydrogen from a pendant allyl group by the uncyclized radical to produce the polymeric allyl radical resonance-stabilized as follows:

$$\sim CH_2-\dot{C}H(CH_2-O-C(=O)-C_6H_4-COOCH_2CH=CH_2) \longrightarrow \sim CH_2-CH_2(CH_2-O-C(=O)-C_6H_4-COO\dot{C}HCH=CH_2) \quad (M\cdot^*)$$

Thus, when $M \cdot ^*$ reinitiates a new chain or couples with the growing polymer radical or another $M \cdot ^*$, the polymer produced must be composed of two or more primary chains separated by a saponifiable branch junction.

These intramolecular chain transfer reactions should have an important bearing on gel-point theory since the branch junction formed thereby does not result in an effective crosslink for gelation. Thus, in the estimation of a theoretical gel point it is $\overline{P}_{n,0}$, which is directly obtained from the conversion-dependence of $\overline{P}_n$ rather than that calculated from $\overline{P}_{n(ch),0}$ and $R_{us,0}$, that should be reasonably employed as the number-average number of divinyl monomer unit per chain.

In this connection, Simpson et al. [29] and Gordon [30] employed a value of 19 for $\overline{P}_{n(ch),0}$ instead of 32 obtained by us in the estimation of a theoretical gel point for the bulk polymerization of DAP at 80 °C. Moreover, they assumed the value $\overline{P}_w = \overline{P}_n$ as being valid for uniform chains or $\overline{P}_w = 2\overline{P}_n$ as being valid for random chains.

Then, the polydispersity coefficient ($\overline{P}_w/\overline{P}_n$) was estimated from an analysis of the SEC curves (Fig. 2) of the prepolymers by using the calibration curve

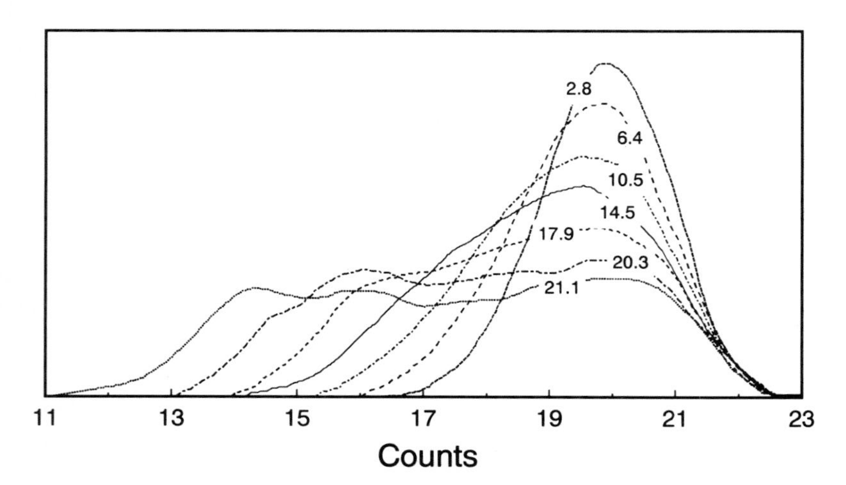

Fig. 2. Variation of SEC curves with conversion (see Fig. 1)

obtained with fractionated DAP prepolymers in place of common standard samples of monodisperse polystyrene. Figure 3 shows the dependence of $\overline{P}_w/\overline{P}_n$ on conversion in the bulk and benzene-solution polymerizations of DAP. From the extrapolation to zero conversion, $\overline{P}_{w,0}/\overline{P}_{n,0}$ was evaluated to be around 2 in both cases, indicating random chains.

Thus, we calculated the theoretical gel points for the polymerizations of DAP, DAI, and DAT from Gordon's equation (Eq. (4)) using $\overline{P}_w = 2\overline{P}_{n,0}$ and $r = 2R_{us,0}$. Table 3 summarizes the results obtained, along with those reported by Simpson et al. [31]; the values estimated by us were lower than those of Simpson.

Now, we can compare actual and theoretical gel points as shown in Table 4. In all cases, gelation occurred at a conversion considerably higher than the predicted value and the discrepancy increased in the order DAT > DAI > DAP.

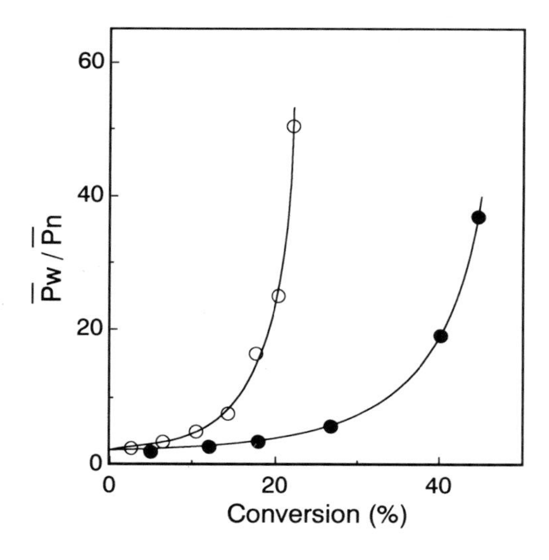

Fig. 3. Dependence of $\overline{P}_w/\overline{P}_n$ on conversion: (○) see Fig. 1; (●) in benzene solution (monomer/benzene = 1/2 by volume)

Table 3. Theoretical gel point

Monomer	[M] mol l^{-1}	r	$\overline{P}_w$	Gel point
DAP	4.55	0.568	64	0.055 (0.140)[a]
	2.27	0.418	48	0.098
	1.52	0.340	40	0.142
DAI	4.56	0.820	62	0.040 (0.073)[a]
	2.28	0.690	48	0.060
	1.52	0.620	44	0.073
DAT	4.56	0.910	64	0.035 (0.069)[a]

[a] Calculated by Simpson and Holt [31].

Table 4. Comparison of actual gel point with theoretical one

Monomer	[M] mol l^{-1}	Gel point	Actual G.P. / Theoretical G.P.
DAP	4.55	0.223 (0.25)[a]	4.1
	2.27	0.370	3.8
	1.52	0.487	3.4
	0.91	0.595	
DAI	4.56	0.218 (0.25)[a]	5.5
	2.28	0.364	6.1
	1.52	0.425	5.8
	0.91	0.530	
DAT	4.56	0.218 (0.26)[a]	6.2
	0.91	0.538	

[a] Data of Simpson and Holt [31].

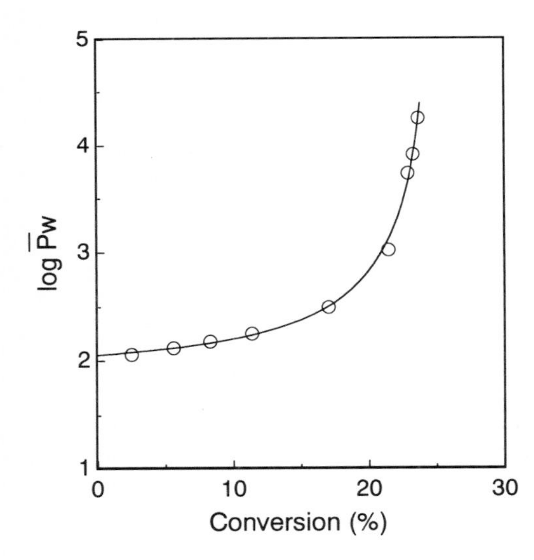

Fig. 4. Dependence of $\overline{P}_w$ on conversion (see Fig. 1)

Quite recently, we rechecked the initial weight-average degree of polymerization $\overline{P}_{w,0}$ as the primary chain length by light scattering. Figure 4 shows the dependence of $\overline{P}_w$ on conversion in the bulk polymerization of DAP. $\overline{P}_{w,0}$ was estimated to be 102, 1.6 times higher than the value in Table 3; this high degree of polymerization, by the consideration of allyl polymerization, may be related to the occurrence of intramolecular chain transfer. Then, the theoretical gel point was calculated to be 3.4% in contrast to 23.5% of actual one, i.e. the gelation was delayed to a 6.9 times higher conversion.

These discussions have been extended to the bulk polymerizations of diallyl adipate (DAA), diethylene glycol bis (allyl carbonate) (DEGBAC), and diallyl diphenate (DAD) accompanied by a greatly delayed gelation: The monomer type, actual gel point (%), theoretical gel point (%) are, respectively, as follows : DAA, 25.6, 3.6; DEGBAC, 23.0, 3.7; DAD, 29.2, 3.4.

2.3 Discussion of the Large Deviation of the Actual Gel Point from Theory

As some of the reasons for the large deviation of the actual gel point from the theoretical one, we may consider the following four factors: intramolecular cyclization, reduced reactivity of pendant vinyl groups of prepolymer, intramolecular crosslinking, and microgelation, which are neglected in ideal network formation by Stockmayer [2]. Thus we will discuss in detail these factors one by one in order to interpret mechanistically the greatly delayed gelation compared to theory in the free-radical polymerization of diallyl dicarboxylates. As a result, these four factors are not significant in the early stages of polymerization, at least up to a theoretical gel point, and finally, a thermodynamic excluded volume effect on the intermolecular crosslinking reaction between growing polymer radical and prepolymer, especially at high-molecular-weight, is discussed as a primary factor.

2.3.1 Intramolecular Cyclization

It is well known that the radical polymerization of diallyl quaternary ammonium salts as carried out by Butler et al. yielded water-soluble linear polymers as opposed to the expected formation of highly crosslinked quaternary ammonium polymers as ion exchange resins. However, quaternary ammonium bromides containing three or more allyl groups were polymerized to yield highly water-swellable but water-insoluble gels suitable as strongly basic ion exchange materials [56, 57]. This behavior was explained by an alternating intramolecular-intermolecular chain propagation mechanism [58], now commonly referred to as cyclopolymerization, through which a wide variety of nonconjugated divinyl compounds have been converted into characteristic cyclopolymers [59, 60].

In fact, the polymerizations of diallyl compounds were by chance at the beginning of the further developed research on cyclopolymerization and crosslinking.

We have investigated the free-radical polymerization of a variety of diallyl dicarboxylates extensively, especially DAP, DAI, and DAT, in terms of allyl polymerization, cyclopolymerization, and gelation [38]. The cyclopolymerization resulted in cyclization constants K_c of 7.5, 1.6, and 0.6 mol/L at 60 °C for DAP, DAI, and DAT, respectively. In this connection, it should be noted that in the polymerizations of DAP, DAI, and DAT, two different modes of intramolecular cyclization should be considered. The cyclization reaction of DAP is predominantly a consecutive intramolecular addition reaction (leading to A-type structures), i.e. addition of an uncyclized radical to the double bond present in the same monomer unit, whereas that of DAT should be considered as a nonconsecutive addition reaction (leading to B-type structures) in which an uncyclized radical adds to any other double bond pendant to the polymer chain to form a loop structure. In the polymerization of DAI, both the consecutive and the nonconsecutive addition takes place, leading to type A and type B:

$C_6H_4(COOCH_2\dot{C}H{-}CH_2R)(COOCH_2CH{=}CH_2) \longrightarrow$ A

$RCH_2{-}CH{-}CH_2 \ldots CHCH_2{-}X{-}CH_2CH{=}CH_2 \ldots CH{=}CH_2 \longrightarrow$ B

$X = {-}OC(O){-}C_6H_4{-}C(O)O{-}$

In this reaction scheme, a preferential occurrence of intramolecular head-to-tail addition is assumed; fortuitously, it was valid in DAP polymerization [61], although many papers have been published reporting the occurrence of intramolecular head-to-head addition especially in the cyclopolymerization of 1, 6-dienes.

These intramolecular cyclization reactions are fruitless as regards the formation of crosslinks. That is, the stronger cyclization is the more gelation behavior should deviate from Stockmayer's theory [2]. In the present discussion concerned with the gelation of diallyl dicarboxylates, however, the intramolecular cyclization forming cyclic structural units in the polymer is omitted as a primary factor for greatly delayed gelation because the theoretical gel point was calculated by using Gordon's equation involving a formal allowance for cyclization as mentioned above.

In addition, the interesting finding that no difference in the actual gel point was substantially observed among three isomeric diallyl phthalates (see Table 4) has been discussed in terms of the correlation between gelation and the difference in cyclization modes, and the difference in reactivity between the uncyclized and cyclized radicals for crosslinking [34]: In the first case, the difference in cyclization modes was reflected in the monomer composition dependence of the residual unsaturation of the resulting copolymer in the copolymerizations of DAP, DAI, and DAT with allyl benzoate (ABz) [62, 63]. In the case of DAP the comonomer acts only as a competitive species according to the cyclopolymerization mechanism. The uncyclized radical reacts intramolecularly with the double bond of the same monomer unit or intermolecularly with another diallyl dicarboxylate monomer or the comonomer. In the case of DAT, the occurrence of a steric suppression of B-type cyclization by the comonomer incorporated into the copolymer chain is additionally expected. These differences in the cyclopolymerization behavior may be reflected in the gelation in the copolymerizations of diallyl aromatic dicarboxylates with ABz because the occurrence of B-type cyclization leading to the formation of loop structure may induce the sterically reduced reactivity of the unreacted pendant allyl groups toward crosslinking. This results in a greatly delayed gelation, while in the copolymerization of DAT with ABz a suppression of B-type cyclization with an increasing mole fraction of ABz in the feed leads to the sterically less reduced reactivity of the unreacted pendant allyl groups, resulting in a less delayed gelation. Figure 5 shows the relationships between the gel points and monomer composition in the bulk copolymerizations of DAP, DAI, and DAT with ABz at 80 °C. The difference in conversion at the gel point tended to increase with increasing molar fraction of ABz in the feed, suggesting a less delayed gelation with higher ABz-feed composition in the DAT-ABz copolymerization. In this connection, a tentative calculation was done to estimate theoretical gel points in the DAT-ABz copolymerization according to Stockmayer's equation (Eq. (3)); the discrepancy of actual gel point from the theoretical one was quite reduced with increasing molar fraction of ABz in the feed, this being attributed

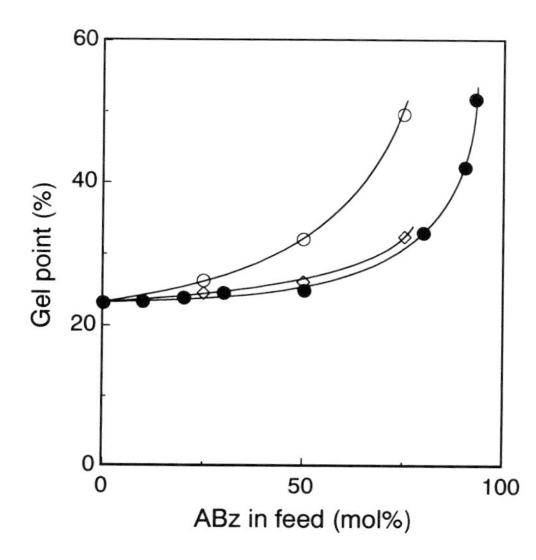

Fig. 5. Relationships between gel point and monomer composition in the bulk copolymerization of (○) DAP, (◇) DAI, and (●) DAT with ABz at 80 °C

to the difference in cyclization modes among DAP, DAI, and DAT as described above.

Secondly, the difference in reactivity between the uncylized and cyclized radicals for crosslinking is discussed as follows. DAP is very much easier to cyclize than DAI and DAT; even in bulk polymerization, 43% of the growing polymer radicals are in a cyclized form [46]. Also, the chain length dependence of the cyclization constant was observed in the cyclopolymerization of DAP [64]; the steric suppression effect of bulky side chains on the intermolecular propagation of uncyclized radical, as expected from the molecular model, is enhanced with increasing chain length by promotion of intramolecular cyclization of the uncyclized radical. In this connection, a cyclized radical may be easily responsible for crosslinking compared with an uncyclized radical promoting the gelation in DAP polymerization.

Uncyclized radical Cyclized radical

Thus, the copolymerizations of the DAP prepolymer were conducted with allyl propyl phthalate (APP) and DAP monomer in dilute solution, corresponding to uncyclized and cyclized radicals, respectively [65]. Figure 6 shows the copoly-

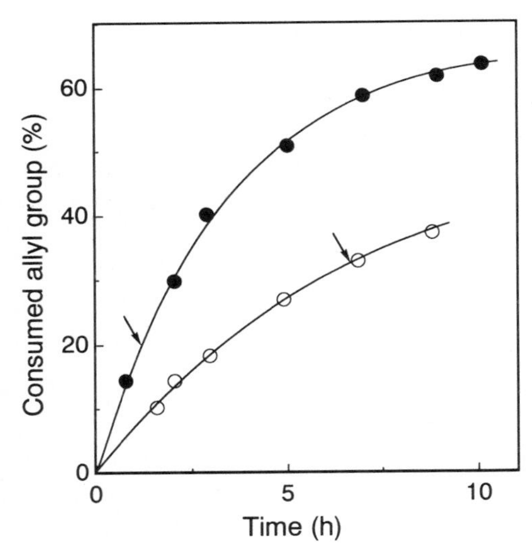

Fig. 6. Solution copolymerization of DAP prepolymer with (○) APP and (●) DAP monomer (molar ratio [APP or DAP]/[allyl groups of prepolymer] = 2.1, [BPO] = 0.1 mol l^{-1}, in benzene, at 80 °C): (↓) gel point

merization results; clearly the rate and gelation were promoted for the copolymerization with DAP monomer, suggesting enhanced reactivity of the cyclized radical for crosslinking.

2.3.2 Reduced Reactivity of the Prepolymer

In the theoretical treatment, an equal reactivity of functional groups belonging to the monomer and polymer is assumed. So, any reduction of the reactivity of pendant allyl groups of the prepolymer compared with the monomer may become one of the reasons for a greatly delayed actual gel point. The reactivity of DAP prepolymer was evaluated kinetically from the post-copolymerization with ABz which has equivalent reactivity to one allyl group of DAP monomer [62].

The DAP prepolymer was prepared by bulk polymerization of DAP, in which conversion was about 10%, and characterized as $R_{us} = 0.275$, $\overline{P}_n = 65.8$, and $\overline{P}_w/\overline{P}_n = 3.71$. That is, the polymer carries, on average, 36 pendant allyl groups and one crosslinkage per prepolymer molecule. It should be noted that under the same polymerization conditions for the preparation of DAP prepolymer the theoretical gel point was estimated to be 3.4%, being quite low compared with 10% conversion.

Figure 7 shows the copolymer composition curve for the post-copolymerization of DAP prepolymer with ABz at 80 °C in which the copolymer composition was determined by extrapolation to zero conversion based on the conversion dependence. The apparent reactivity of pendant allyl groups of the prepolymer proved to be significantly higher than that of the monomer and to be strongly influenced by the concentration as was the case in the post-copolymerization of DAP prepolymer with vinyl acetate [66]. These interesting results at an early stage of the post-copolymerization may be interpreted by considering the following reaction

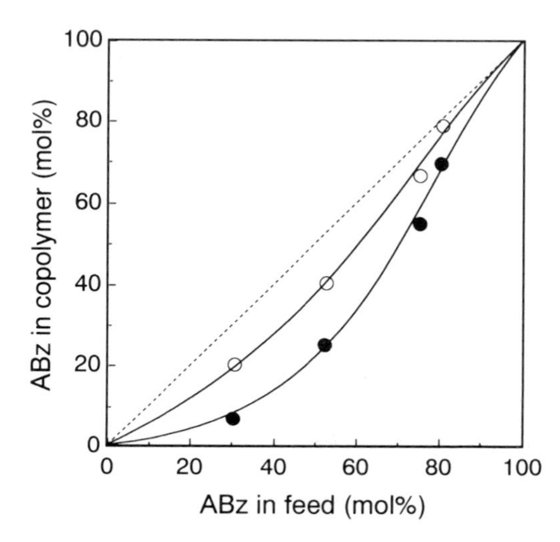

Fig. 7. Copolymerization of DAP prepolymer with ABz ([allyl groups of prepolymer] = (○) 0.6 and (●) 0.2 mol l^{-1}, in benzene, at 80 °C)

mechanism, in which the intramolecular cyclization reaction is taken as significant [38].

$$
\begin{array}{llll}
M_1\cdot + M_1 & \rightarrow M_1\cdot & k_{11}[M_1\cdot][M_1] \\
M_1\cdot + M_2 & \rightarrow M_2\cdot & k_{12}[M_1\cdot][M_2] \\
M_1\cdot & \rightarrow M_2\cdot & k_{1c}[M_1\cdot] \\
M_2\cdot + M_1 & \rightarrow M_1\cdot & k_{21}[M_2\cdot][M_1] \\
M_2\cdot + M_2 & \rightarrow M_2\cdot & k_{22}[M_2\cdot][M_2] \\
M_2\cdot & \rightarrow M_2\cdot & k_{2c}[M_2\cdot]
\end{array}
$$

where M_1 denotes ABz and M_2, pendant allyl group of the DAP prepolymer.

The kinetic treatment revealed that the reactivity of pendant allyl groups of the prepolymer is approximately equal to that of the monomer. The ratios $r_1 = k_{11}/k_{12}$ and $r_2 = k_{22}/k_{21}$ have been estimated to be 1.0 and 0.9, respectively [67]. A similar result was also observed for the post-copolymerization of DAI prepolymer with ABz [68]. Now, we can conclude that the concept of equal reactivity of functional groups belonging to the monomer and polymer is valid for the radical polymerization of diallyl aromatic dicarboxylates at an early stage of polymerization, at least up to the theoretical gel point. However, in the case of the highly branched prepolymer formed at a late stage of polymerization, i.e. far beyond the theoretical gel point, the reactivity of pendant allyl groups may be reduced by steric hindrance.

Thus, the reduced reactivity of the prepolymer, so far referred to as one of the reasons for the greatly delayed gelation, is not a primary factor when one considers that no reduction of reactivity of the prepolymer at about 10% conversion, which is 3 times higher than the theoretical gel point, was observed. Here, this unreduced reactivity may be easily understood from the fact that the kinetic treatment of the reactivity is based on the number-average values and the number-average molecular weight of the prepolymer is quite low in the polymerization of diallyl dicarboxylates. The discussion on gelation is based on the weight-average and, therefore, the reactivity of pendant double bonds belonging to the high-molecular-weight prepolymer plays an important role, regardless of the amount. This is relevant to a thermodynamic excluded volume effect on the intermolecular crosslinking reaction between the growing radical and prepolymer, especially at high molecular weights, taken as the primary factor and intramolecular crosslinking and microgelation as secondary factors.

In this connection, the variation of the shape of molecular-weight-distribution (MWD) curves with conversion is theoretically expected to become broader towards the high-molecular-weight side [69]. In contrast, the observed MWD curves did not become broader, as shown in Fig. 2.

2.3.3 Intramolecular Crosslinking

As polymerization proceeds beyond the theoretical gel point and the resulting prepolymer concentration becomes higher, entanglements of polymer chains are formed, the occurrence of intermolecular crosslinking reaction between growing polymer radical and prepolymer being enhanced; this results in the formation of highly branched prepolymer having abundant pendant double bonds. Concurrently, the occurrence of the intramolecular crosslinking as the propagation reaction of the growing C-polymer chain radical with the pendant double bond belonging to the primary A-polymer chain preceded by the intermolecular crosslinking reaction with the primary B-polymer chain, will be enhanced.

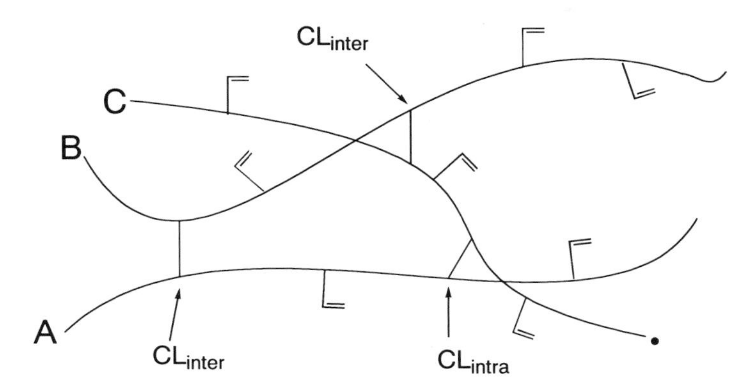

This kind of intramolecular crosslinking wastes the pendant double bond without contributing to the increase of molecular weight and, moreover, leads to the formation of multiple crosslinkages.

The significance of intramolecular crosslinking was clearly demonstrated in the post-polymerization of DAP prepolymer (Fig. 8) and its post-copolymerization with ABz (Fig. 7). This was also checked by SEC-LALLS; the occurrence of

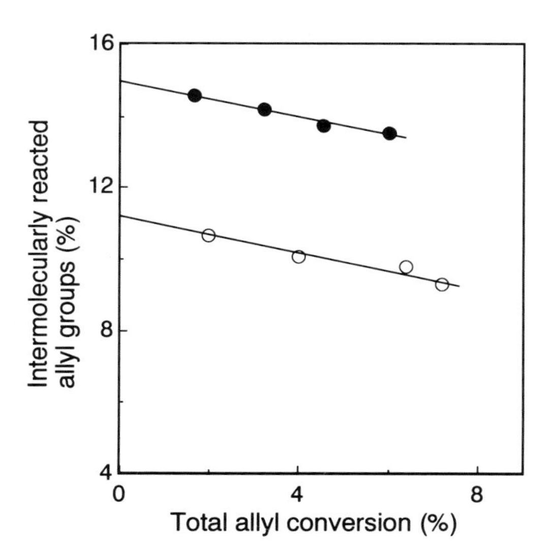

Fig. 8. Dependence of percentage of intermolecularly reacted allyl groups on total conversion of intramolecular and intermolecular reactions in the solution post-polymerization of DAP prepolymer in (○) benzene and (●) EtAc ([allyl groups of prepolymer] = 0.43 mol l^{-1}, at 80 °C)

intramolecular crosslinking induces a shrinkage of molecular size of the resulting polymer and thus, the molecular weight determined by SEC deviates toward the low-molecular-weight side compared with the one determined by LALLS [70].

It should be noted that the intramolecular crosslinking preceded by the intermolecular crosslinking becomes significant at a rather high conversion, i.e. far beyond the theoretical gel point as is evident from the variation of SEC curves with conversion (Fig. 2). So, the occurrence of intramolecular crosslinking is not a primary factor for the greatly delayed gelation although it becomes more and more important with the progress of polymerization, i.e. with the enhanced occurrence of intermolecular crosslinking leading to the formation of a highly branched polymer.

2.3.4 Microgelation

The intramolecular crosslinking may tend to occur locally because the formation of the crosslinked unit induces the decrease of the interaction between polymer segment and solvent or the increase of the interaction between polymer chains leading to an enhanced occurrence of intramolecular crosslinking. Thus, the locally enhanced occurrence of intramolecular crosslinking accompanied by microsyneresis could lead to the formation of a microgel having a highly crosslinked microdomain which may induce microphase inversion. Here, the microgel is conceived as consisting of both core and shell parts of high and low crosslinking densities, respectively, although it is soluble due to a strong interaction of the shell part with solvent overcoming the presence of the core part, just like a microsolid, having quite a weak interaction with the solvent. That is, it is a highly shrunken molecule having a much lower interaction with

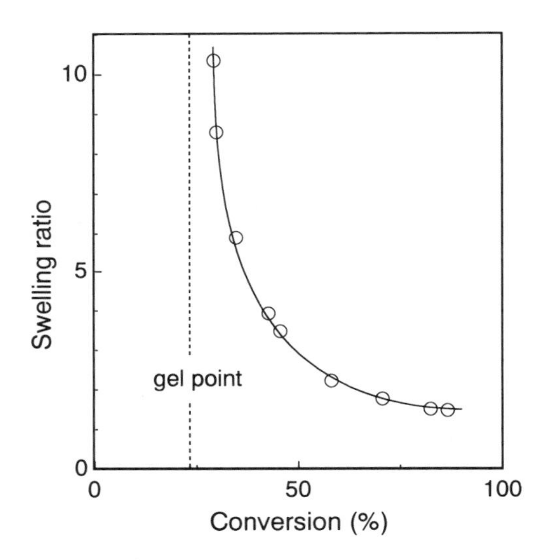

Fig. 9. Decrease in the swelling ratio of resulting gel with the progress of polymerization beyond the gel point (see Fig. 1)

the solvent as compared with a linear polymer of the same molecular weight. This molecular size shrinkage and lowered interaction force will be reflected as the lowering of the r.m.s. radius of gyration and second virial coefficient, respectively.

Thus the bulk polymerization process of DAP was followed in detail until the gel point by light scattering and any abnormality due to the presence of microgel was not observed [71], and moreover the swelling ratio of the gel obtained just beyond the gel point was very high (Fig. 9) [72]; this strongly suggests that no microgelation was observed up to the gel point in DAP polymerization. In conclusion, microgelation is not a factor for the gelation being greatly delayed.

2.3.5 Thermodynamic Excluded Volume Effect

As a summary of the above discussion, four factors including intramolecular cyclization, reduced reactivity of prepolymer, intramolecular crosslinking, and microgelation are not primary factors for the greatly delayed gelation in the radical polymerization of diallyl dicarboxylates. So far, these four factors have been discussed quite often, but we have to look for other reasons for delayed gelation from ideal network formation, especially at an early stage of polymerization, i.e. at least up to the theoretical gel point.

So, we proposed the significance of the thermodynamic excluded volume effect [73] for the intermolecular crosslinking between high-molecular-weight prepolymers [74]: In the post-polymerization of DAP prepolymer, the intermolecular crosslinking reaction, reflected as the increment of polydispersity coefficient $\overline{M}_w/\overline{M}_n$, was quite reduced in a good solvent (Fig. 10). This suggests the significance of the excluded volume effect, although a remarkable effect was observed

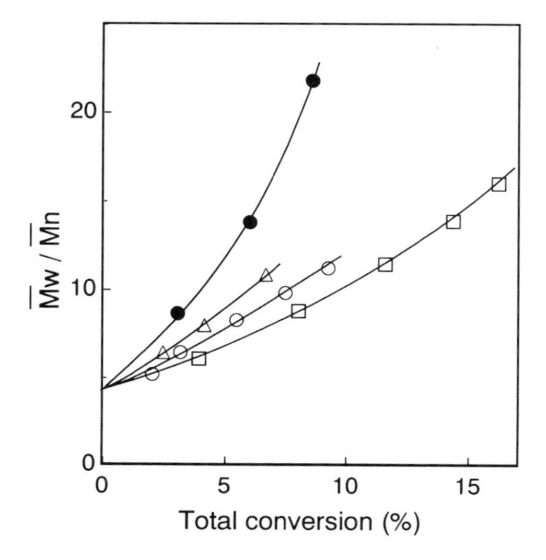

Fig. 10. Dependence of $\overline{M}_w/\overline{M}_n$ on total conversion (see Fig. 8): (•) in EtAc, (○) in benzene, (△) in methyl ethyl ketone, and (□) in dimethyl phthalate

for the high-molecular-weight prepolymer (> 100 000). The excluded volume effect should be important even in the bulk polymerization of DAP, DAI, and DAT by considering that each monomer is a good solvent for the corresponding prepolymers produced. Thus the intermolecular crosslinking reaction between the growing polymer radical and prepolymer, especially at high-molecular-weight, is suppressed at an early stage of polymerization such as at a theoretical gel point conversion, since the high-molecular-weight prepolymer concentration is quite low. With the progress of polymerization the prepolymer concentration increases and inversely, the significance of excluded volume effect on intermolecular crosslinking is reduced.

In conclusion, the thermodynamic excluded volume effect is a primary factor for the large deviation of the actual gel point from the theoretical one at an early stage of polymerization, especially up to the theoretical gel point. Its significance should be reduced with conversion as a result of increased prepolymer concentration and concurrently, at a stage of polymerization beyond the theoretical gel point, the intramolecular crosslinking preceded by the intermolecular crosslinking, as a secondary factor, becomes more and more important with the progress of polymerization. The intramolecular crosslinking leads to the restriction of segmental motion of prepolymer and moreover, imposes the steric hindrance, inducing the significance of the reduced reactivity of prepolymer as a tertiary factor.

The thermodynamic excluded volume effect should be also reflected as a solvent effect on gelation. That is, a more delayed gelation is expected in a good solvent than in a poor solvent; this was the case in the solution polymerization of DAP in various solvents [75, 76]. In this connection, Walling [3] has reported the solvent effect on the gelation in the copolymerization of MMA with EDMA, but the result was contrary to our expectations. So, we will discuss this subject in more detail later.

2.4 Microheterogenization of the Gel During Polymerization Beyond the Gel Point

One of the reasons we have taken up the gelation of diallyl dicarboxylates first is that, as we pointed out in the Introduction, even beyond the gel point we can follow the conversion dependence of network formation until a completely cured resin is obtained. Thus we have investigated in detail the process of growth of the gel with the progress of polymerization beyond the gel point in the bulk polymerization of DAP, especially in terms of microheterogenization [72]: Although Erath and Robinson [77] observed the DAP cured resin directly as the agglomerate of colloidal particles by electron microscopy, our experimental results clearly demonstrated that no microgelation occurred up to the gel point in the bulk polymerization of DAP. Thus the formation of colloidal particles should be induced during the polymerization process beyond the gel point.

The gel fraction in the polymer obtained beyond the gel point increased rapidly with conversion (Fig. 1), and thus the polymerization system changed toward the copolymerization of the gel with unreacted monomer. For example, the gel fraction in the polymer obtained at 50% conversion was more than 90%, that is, the resulting gel was swollen by an approximately equal amount of the unreacted monomer although the amount of monomer as solvent was quite insufficient for a full swelling of the gel produced in the polymerization system (Fig. 9). Thus, the polymerization proceeded heterogeneously in an insufficiently swollen gel at a late stage of polymerization, but no abnormality of the polymerization behavior was observed. This was compared with an early stage of polymerization, in which both the prepolymer and the gel were derived to poly(allyl acetate)s which were subjected to IR, ^{1}H- and ^{13}C-NMR, and SEC measurements. These results may suggest that the microheterogeneity of polymerization system; the microheterogenization of the resulting gel, i.e. quasi-microgelation, proceeds rapidly beyond the gel point and at a late stage of polymerization, the unreacted monomer mainly polymerizes in microspaces among quasi-microgels. In other words, the occurrence of intramolecular crosslinking inside the highly swollen gel produced just beyond the gel point would be enhanced with the progress of gelation. Moreover, the intramolecular crosslinking may tend to occur locally as described in Sect. 2.3.4. Thus the segmental density of the interface among quasi-microgels would be low and unreacted monomer molecules can move freely without restriction from the swollen gel. This is supported by the drastic decrease in swelling ratio of the resulting gel with the progress of polymerization beyond the gel point (Fig. 9), the conversion dependence of residual unsaturation (Fig. 11) and reduced viscosity of the finely pulverized gel, and the permeability of solvent into the gelled polymerization system.

This investigation of the microheterogenization of the gel during polymerization beyond the gel point was applied to the microheterogeneous copolymerization of DAP with comonomers accompanied by the formation of quasi-microgel and moreover, macrogel with the intention of improving the mechanical properties of DAP resins [80, 81]: First, three kinds of glycol bis(allyl phthalate)s, including ethylene glycol bis(allyl phthalate), tetraethylene glycol bis(allyl phthalate), and

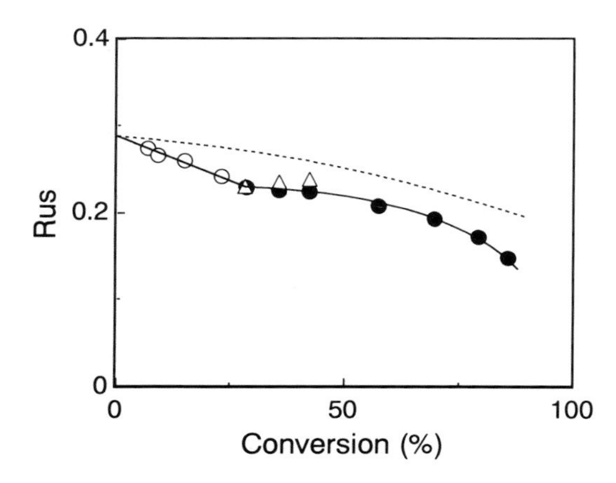

Fig. 11. Dependence of residual unsaturation R_{us} of (○) prepolymers, (△) sols, and (●) gels on conversion (see Fig. 1). *Dotted line* corresponds to the conversion dependence of R_{us} calculated by assuming only the occurrence of intramolecular cyclization

tridecaethylene glycol bis(allyl phthalate) were crosslinked with DAP and the resultant crosslinked products were evaluated for the elucidation of correlations between mechanical properties and crosslinkers [80]. Thus DAP resins with improved flexibility were successfully obtained by introducing flexible crosslinked units on the surface of the microgel, i.e. by flexibly crosslinking microgels to form the macrogel.

Second, DAP was cocured in bulk with those vinyl monomers having long-chain alkyl groups, including vinyl laurate, dioctyl fumarate, lauryl methacrylate (LMA), and stearyl methacrylate (SMA) [81]. Thus, for the DAP-LMA copolymerization process leading to the microgel formation, the initially obtained copolymers or precopolymers of high LMA content as a reflection of a high polymerizability of LMA compared with DAP can not be compatible with DAP-enriched polymer chains which increase rapidly with the progress of polymerization. Therefore they may exist predominantly in the spaces between microgels and act as flexible crosslinkers of microgels to form the macrogels.

This type of microheterogeneous copolymerization of DAP with vinyl monomers having long-chain alkyl groups was applied further for the bulk copolymerization systems to obtain direct evidence to support the idea of the microheterogeneity of the systems beyond the gel-point conversion [82]. Also, the solution copolymerization of DAT was explored to demonstrate the incompatibility of the initially obtained precopolymer with a high content of comonomer units with DAT-enriched polymer chains [83].

3 Crosslinking Polymerization and Copolymerization of Common Multivinyl Compounds

Here we will extend the above discussion of network formation in allyl resins to the polymerization of common multivinyl compounds accompanied by gelation since allyl compounds are rather a special type vinyl compound and our findings are not always in conformity with the results reported by other workers on network formation in common multivinyl polymerizations.

3.1 Comparison of Gelation Behavior in Allyl and Vinyl Polymerizations

The most significant difference between allyl and vinyl polymerizations is in the length of their primary chains which has a predominant influence on gelation, since in the allyl polymerization an occurrence of monomer chain transfer is quite remarkable and only the oligomer is formed. Therefore, we tried to carry out the telomerization of EDMA in the presence of CBr_4 in order to reduce the primary chain length [84] down to a comparable order in allyl polymerization; otherwise, a direct comparison of both DAP and EDMA polymerizations can not be made easily since we do not have an alternative way of enlarging the primary chain length in the allyl polymerization. Figure 12 shows the conversion-time curve in the bulk polymerization of EDMA in the presence of CBr_4 as one of the typical data; the gel formation curve was similar to the one in Fig. 1 and no gel effect was observed even beyond the gel point conversion. The variation of SEC curves of the resulting prepolymers with conversion was also quite similar to DAP polymerization (Fig. 2). The gel point was obtained as 18.5% and the primary chain length $\overline{P}_{w,0}$ was estimated, from the extrapolation of the conversion-dependence of weight-average degree of polymerization of the prepolymer determined by LS to zero conversion, to be 140, enabling us to calculate the theoretical gel point as 1.46% according to Gordon's equation, Eq. (4) in which $r = 0.98$; no substantial difference was observed between allyl and vinyl polymerizations in the case where the primary chain lengths were adjusted to be comparable.

In this connection, many homopolymerizations and copolymerizations of multivinyl compounds were carried out to compare the delay of actual gelation with that of theory [13, 85–102], and we reached the conclusion that in the radical polymerizations of multivinyl compounds, the deviation of actual gel point from theoretical one become greater with an increase in the primary chain length, an increase in the content of pendant vinyl groups of the primary chain, and a decrease in the monomer concentration. Its range became quite wide to almost 10^3. The enhancement of the significance of thermodynamic excluded volume effect, the occurrence of intramolecular crosslinking, and the reduced reactivity of prepolymer should lead to a greater deviation from theory.

An abundance of instances of the large deviation prompted us to check the validity of Flory and Stockmayer's gelation theory (F-S theory) as a next step.

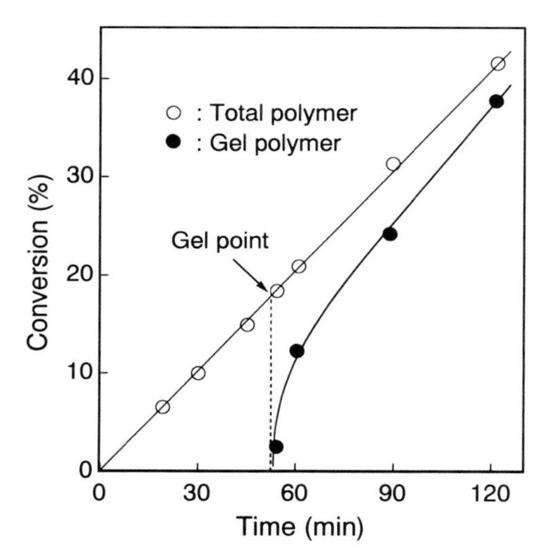

Fig. 12. Conversion-time curve in the bulk polymerization of EDMA in the presence of CBr_4 (molar ratio $[CBr_4]/[EDMA] = 1/5$, $[AIBN] = 0.005$ mol l^{-1}, at 50 °C)

3.2 *Validity of F-S Theory*

Here it should be noted that our discussion so far has been based on F-S theory. This classical theory has been tested for many systems including homopolymerizations and copolymerizations of a variety of multivinyl compounds by many workers; gelation always occurred 1–2 orders of magnitude later than predicted as mentioned above. In all cases, the prepolymer concentration was too low at the theoretical gel point to affect significantly the thermodynamic excluded volume effect on the intermolecular crosslinking between high-molecular-weight prepolymers. Moreover, in solution polymerizations, the occurrence of the intramolecular crosslinking would be enhanced with conversion.

Our interpretation of the greatly delayed gelation strongly suggests that the actual gel point should be close to the theoretical one if the experiment were done under the polymerization conditions in which the polymer concentration at the theoretical gel point is high enough to reduce the significance of a thermodynamic excluded volume effect as a primary factor and in addition, the intramolecular crosslinking preceded by the intermolecular crosslinking would be suppressed.

Thus, MMA-EDMA copolymerizations involving different amounts of EDMA were carried out in the presence of lauryl mercaptan as a chain transfer reagent in order to reduce the primary chain length and keep it constant. Table 5 summarizes the results obtained. The higher the theoretical gel point, the smaller the ratio of the actual gel point to the theoretical one. In the presence of only 0.03 mol% of EDMA it reached 1.3, very close to unity, supporting the validity of F-S theory, although Dotson [36] and Hamielec [37] strongly demonstrated in their recent reviews that the classical theory was inapplicable and, moreover, an alternative theory should be found. The theory of ideal network formation also suggests that the MWD curves should be rapidly broadened with conversion

Table 5. Deviation of actual gel points from theoretical ones in MMA-EDMA copolymerizations[a]

EDMA mol%	$\bar{P}_{w,0}$[b] $\times 10^{-3}$	Gel point (%)		A.G.P.[e] / T.G.P.[c]
		Theoretical[c]	Actual[d]	
0.03	2.94	56.6	72.5 (72.4)[e]	1.3
0.05	3.08	32.4	68.7 (68.6)[e]	2.1
0.1	3.80	13.2	35.8 (35.7)[e]	2.7
0.5	4.24	2.37	12.0 (11.9)[e]	5.0
1	4.85	1.05	8.7 (8.5)[e]	8.1

[a] In dioxane, dilution 2/3, [AIBN] = 0.04 mol l^{-1}, 50 °C, [LM] = 1.68×10^{-3} mol l^{-1}.
[b] Estimated by SEC-MALLS.
[c] Theoretical gel point $\alpha_c = (1/\rho)(\bar{P}_{w,0} - 1)^{-1}$.
[d] Obtained as monomer basis.
[e] Obtained as the vinyl group conversion calculated by assuming an equal reactivity of vinyl groups belonging to MMA and EDMA.

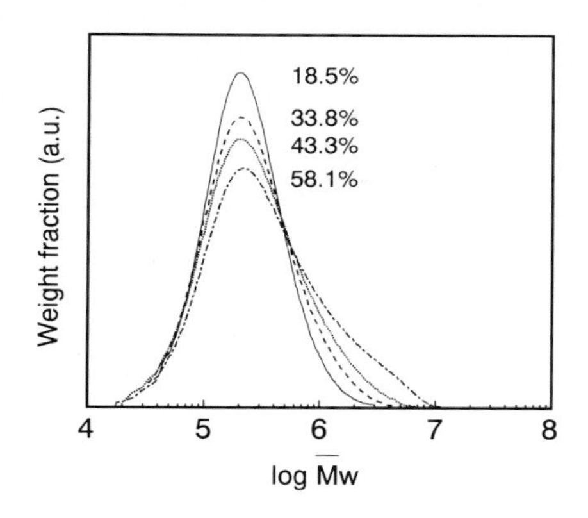

Fig. 13. Variation of MWD curves with conversion in MMA-EDMA copolymerization in the presence of 0.03 mol% of EDMA (see Table 5)

1) Intermolecular propagation with monomer:

2) Intramolecular cyclization:

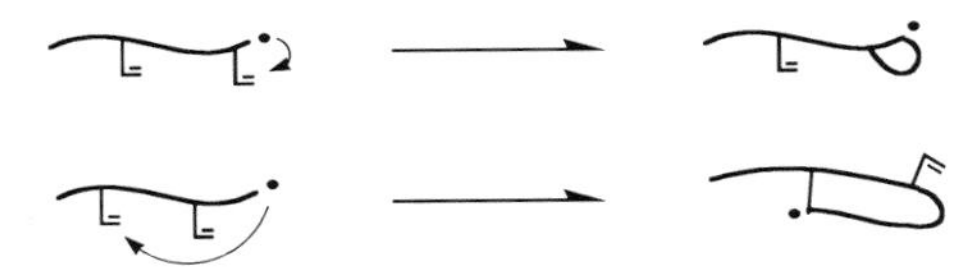

3) Intermolecular crosslinking with prepolymer:

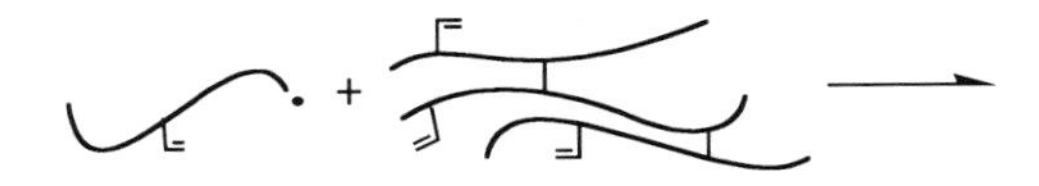

4) Intramolecular crosslinking:

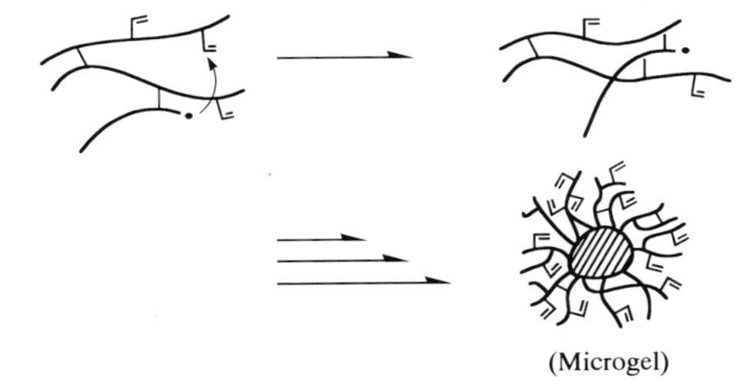

Fig. 14. Reaction scheme for the network formation processes in the free-radical monovinyl-divinyl copolymerization

as a result of enhanced intermolecular crosslinking reaction of higher-molecular-weight prepolymers [69]. The results shown in Fig. 13 were in a good agreement with the ones expected, and quite different from those in Fig. 2 in which the gelation was greatly delayed from theory.

Next, we will discuss some problems involved in the gelation of multivinyl compounds. Figure 14 shows the reaction scheme for the network formation processes in a free-radical monovinyl-divinyl copolymerization, essentially involving four reactions of a growing polymer radical: (1) intermolecular propagation with two types of monomer, (2) intramolecular cyclization leading to the formation of ring or loop structures, (3) intermolecular crosslinking with the prepolymer to form effective crosslinkage, eventually leading to the gel, and (4) intramolecular crosslinking leading to the formation of multiple crosslinkages, the locally extensive occurrence of which induces microgelation. Thus the subject of the molecular design of a vinyl-type three-dimensional polymer is reduced to the control of the elementary reactions shown in Fig. 14. In this connection, the detailed discussion of each elementary reaction should be required.

3.3 Intramolecular Cyclization

As described in the Introduction, Dusek [35] emphasized the importance of cyclization and quite recently, Dotson [36] has argued it again. Here it should be noted that they have treated indiscriminately both intramolecular cyclization (2) and intramolecular crosslinking (4) as cyclization and strongly emphasized the marked occurrence of such cyclization at a quite early stage of polymerization where no intermolecular crosslinking occurs. Eventually, the enhanced reactivity of pendant double bonds because of their spacial proximity to the terminal radical site would lead to the formation of microgel at a low conversion.

However, we insist that both reactions (2) and (4) (Fig. 14) should be clearly discriminated for the discussion of gelation: The intramolecular cyclization leads to only the incorporation of ring or loop structures into the primary chain, never forming crosslinkages between primary chains, while the intramolecular crosslinking leads not only to the formation of loop structures, but also of crosslinkages resulting in multiple crosslinkage, influencing the crosslinking density of the gel. Moreover, the intramolecular cyclization always occurs competitively with the intermolecular propagation (1) during the formation of each primary chain and it is accelerated with conversion as a reflection of reduced monomer concentration, although it is not a primarily important factor for the greatly delayed gelation as discussed in Sect. 2.3.1 except for the case where cyclization occurs preferentially to form a linear, cyclic polymer [59, 60]. On the other hand, the intramolecular crosslinking should be preceded by the intermolecular crosslinking (3) and its situation is more complicated as will be discussed in a subsequent section. Anyhow, the discussions of Dusek [35] and Dotson [36] are based on the experiments [7, 17] in which the gel point conversions were quite low, although we tried to set up the polymerization conditions under which gelation occurred at more than 20% conversion, in order to ensure the detailed pursuit of the network formation processes in terms of the reaction.

In addition to the above discussion (Sect. 2.3.1) on intramolecular cyclization, we will deal with the large cyclic ring formation in the common monovinyl-multivinyl copolymerizations in connection with gelation. Thus the chain length of the resulting copolymers is remarkably high compared with allyl polymerization to form a large cyclic ring as follows:

$$\sim CH_2-\underset{\underset{CH=CH_2}{|}}{\underset{X}{|}}{CH}\!-\!(CH_2-\underset{Y}{\underset{|}{CH}})_n-CH_2-\dot{C}H-Y \longrightarrow \sim CH_2-\underset{\underset{\bullet CH-CH_2}{|}}{\underset{X}{|}}{CH}\!-\!(CH_2-\underset{Y}{\underset{|}{CH}})_n-CH_2-CH(-CH_2)-Y$$

Soper et al. have investigated the copolymerization of St with DVB [21]; the intrinsic viscosity [η] of the resulting polymer was measured in order to explore the occurrence of intramolecular cyclization leading to the formation of loop structure and thus, resulting in the reduced [η] of the polymer. As a matter of course, this intramolecular cyclization is influenced by the flexibility of the polymer chain, the content of pendant double bonds, and the primary chain length.

Also, this kind of intramolecular cyclization led to increased primary chain length in the copolymerization of MMA with several polyethylene glycol dimethacrylates (PEGDMA) having different numbers of oxyethylene units: Firstly, gelation in these copolymerizations has been explored [13]; as opposed to the expectation [103], gelation was accelerated with an increase in the oxyethylene unit number beyond two. This was attributed to the increased primary chain length estimated by the extrapolation of the conversion dependence of weight-average molecular weight (Fig. 15) to zero conversion. Secondly, we tried to

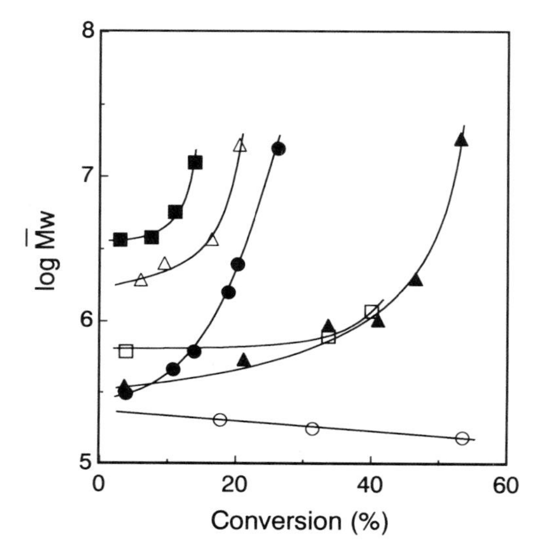

Fig. 15. Dependence of $\overline{M}_w$ on conversion in the solution copolymerizations of MMA with 1 mol% of (•) EDMA, (▲) PEGDMA-2, (□) PEGDMA-3, (△) PEGDMA-9, and (■) PEGDMA-23, along with (○) MMA homopolymerization (volume ratio total monomer/1,4-dioxane = 1/4, [AIBN] = 0.04 mol l^{-1}, at 50 °C)

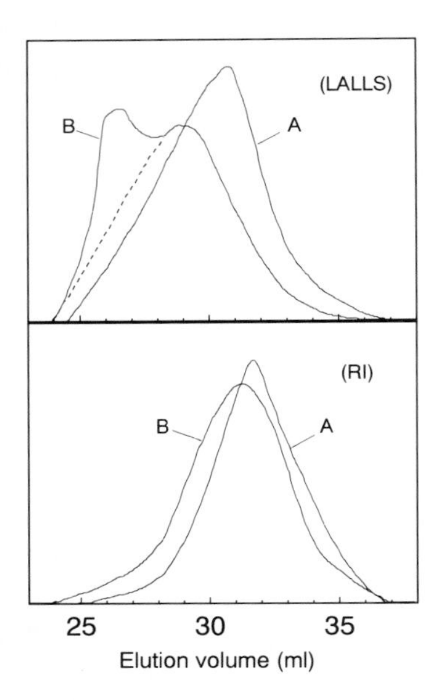

Fig. 16. SEC-LALLS curves for (*A*) MMA-EDMA and (*B*) MMA-TMPTMA copolymers (see Fig. 15)

elucidate the interesting enhancement of primary chain length by increasing the number of oxyethylene units of PEGDMA [39]. In conclusion, this is ascribable to the suppression of the intermolecular termination between growing polymer radicals having a loop structure formed through intramolecular cyclization and moreover, the molecular-weight distribution of the polymer obtained at an early stage of polymerization was broad. Additional supporting evidence was obtained by the copolymerizations of St with PEGDMAs [89] and of MMA with trimethylolpropane trimethacrylate (TMPTMA) [90]. Figure 16 shows the SEC-LALLS curve of the MMA-TMPTMA copolymer obtained at 7% conversion along with that of the MMA-EDMA copolymer. The intensity of light scattering at a low elution volume for the MMA-TMPTMA copolymer was enhanced compared with the calculated value (dotted line) assuming a linear polymer. This is in line with the view that bimolecular termination between growing polymer radicals having a loop structure is sterically suppressed to yield a high-molecular-weight polymer. However, no intensified light scattering was observed for the MMA-EDMA copolymer having a linear primary chain ($[\eta]_0/[\eta]_L = 0.98$) [90] as expected, although, even in MMA-EDMA copolymerization, a loop structure formation was observed in a very dilute solution [92]. Also, the results shown in Fig. 15 suggest that it is not always reasonable to employ the $P_{w,0}$ value from the homopolymerization of monovinyl monomer as the primary chain length of a monovinyl-divinyl copolymerization system for the calculation of the theoretical gel point. This kind of suppressed occurrence of intermolecular termination due to the steric hindrance and/or reduced segmental motion that arise from the formation of a loop structure may be related to the occurrence of intramolecular crosslinking and microgelation as will be discussed below.

3.4 Intramolecular Crosslinking

As discussed in Sects. 2.3.3 and 3.3, the intramolecular crosslinking preceded by the intermolecular crosslinking is not a primary factor for greatly delayed gelation, but its significance becomes important with the progress of polymerization. That is, at an early stage of polymerization in which the prepolymer concentration is quite low, the occurrence of intermolecular crosslinking is negligible; in this connection, the possibility of intermolecular crosslinking between growing polymer radicals having pendant double bonds, concurrently occurring along with bimolecular termination, will also be negligible when one considers the higher activation energy of the former ones. In addition, the occurrence of intermolecular crosslinking is delayed more with higher conversion in the polymerization systems where a thermodynamic excluded volume effect becomes more significant such as with dilution. Thus, in the case where the theoretical gel point is quite low, i.e. at a quite early stage of polymerization, the occurrence of intermolecular crosslinking is negligible, at least up to the theoretical gel point, and under these circumstances no contribution of intramolecular crosslinking is expected for the greatly delayed gelation. Inevitably, the prepolymer concentration becomes higher with conversion and thus, the occurrence of intermolecular crosslinking is enhanced, inducing a subsequent intramolecular crosslinking. Concurrently, the intramolecular crosslinking preceded by the physical intermolecular crosslinking such as an entanglement of polymer chains may act as an enhancement of intermolecular crosslinking. The extent of intramolecular crosslinking becomes higher with increasing numbers of primary chains in the prepolymer or an increased degree of branching of the prepolymer. In solution polymerization it depends on monomer concentration and furthermore, the solvent effect would be significant in connection with the conformational change of the prepolymer chains [91].

The occurrence of intramolecular crosslinking leads to the complication of the prepolymer structure sterically influencing the intermolecular crosslinkability of pendant vinyl groups; the steric effect is more remarkable for the vinyl groups located at the inner part of branched prepolymer molecule. Theoretically, it is assumed that all double bonds residing in the polymerization system should have an equal reactivity, and thus the sterically suppressed reactivity of pendant vinyl groups significantly contributes to the greatly delayed gelation as a shielding effect [19] or a steric excluded volume effect. This kind of complication of prepolymer structure induced by intramolecular crosslinking influences not only the intermolecular crosslinking, but also the bimolecular termination between growing polymer radicals [14, 15]. The latter living-type radical formation leads to the enlargement of the primary chain length and furthermore, may play an important role for microgelation.

In addition, the occurrence of intramolecular crosslinking, which does not affect molecular weight, leads to the shrinkage of molecular size in solution; the latter is reflected in the correlation of intrinsic viscosity, r.m.s. radius of gyration, or the second virial coefficient with molecular weight [90–93, 96, 98, 101, 105], and also shown by SEC-LALLS [90, 91, 104] or SEC-MALLS [99, 102]. The formation of crosslinkages between primary chains may accelerate the intermolecular interaction between polymer chains as a sign of decreased interaction of the polymer segment with the solvent and/or unreacted monomer around the crosslinked unit; this may be accompanied by an enhanced occurrence of intramolecular crosslinking. The intensified intramolecular crosslinking induces a local microsyneresis leading to the formation of microgel (Fig. 14). More detailed discussion of microgelation will be described below.

3.5 *Microgelation*

In 1935, Staudinger et al. [106] reported the formation of St-DVB microgel. Thereafter, numerous reports on microgel formation were published in the homopolymerizations of multivinyl compounds and their copolymerizations with monovinyl monomers [6–8, 11, 24]. However, no microgelation was observed in the bulk polymerization of DAP giving a quite low primary chain length as described above and also in the monovinyl-divinyl copolymerizations with only a small amount of divinyl monomer [26]. Thus a detailed study on microgelation, especially up to the actual gel point conversion, is required because microgelation leads not only to delayed gelation, but also to the microheterogeneity of network structures of three-dimensional polymers as closely related to their properties.

First, the radical solution copolymerizations of MMA with EDMA in dioxane were carried out by changing volume ratio monomer : solvent and feed molar ratio MMA : EDMA, although the polymerization conditions were adjusted so as to obtain the prepolymer of long primary chain length and to observe the gel point conversion of more than 20%. This was because no microgelation had been reported previously when the primary chain length was short or the content of pendant vinyl groups useful for crosslinking was small. Thus a critical condition under which microgelation could occur up to a gel-point conversion was sought for by investigating the conversion dependencies of the solution properties of resulting prepolymers by light scattering (LS) and viscosity measurements. Figure 17 shows double logarithmic plots of the r.m.s. radius of gyration versus the weight-average molecular weight of prepolymers as a typical example of the results. In conclusion, the formation of microgel-like molecules was suggested as taking place under the polymerization conditions of higher dilution and increased EDMA contents; for example, microgelation occurred at a dilution of 1/5 and 1 mol% of EDMA [93]. This procedure was extended to the copolymerization of St with p-DVB; clearly, the formation of microgel-like molecules was observed under the polymerization conditions of a 1/5 dilution and 20 mol% of p-DVB. In addition, it was suggested that microgelation depends not only on the amount of crosslinker added and the degree of dilution, but also on the primary chain length and the rigidity of the polymer chain [98].

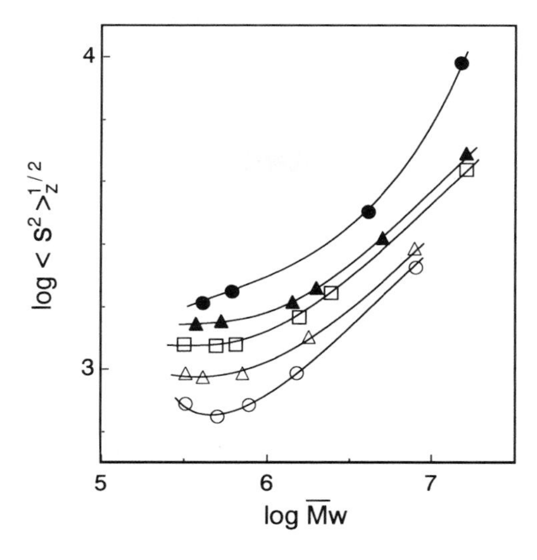

Fig. 17. Double logarithmic plots of $\langle S^2 \rangle_z^{1/2}$ versus $\overline{M}_w$ in MMA-EDMA solution copolymerization under various conditions: volume ratio total monomer/1,4-dioxane, EDMA mol%; (●) 1/2, 0.5; (▲) 1/3, 0.8; (□) 1/4, 1; (△) 1/5, 1.5; (○) 1/7, 1.5 (see Fig. 15 for other polymerization conditions)

Second, the process of microgel formation was followed directly by using our new technique of LS measurement: The polymerization solution adjusted to a required concentration by adding solvent was subjected to LS measurement without isolating the resulting prepolymer because the microgel or microgel-like prepolymer cannot pass through the micropore filter used for optical clarification and also, the possibility of post-polymerization of the isolated prepolymer is quite high [92, 96, 105]. Thus a significant decrease in the radius of gyration and the second virial coefficient was observed under the polymerization conditions where an occurrence of intramolecular crosslinking leading to microgel formation is expected to be enhanced. The microgelation processes in the polymerization of EDMA and its copolymerization with MMA were followed successfully and then, the correlations between microgelation and the primary chain length, the content of pendant vinyl groups, monomer concentration, or solvent were clarified; the microgelation became noticeable with an increase in the primary chain length, an increase in the pendant vinyl content, a decrease in the monomer concentration, and in the good solvent. In the polymerizations of allyl methacrylate and vinyl methacrylate, there are two types of vinyl groups in which the markedly high polymerizability of methacryloyl group compared with allyl or vinyl group will lead to a low crosslinking density and no microgelation was observed up to the gel point.

Furthermore, alkyl methacrylate (RMA) having a long-chain alkyl group such as LMA or SMA was added to a MMA-EDMA copolymerization system in dioxane as a good solvent or in butyl acetate (BuAc) as a poor solvent [107]: The addition of LMA reduced the occurrence of intramolecular crosslinking by the steric hindrance of bulky long-chain alkyl groups leading to the formation of a less densely crosslinked core of microgel. The growth of microgel to macrogel was also reduced, especially in BuAc as a result of the interaction between LMA and BuAc which favors the introduction of a lauryl group on the surface of the microgel leading to a suppressed occurrence of the collision between core parts of microgels having pendant double bonds responsible for intermolecular crosslinking.

3.6 Solvent Effect on Gelation

The first report of the solvent effect on gelation in the free-radical polymerization of multivinyl compounds was published by Walling [3]. Thus the solution copolymerizations of MMA with EDMA were carried out in benzene and ethyl acetate (EtAc), and the gelation was clearly delayed in EtAc as a poor solvent. This was interpreted as follows: The reaction mixture is pictured as consisting of polymer molecules as discrete masses or spheres, highly swollen by, and floating in, the solvent, the rate of diffusion of which is slow compared with the rate of polymer chain growth. As polymerization proceeds, new spheres are continually generated. Crosslinking of polymer molecules generally occurs by a growing sphere becoming attached to one which is already formed. Consequently, the swelling factor of the sphere significantly influences the occurrence of crosslinking, thus leading to the delayed gelation in a poor solvent as a result of the reduction in the swelling factor.

In contrast, our previous discussion of the significance of the thermodynamic excluded volume effect on the intermolecular crosslinking reaction between growing polymer radical and prepolymer, especially at high-molecular-weight, will be extended to the expectation that the gelation should be delayed in a good solvent as opposed to Walling's result. Here it should be noted that in Walling's experiments the gel point was taken as the point when a bubble would no longer rise in the solution and the primary chain length, that of poly(MMA) obtained under the same polymerization conditions although the molecular weight was determined viscometrically. In this connection, Flory pointed out that the values used by Walling for the weight-average degree of polymerization are unfortunately unreliable [108].

So we rechecked the gelation in the same copolymerization system: The gel point was determined according to our method as shown in Fig. 1 and the primary chain length was estimated by extrapolating the conversion dependence of the weight-average molecular weight, measured by light scattering, of prepolymer to zero conversion. Finally, we reached the conclusion that the gelation was clearly delayed in a good solvent as a result of the thermodynamic excluded volume effect [76, 87]. EDMA homopolymerization in the presence of chain transfer reagent [76, 86] and St-DVB copolymerization [88] fell into the same category.

A similar solvent effect was also observed in the gelation behavior of the copolymerization of MMA with nonaethylene glycol dimethacrylate [104]: The delay between the actual gel point and the theoretical one was much larger in dioxane as a good solvent. The primary chain length was relatively shortened for the polymer obtained in BuAc as a poor solvent. The occurrence of intramolecular cyclization, leading to the formation of loop structures, was reduced in a poor solvent as shown by viscometry and SEC-LALLS. The relationships between intrinsic viscosity or root-mean square radius of gyration and weight-average molecular weight are discussed in terms of the solvent effect on intramolecular cyclization and crosslinking.

The solvent effect on the gelation in the copolymerization of MMA with TMPTMA suggested a significant influence of the thermodynamic excluded volume effect on the intermolecular crosslinking reaction. On the other hand, the delay of the actual gel point from the theoretical one was larger in a poor solvent for the copolymerization in a high dilution with a small amount of TMPTMA; this may be ascribed to a relatively reduced number of pendant vinyl groups responsible for crosslinking in the prepolymer as a reflection of the enhanced occurrence of intramolecular crosslinking [91].

3.7 Some Attempts to Control Network Formation

As an extension of our research program concerned with the elucidation of the mechanism of three-dimensional network formation in the radical polymerization of multivinyl compounds, we attempted to control network formation with the intention of collecting the basic data for the molecular design of three-dimensional vinyl-type polymers with high performance and high functionality.

First, the steric effects of long-chain alkyl groups on free-radical vinyl polymerization including a multivinyl compound were discussed in connection with the control of gelation [94]: 1) Three kinds of RMA including butyl methacrylate, LMA, and SMA, were added to a MMA-EDMA copolymerization system in order to examine the steric effect of the long-chain alkyl group of RMA on the gelation, 2) the gelation in the copolymerization of LMA with several PEGDMAs depended on the number of oxyethylene units (n) of PEGDMA as a crosslinker although no gelation was observed up to $n = 4$, and 3) LMA-EDMA copolymerization behavior was examined in detail by changing monomer concentration and the amount of EDMA, the occurrence of intermolecular crosslinking leading to gelation being obviously suppressed by the steric effect of the long-chain alkyl group, providing a new way for the preparation of novel self-crosslinkable prepolymers having pendant vinyl groups.

This was further extended to the copolymerization of RMA with TMPTMA [97, 102], in which trivinyl monomer was employed in place of the divinyl monomer as crosslinker because more than one methacryloyl groups, i.e. uncyclized structure units having two methacryloyl groups or monocyclic structure units having one methacryloyl group, will be introduced by the incorporation of one TMPTMA unit into the polymer chain [90]: The primary chain length increased with an increase in the size of the alkyl group in RMA and concurrently, the deviation of the actual gel point from the theoretical one became greater because of the reduced occurrence of bimolecular termination and intermolecular crosslinking caused by the steric effect of a long-chain alkyl group. The effects of primary chain length, dilution, TMPTMA content, and solvent on the gelation were also investigated.

Second, we attempted to control the network formation by means of the polar effect. Thus the gelation in the terpolymerization among benzyl methacrylate (BzMA), methoxytriicosaethylene glycol methacrylate (MPEGMA-23), and EDMA was examined in terms of the solvent effect because BzMA and EDMA lead to the formation of nonpolar main-chain and MPEGMA-23, polar side-chains, and therefore the difference in the polar nature of main-chain and side-chains induces the conformational change of the prepolymer in various solvents giving a significant influence on the gelation behavior. For example, in methanol the shrinkage of the main-chain and the expansion of side-chains should promote the microgelation; the resulting prepolymers were characterized as microgel-like polymers from the data of radius of gyration, second virial coefficient, and intrinsic viscosity [95]. As another example [101], the terpolymerization among 2-hydroxyethyl methacrylate (HEMA), docosyl methacrylate (DMA), and EDMA was carried out in benzene/methanol mixed solvents because HEMA leads to the formation of polar main-chain and DMA, nonpolar side-chains.

Third, we tried to elucidate the relationship between the rigidity of the main-chain and the gelation behavior; *tert*-butyl methacrylate and d-bornyl methacrylate were chosen and their copolymerizations with dimethacrylates were compared with MMA-systems [99, 100]. Thus the occurrence of intramolecular cyclization accompanied by the loop structure formation, intermolecular crosslinking, and intramolecular crosslinking was suppressed as a result of reduced flexibility of the polymer chain and a steric effect due to the bulky *tert*-butyl and d-bornyl groups. Supporting evidence is obtained from the correlation curve between molecular weight and elution volume and the conversion dependence of MWD curves determined by SEC-MALLS, and also the correlations of both the r.m.s. radius of gyration and second virial coefficient with molecular weight by LS measurements.

4 Concluding Remarks

The mechanistic elucidation of network formation in free-radical crosslinking polymerization and copolymerization of multivinyl compounds – a long-term controversial problem because of the complexity of the reactions involved – has been discussed on the basis of the comparison of the experimental results obtained mainly in our laboratory with F-S theory and assuming ideal network formation.

Thus, in the bulk polymerization of DAP, the actual gel point was found to be 6.9 times higher than the theoretical one. In common multivinyl polymerization systems, the discrepancy was more than 10 times and sometimes, more than 10^2. Moreover, the extent of deviation was greater with increasing the primary chain length, the content of pendant double bonds in the prepolymer, and dilution. In order to interpret the greatly delayed gelation reasonably, the mechanistic discussion of intramolecular cyclization, the reduced reactivity of the prepolymer, the intramolecular crosslinking, and microgelation, research into factors has been done in detail. Finally, we reached the following conclusions: The primary factor is the thermodynamic excluded volume effect on the intermolecular crosslinking reaction between growing polymer radical and prepolymer, especially at high molecular weight. Beyond the theoretical gel point, a secondary factor is the intramolecular crosslinking which becomes progressively important with conversion. The latter leads to the restriction of segmental motion of the prepolymer and, moreover, imposes steric hindrance, i.e. shielding effect or steric excluded volume effect, inducing the significance of the reduced reactivity as a tertiary factor.

Along with the above discussion, some problems involved in the gelation of multivinyl compounds have been discussed extensively. Thus, the validity of F-S theory was first confirmed by copolymerizing MMA with EDMA under the conditions in which the polymer concentration at the theoretical gel point is high enough to reduce the significance of a thermodynamic excluded volume effect and in addition, the intramolecular crosslinking is suppressed. F-S theory is quite simple, but it works effectively for the present mechanistic discussion to interpret the greatly delayed gelation. Then, the intramolecular cyclization, discriminated from intramolecular crosslinking, leading to the formation of loop structure was explored in detail by emphasizing the importance of the suppression of bimolecular termination which leads to the enlargement of primary chain length, although the correlation between gelation and the difference in cyclization modes, and the difference in reactivity between the uncyclized and cyclized radicals for crosslinking were also significant. Next, the intramolecular crosslinking was discussed mainly from the standpoint of the complication of the resulting prepolymer structure. The formation of microgel as an extremely complicated prepolymer was then investigated by LS measurements. The significance of the thermodynamic excluded volume effect proposed by us was reflected in the solvent effect in contrast to Walling's results [3]; gelation was delayed in a good solvent. Finally, our preliminary attempts to control the network formation have been described by means of the steric effect of long-chain alkyl groups, the polar effects such as the difference in the polar nature of main-chain and side-chains, or the flexibility of the main-chain with the intention of collecting the basic data for the molecular design of three-dimensional vinyl-type polymers with high performance and high functionality.

As is evident from the above mechanistic discussion, the problems involved in the network formation of free-radical multivinyl polymerization are quite complicated and many factors are relevant to the network formation processes and their contributions depend on the polymerization conditions. Thus, finding an appropriate model for a comprehensive gelation theory and, moreover, controlling the network structure are very important but they are quite difficult subjects. In the polymerization of multivinyl compounds having different types of vinyl groups and the monovinyl-multivinyl copolymerization between different vinyl groups, the factors contributing to the network formation are much more complicated. Electrostatic repulsion for intramolecular and intermolecular crosslinkings also becomes important in aqueous solution polymerizations of ionic multivinyl compounds [109]. Further studies are still required for a full understanding of network formation of the free-radical multivinyl polymerization.

Acknowledgements. The author wishes to express his sincere appreciation to the late Professor Masayoshi Oiwa for his helpful discussions and continuous encouragement. Our research group members are thanked for their contributions.

5 References

1. Flory PJ (1941) J Am Chem Soc 63: 3083, 3091, 3096
2. Stockmayer WH (1943, 1944) J Chem Phys 11: 45; 12: 125
3. Walling C (1945) J Am Chem Soc 67: 441
4. Gordon M, Roe RJ (1956) J Polym Sci 21: 27, 39, 75
5. Jokl J, Kopecek J, Lim D (1968) J Polym Sci A-1 6: 3041
6. Horie K, Otagawa A, Muraoka M, Mita I (1975) J Polym Sci Polym Chem Ed 13: 445
7. Galina H, Dusek K, Tuzar Z, Bohdanecky M, Sokr J (1980) Eur Polym J 16: 1043
8. Spevacek J, Dusek K (1980) J Polym Sci Polym Phys Ed 18: 2027
9. Dusek K, Spevacek J (1980) Polymer 21: 750
10. Whitney RS, Burchard W (1980) Makromol Chem 181: 869
11. Shah AC, Parsons IW, Haward RN (1980) Polymer 21: 825
12. Hild G, Okasha R (1985) Makromol Chem 186: 389
13. Matsumoto A, Matsuo H, Oiwa M (1987) Makromol Chem Rapid Commun 8: 373
14. Landin DT, Macosko CW (1988) Macromolecules 21: 846
15. Zhu S, Hamielec AE (1989) Macromolecules 22: 3093
16. Zhu S, Tian Y, Hamielec AE (1990) Polymer 31: 154
17. Dotson NA, Diekmann T, Macosko CW, Tirrell M (1992) Macromolecules 25: 4490
18. Matsumoto A (1993) Makromol Chem Macromol Symp 76: 33
19. Storey BT (1965) J Polym Sci A3: 265
20. Malinsky J, Klaban J, Dusek K (1971) J Macromol Sci Chem A5: 1071
21. Soper B, Haward RN, White EFT (1972) J Polym Sci A-1 10: 2545
22. Kwant PW (1979) J Polym Sci Polym Chem Ed 17: 1331
23. Okasha R, Hild G, Rempp P (1979) Eur Polym J 15: 975
24. Leicht R, Fuhrmann J (1981) Polym Bull 4: 141
25. Fink JK (1981) J Polym Sci Polym Chem Ed 18: 195
26. Hild G, Rempp P (1981) Pure Appl Chem 53: 1541
27. Hild G, Okasha R (1985) Makromol Chem 186: 93
28. Antonietti M, Rosenauer C (1991) Macromolecules 24: 3434
29. Simpson W, Holt T, Zetie RJ (1953) J Polym Sci 10: 489
30. Gordon M (1954) J Chem Phys 22: 610
31. Simpson W, Holt T (1955) J Polym Sci 18: 335
32. Oiwa M, Ogata Y (1958) Nippon Kagaku Zasshi 79: 1506
33. Matsumoto A, Yokoyama S, Khono T, Oiwa M (1977) J Polym Sci Polym Phys Ed 15: 127
34. Matsumoto A, Ogasawara Y, Nishikawa S, Aso T, Oiwa M (1989) J Polym Sci Part A Polym Chem 27: 839
35. Dusek K (1982) Developments in polymerization-3. Appl Sci Pub, p 143
36. Dotson NA, Macosko CW, Tirrell M (1992) Synthesis, characterization, and theory of polymeric networks and gels. Plenum, p 319

37. Zhu S, Hamielec AE (1992, 1993) Makromol Chem Macromol Symp 63: 135; 69: 247
38. Oiwa M, Matsumoto A (1974) Progress in polymer science, Japan vol 7. Kodansha, p 107
39. Matsumoto A, Matsuo H, Oiwa M (1988) J Polym Sci Part C Polym Lett 26: 287
40. Matsumoto A, Murata H, Yamane H, Oiwa M, Murata Y (1988) J Polym Sci Polym Chem Ed 26: 1975
41. Matsumoto A, Ishido H, Urushido K, Oiwa M (1982) J Polym Sci Polym Chem Ed 20: 3207
42. Bartlett PD, Altshul R (1945) J Am Chem Soc 67: 812, 816
43. Butler GB (1992) Cyclopolymerization and cyclocopolymerization, Marcel Dekker
44. Carothers WH (1931) Chem Rev 8: 402
45. Laible RC (1958) Chem Rev 58: 807
46. Matsumoto A, Asano K, Oiwa M (1969) Nippon Kagaku Zasshi 90: 290
47. Gordon M, Ross-Murphy SB (1975) Pure Appl Chem 43: 1
48. Boots HMJ, Pandey RB (1984) Polym Bull 11: 415
49. Dusek K (1986) Adv Polym Sci 78: 1
50. Mikos AG, Tsakoudis CG, Peppas NA (1986) Macromolecules 19: 2174
51. Dotson NA, Galvan R, Macosko CW (1988) Macromolecules 21: 2560
52. Tobita H, Hamielec AE (1989) Macromolecules 22: 3098
53. Scranton AB, Peppas NA (1990) J Polym Sci Part A Polym Chem 28: 39
54. Imoto T, Nakajima T (1966) Kogyo Kagaku Zasshi 69: 520
55. Matsumoto A, Inoue I, Oiwa M (1976) J Polym Sci Polym Chem Ed 14: 2383
56. Butler GB, Bunch RL (1949) J Am Chem Soc 71: 3120
57. Butler GB, Ingley FL (1951) J Am Chem Soc 73: 895
58. Butler GB, Angelo RJ (1957) J Am Chem Soc 79: 3128
59. Butler GB (1982) Acc Chem Res 15: 370
60. Butler GB (1986) Encyclopedia of polymer science and engineering, 2nd edn, vol 4. John Wiley, p 577
61. Matsumoto A, Iwanami K, Oiwa M (1980) J Polym Sci Polym Lett Ed 18: 307
62. Matsumoto A, Oiwa M (1969) Kogyo Kagaku Zasshi 72: 2127
63. Matsumoto A, Sasaki H, Oiwa M (1972) Nippon Kagaku Zasshi 1972: 166
64. Matsumoto A, Nakane T, Oiwa M (1983) J Polym Sci Polym Lett Ed 21: 699
65. Matsumoto A, Aso T, Oiwa M (1973) J Polym Sci Polym Chem Ed 11: 2357
66. Matsumoto A, Tachibana S, Oiwa M (1972) Polym Prep Jpn 21: 470
67. Matsumoto A, Ohtsuka S, Hirota Y, Ohara Y, Oiwa M (1978) Polym Prep Jpn 27: 164, 1060
68. Matsumoto A, Nakajima H, Oiwa M (1981) CSJ Prep 43: 1105
69. Oiwa M (1955) Nippon Kagaku Zasshi 76: 684
70. Matsumoto A et al. (unpublished results)
71. Matsumoto A, Morita T, Oiwa M(1990) CSJ Prep 59: 1012
72. Matsumoto A, Nakajima H, Oiwa M (1988) Netsukokasei Jushi (J Thermoset Plast Jpn) 9: 141
73. Morawetz H, Cho JR, Gans PJ (1973) Macromolecules 6: 624
74. Matsumoto A, Ohtsuka S, Ohara Y, Oiwa M (1979) ACS Polym Prep 20: 921
75. Ohara Y, Honda K, Matsumoto A, Oiwa M (1980) Polym Prep Jpn 29: 191
76. Matsumoto A, Oiwa M (1984) 3rd Japan-China Symposium on Radical Polymerization, p 29
77. Erath EH, Robinson M (1963) J Polym Sci C3: 65
78. Bobalek EG, Moore ER, Levy SS, Lee CC (1964) J Appl Polym Sci 8: 625
79. de Boer JH (1936) Trans Faraday Soc 32: 10
80. Matsumoto A, Aoki K, Kukimoto Y, Oiwa M, Ochi M, Shimbo M (1983) J Polym Sci Polym Lett Ed 21: 837
81. Matsumoto A, Aoki K, Oiwa M, Ochi M, Shimbo M (1983) Polym Bull 10: 438
82. Matsumoto A, Kurokawa M, Oiwa M (1988) Netsukokasei Jushi (J Thermoset Plast Jpn) 9: 85
83. Matsumoto A, Kurokawa M, Oiwa M (1989) Eur Polym J 25: 207
84. Matsumoto A, Sumi H, Oiwa M (1980) CSJ Prep 41: 523
85. Sumi H, Matsumoto A, Oiwa M (1981) Polym Prep Jpn 30: 53, 804
86. Matsumoto A, Noguchi A, Oiwa M (1982) Polym Prep Jpn 31: 348, 1109
87. Matsumoto A, Sumi H, Oiwa M (1982) Symp Thermoset Plast 32: 96
88. Matsumoto A, Naito Y, Oiwa M (1984) Polym Prep Jpn 33: 329, 1347
89. Matsumoto A, Yonezawa S, Oiwa M (1988) Eur Polym J 24: 703
90. Matsumoto A, Ando H, Oiwa M (1989) Eur Polym J 25: 385
91. Matsumoto A, Ando H, Oiwa M (1989) Kobunshi Ronbunshu 46: 583
92. Matsumoto A, Takahashi S, Oiwa M (1989) Polym Prep Jpn 38: 261, 1787

93. Matsumoto A, Yamashita Y, Oiwa M (1991) Netsukokasei Jushi (J Thermoset Plast Jpn) 12: 135
94. Matsumoto A, Nishi E, Oiwa M, Ikeda J (1991) Eur Polym J 27: 1417
95. Matsumoto A, Terada S, Nishi E, Oiwa M (1991) Polym Prep Jpn 40: 179
96. Matsumoto A (1992) Rep Asahi Glass Found 61: 187
97. Matsumoto A, Fujise K, Minamino H (1992) Polym Prep Jpn 41: 210
98. Matsumoto A, Yamashita Y, Oiwa M (1993) Netsukokasei Jushi (J Thermoset Plast Jpn) 14: 139
99. Hasei Y, Aota H, Matsumoto A (1993) Polym Prep Jpn 42: 203, 2964
100. Matsumoto A, Hasei Y, Aota H (1994) Netsukokasei Jushi (J Thermoset Plast Jpn) 15: 117
101. Matsumoto A, Yamamoto Y, Terada S, Aota H (1993) Polym Prep Jpn 42: 204
102. Fujise K, Aota H, Matsumoto A, Yoneno H, Ikeda J (1993) Polym Prep Jpn 42: 205, 2958
103. Rabadeux JC, Durand D, Bruneau C (1984) Makromol Chem Rapid Commun 5: 191
104. Matsumoto A, Matsuo H, Ando H, Oiwa M (1989) Eur Polym J 25: 237
105. Matsumoto A, Takahashi S, Oiwa M (1990) ACS Polym Prep 31: 149
106. Staudinger H, Husemann E (1935) Chem Ber 68: 1618
107. Matsumoto A, Morita T, Aota H (1992) Polym Prep Jpn 41: 3062
108. Flory PJ (1953) Principles of polymer chemistry. Cornell University Press, Ithaca NY, p 391
109. Matsumoto A, Kohama Y, Oiwa M (1990) Polymer 31: 2141

Editor: Prof. K. Dušek

Received: 30 November 1994

Imprinting of Synthetic Polymers Using Molecular Templates

J. Steinke, D.C. Sherrington and I.R. Dunkin
Department of Pure & Applied Chemistry, University of Strathclyde, 295, Cathedral Street, Glasgow G1 1XL, UK

The concept of using a molecular template to generate recognition sites for selective separations or reactions within a polymeric network is an exciting, challenging and far-reaching one. This review seeks to explain and define the concept and its components. It traces the early developments by the pioneers in the field and highlights the important advances made, or the key crossroads passed, in reaching the current state-of-the-art. The various types of templates and template binding which have been explored are reviewed and achievements in re-binding or recognition discussed. Finally, the practicalities of preparing polymer networks imprinted in this manner are dealt with and guidance given in what is currently achievable and what limitations still exist.

Advances in Polymer Science, Vol. 123

1 Introduction

This review seeks to gather together and distil the published work involving the synthesis and exploitation of molecularly imprinted synthetic polymers. The latter are polymeric networks synthesised in the presence of a template molecule as pioneered and first reviewed by Wulff [1]. The use of a template species is therefore crucial and, of course, the origin of molecular templating dates from far earlier. Thus following the discovery of the structure of DNA double helix by Watson and Crick [2], and that replication must involve templating, Todd suggested in 1956 [3] that organic template molecules might be used to control the laboratory synthesis of organic molecules in a similar way. In the intervening period the exploitation of this concept has been achieved in a number of areas.

This review will focus on the situation where a template molecule interacts either with a performed polymer or with the constituents of a polymerisable mixture. The interactions exploited have to be reversible and can either be of a well-defined nature, such as in the formation of a covalent bond, or rather ill-defined as in the case of hydrophobic interactions. The preformed polymer undergoes conformational changes in the presence of the template, whereas in the case of the polymerisable mixture an entirely new polymer is formed around the template. In both instances the template is incorporated into the polymer network and subsequently extracted. The template leaves behind spaces or cavities, which to varying extents are complementary to the shape and/or electronic features of the template. The properties of the polymers obtained and their potential applications depend on the characteristics of the template and its mode of interaction with the precursor polymer or polymerisable mixture, but also on the final polymer structure itself.

2 Templating – A Wider Perspective

Before looking in detail at molecular imprinting in synthetic polymers it is perhaps appropriate to mention some of the important and closely related areas where the use of templates has brought scientific advance. The in vivo transcription of the information coded in DNA involves the production of messenger RNA, which itself functions as a template, and eventually stimulates the synthesis of specific proteins within cells. The development of the polymerase chain reaction (PCR) [201] is perhaps the most important example of a natural templating process which has been exploited in numerous in vitro applications [202]. The PCR enables the exponential scaling up of minute quantities of a DNA sequence to yield sufficient material for exploitation in e.g. genetic engineering and diagnostic and forensic applications. DNA polymerase enzymes had been isolated, identified and utilised prior to 1985, but at that time the strategy and methodology for using these enzymes for exponential amplification of DNA samples had not been conceived. This group of enzymes are able to take a single strand of DNA and with a short complementary primer (DNA oligomer) bound at one end, extend the primer to produce the double stranded DNA. Mullis [201] then made the conceptually simple, but far reaching proposal that the new double stranded DNA might be separated or "denatured" by heating, and that combination of the two isolated strands of DNA with appropriate primers should allow *two* double-stranded DNA macromolecules to be produced. Repetition of separation followed by replication in principle might allow exponential growth of DNA. In a few short years this concept was not only reduced to practice, but also the enormous significance of its application was realised. Early problems involving the limited stability of the *Escherichia coli* DNA polymerase were removed with the introduction of the robust Taq DNA polymerase isolated from the thermophilic bacterium *Thermus aquaticus* [203]. The availability of this thermostable enzyme has itself transformed the PCR by allowing the development of simple automated thermal cycling devices for carrying out the amplification reaction in a single tube containing the necessary reagents [204]. Nowadays the methodology is capable of utilising such small initial levels of DNA that great care is required to avoid contamination by stray DNA sequences.

Many of the proteins formed by expression of DNA genes also perform as templates. Enzymes, for example, are in fact "concave" templates (see later) and provide a complementary, enantioselective substrate environment or "template cavity" within which a chemical reaction takes place. Antibodies likewise bind to antigens selectively. Fidelity, selectivity and specificity of templating in biological systems are so far unsurpassed by man-made ones. Nevertheless very recently Mosbach et al. [4] were able to show that an immunoassay based on antibody recognition could equally well be conducted with an imprinted polymer.

There are now many examples in synthetic chemistry where attempts have been made, with varying degrees of success, to emulate and mimic these concepts of templating in biological systems. Topics where templating contributes include: enzyme mimics, modified enzymes, template assembled synthetic proteins (TASP), enzymes in non-aqueous environments, catalytic antibodies, synthetic bilayer membranes, micellar systems, cyclisation reactions, zeolite synthesis, ionophores, ion- and enantioselective receptors, light-harvesting systems, molecular imprinting in polymers, self-assembling molecules, supramolecular assemblies, self-replicating molecules, catalytic metal complexes, template-assisted crystallisation and crystal habit modifiers, and template polymerisation. Even some uses of sophisticated chiral auxiliaries in enantioselective syntheses, and the stereoregulation polymerisations observed with Ziegler-Natta catalysts could be argued to involve templating.

The recent work on template assembled synthetic proteins (TASP) is worthy of further mention [5, 6]. These are synthetic proteins in which short characteristic structural elements (e.g. α-helices, β-sheets) are brought together in a predetermined, often computer simulated, array by attachment to a tailor-made template molecule (Fig. 1). This directs the component chains into a protein-like packing arrangement, e.g. a bundle of four α-helices [7, 8]. The structural elements are amphiphilic secondary-structure-forming peptide blocks, and the template is also an oligopeptide. Cyclic species have great potential in this role, and these and the structural elements can be readily synthesied using solid phase synthesis methodology [9]. Likewise, the expanding area of catalytic antibodies or synzymes [10, 11] is very important in the context of this review, since the attempts to produce molecularly imprinted polymers which perform similar chemical catalyses to catalytic antibodies are slowly producing interesting results (see later).

A central conceptual issue with all templating work is whether the template molecule or macromolecule becomes an integral part of the final structure or is ultimately released, hopefully to be re-used. Lehn demonstrated that Cu(I) can be used as a template to assemble double-stranded helices or "helicates" with up to five Cu(I) centres coordinated by oligomers of 2,2'-bipyridine [12, 13]. Similarly Sauvage's synthesis of catenanes uses Cu(I) as a template to induce two phenanthroline ligands to adopt a tetrahedral conformation around the metal centre. This approach has been extended to achieve the synthesis of a trefoil knot [14, 15] (Fig. 2). In the latter case full demetallation of the product was achieved, i.e. complete removal of the template. In contrast the remarkable syntheses of catenanes and rotaxanes achieved by Stoddart et al. exploit π–π interactions to enhance both cyclisation and interweaving, where in this instance the templating components become ,an integral part of the final product [16]. Supramolecular self-assembly is probably therefore a better description for this process. A recent collection of papers on "topology in molecular chemistry" edited by Sauvage reviews the remarkable progress made in this area [17].

The use of metal ion templates to achieve selective cyclisation reactions is one of the most successful applications of templating [18] and the concept has been extended to other cyclisations. In this context Sanders et al. [19] have described a "negative" template, a tetrapyridyl porphyrin. They have used this template to suppress formation of the otherwise favoured cyclic dimer por-

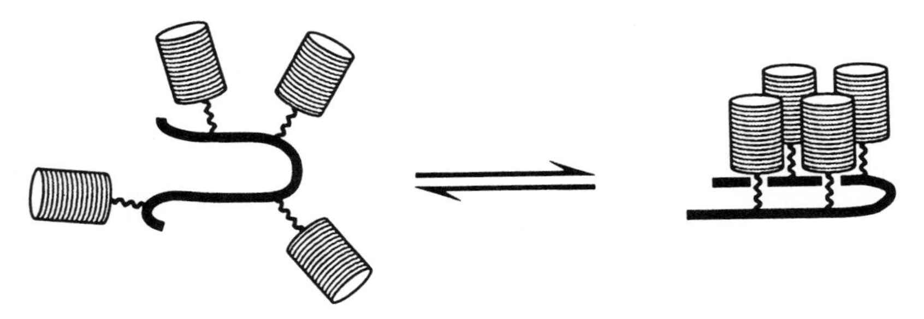

Fig. 1. Template assembled synthetic proteins (TASP); hypothetical folding path. (Adapted from [5])

Fig. 2. Trefoil knots; an enantiomeric pair

phyrin from a linear dimer, and hence favour the cyclic tetrameric porphyrin product.

The question of the definition of a molecular template is therefore an important one, and one which needs consideration in any particular area in which templates might be used.

3 Definition of a Template

Busch offered the following structural definition of a chemical template: "a chemical template organises an assembly of atoms with respect to one or more geometric loci in order to achieve a particular linking of atom" [20]. Noticeably absent from this definition is any mention of whether the template is subsequently removed from the structure achieved, or becomes part of that structure. In the present work the intention is always to remove the template and so a more appropriate definition might be: "a template induces the self-assembly of molecular components to allow reaction or transformation to take place in forming a novel molecular structure; thereafter the template is separated from the novel structure". Ideally the template should be recyclable (Fig. 3). Busch neglects the aspect of recycling in his definition and of course this can be pivotal to the application or properties being sought. For example, spherands can be synthesised in the presence of an ion which acts as a template, but in some cases is it impossible to retrieve the metal-ion from the macrocycle obtained. Instead, it can be found inside the "cage" where it stays trapped [21]. The properties of spherands with and without permanently trapped ions are of course quite different.

In some instances recycling of a template may be feasible but not straightforward. For example, a chiral auxiliary may involve some templating aspects, and by definition is always separated from the final product. However, the form in which the auxiliary appears at the end of the reaction may be very different to its original structure, and elaborate chemistry may be required to re-establish that structure, far more elaborate than worthwhile.

In the context of this review a very apt definition, which focuses more on the process of self-assembly, has been given by Lindsey [22] in an excellent review on self-assembling systems. He distinguishes very clearly and convincingly seven

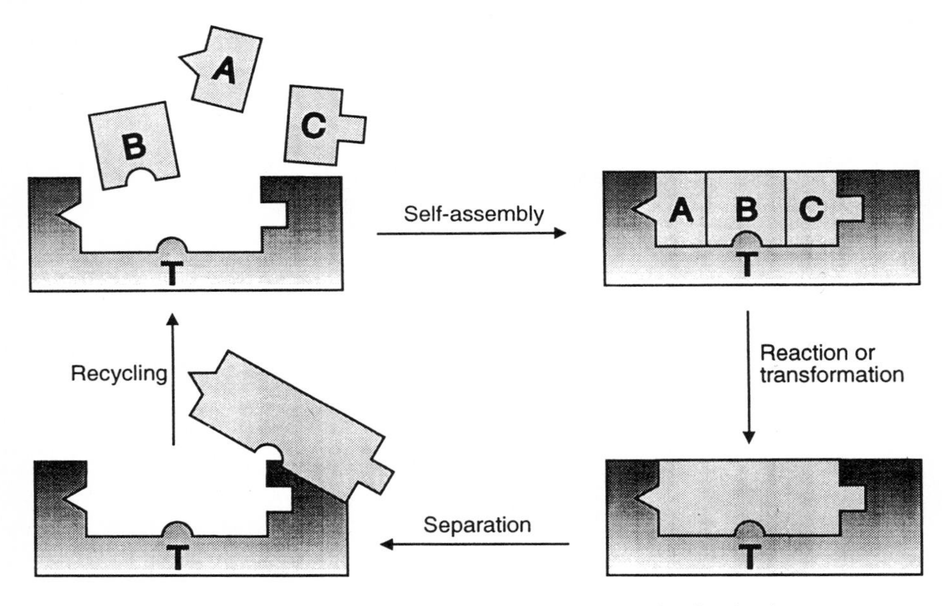

Fig. 3. The concept of a recyclable template, T, inducing self-assembly of molecular components

classes of self-assembly. One of them is termed "directed self-assembly" and is defined as follows: "In this case a temporary scaffold agent, jig or template participates as a structural element in the assembly process, but does not itself appear in the final assembled product". He also touches on the different modes in which the "external element" is able to take part in the process. He writes: "The external element can play a thermodynamic role by stabilising an association of subunits (or by destabilising an undesired aggregate), and/or a kinetic role by channelling the direction of assembly along a particular reaction pathway". Although Lindsey's definition is more elaborate and specific about the self-assembling process, in essence it is the same as Busch's definition of templating apart from the word "recycling". It could be argued that it is this particular aspect of the definition of a template which makes it possible to distinguish between a directed self-assembly process on the one hand and a template process on the other.

It is possible to conceive of situations where the chemical linking of molecular components around a template is not as crucial as the formation of defined, non-covalent interactions during templating. This may be exemplified by the polymerisation of a nematic liquid crystalline crosslinker in the presence of a template, a non-polymerisable cholesteric mesogen [23]. The chiral dopant forces the crosslinker to form a cholesteric phase. After polymerisation of the crosslinker, the polymer still exhibits a helicoidal structure which is stable over a wider temperature range than the initial cholesteric phase. It is not reported in this work whether extraction of the chiral mesogen has been attempted or not, but it is reasonable to assume that this is possible. If the helicoidal structure were to be retained and were still to possess some flexibility, then this would suggest that the helicity is due to a conformational interlocking of the crosslinker mesogens. The template would therefore cause the mesogens to favour a certain conformation, rather than allow a particular linking of atoms and so the *ordering* induced by the template would be the crucial feature.

4 Classification of Templates

As far as we are aware there has been no previous attempts to classify templates systematically. Recently in the context of a reaction governed by strict self-assembly [24], Percec et al. did distinguish between "endo" and "exo" templates. In this review the parameters "shape" and "size" are identified as fundamental, with shapes classified as "convex", "concave" and "linear"; and sizes as "low molecular weight compound or ion", "oligomer" and "polymer". Strictly speaking templates should also be sub-divided as "reactive or catalytic" or "non-reactive or non-catalytic", to distinguish those situations where the presence of the template accelerates the rate of the chemical linking process occurring and those where the reaction rate is unaffected. Since little quantitative data is available and indeed since such *absolute* data might be difficult to obtain in many instances, this sub-classification will not be pursued here.

The sub-divisions of "convex", "concave" and "linear" are more straightforward. The use of a metal-ion as a template in cyclisations [18] is an example of a "convex" template (Fig. 4). An example of a "concave" template is an enzyme. The pocket in which the substrate has to fit to undergo a chemical transformation has a predominantly concave character (Fig. 5). Finally, a "linear" template is typified by the use of a synthetic macromolecule in template polymerisation [25, 26], e.g. the polymerisation of methacrylic acid bound along the backbone of poly(4-vinylpyridine) (Fig. 6). One might argue that synthetic linear polymers are coils, and therefore cannot be classified properly as "linear". However, although they form random coils in solution the part of the polymer where the polymerisation step takes place must be molecularly linear, i.e. there is persistence length, otherwise the growing macroradial would not be able to attack the neighbouring double bond. Hence in this case the template is indeed molecularly "linear".

The sub-division "size" ("low molecular weight compound", "oligomer" and "polymer") is self-explanatory but also not without difficulties. This arises primarily from the uncertainty associated with the term "oligomer" and indeed it is perhaps not appropriate to be too specific here. In general however a "low molecular weight compound" is a molecule without any repetitive structural aspects, an "oligomer" is a molecule where certain structural components are repeated a number of times, e.g. oligopeptide, tetraporphyrin etc. A "polymer" is

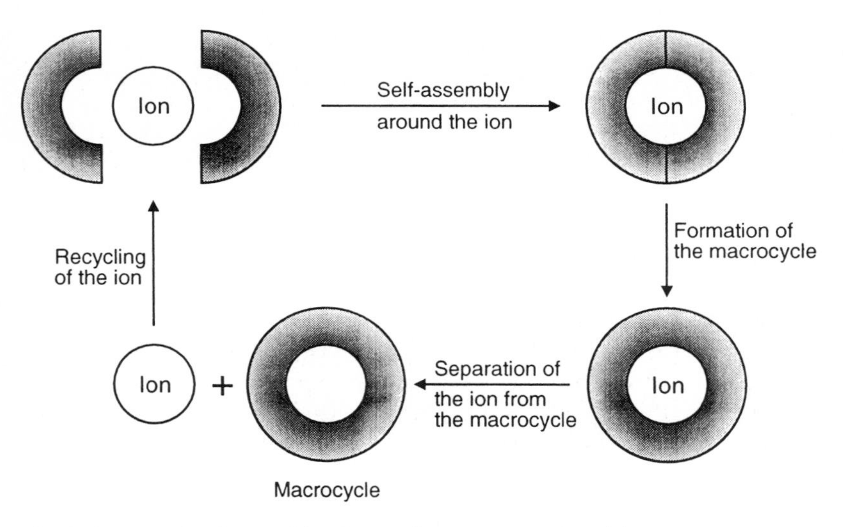

Fig. 4. An example of a metal ion functioning as a convex template

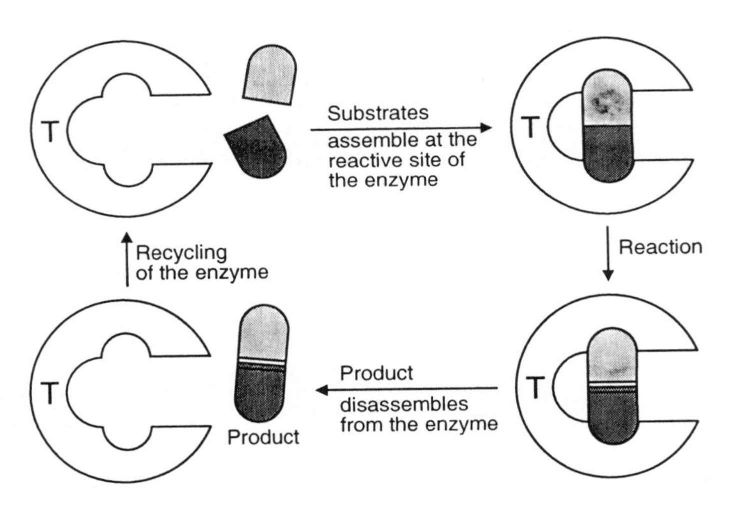

Fig. 5. An example of an enzyme as a concave template, T

a molecule where some structural component is repeated a larger number of times, generating a macromolecule, e.g. polystyrene, a protein etc.

The parameters "shape" and "size" might also be applied to the product arising from the templating process. Pushing all of these classifications too far simply as an exercise in its own right is not valuable. However, in the course of preparing this review, thinking in these terms has proved of value to us in identifying research areas in templating where as yet there has been no exploitation. Some of these will emerge in due course.

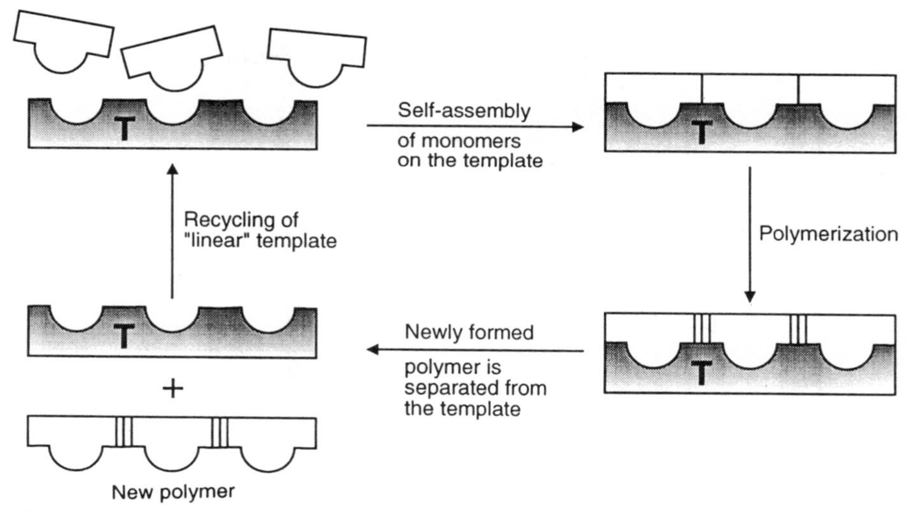

Fig. 6. Application of a "linear" template, T, directing the assembly of a daughter polymer

In terms of imprinting in polymers essentially the only class of templates used to date are "low molecular weight compounds" which are predominantly "convex" in shape. Clearly other opportunities remain available. Fig. 7 summaries the classification scheme for templates.

5 Molecular Imprinting – The Principle

More often than not, molecular imprinting in polymers has involved the synthesis of a crosslinked polymeric network around a template molecule [27,28]. Generally an appropriate template molecule, T, is identified or synthesised in order to complex with suitable polymerisable binding sites in a solvent (Fig. 8). A crosslinking comonomer and a free radical initiator are added, and a radical chain polymerisation initiated thermally.

Polymerisation and network formation is usually accompanied by phase separation of the polymer matrix as it forms. The template molecule is engulfed by polymer chains, within which the binding sites are copolymerised as an integral part of the polymer backbone. The template molecules are then removed to varying levels by a washing or extraction procedure, sometimes preceded by a chemical cleavage reaction depending upon how the template is bound to the polymerising groups. Usually some template molecules remain entrapped in the network. The rigidity of the latter ensures that at least some of

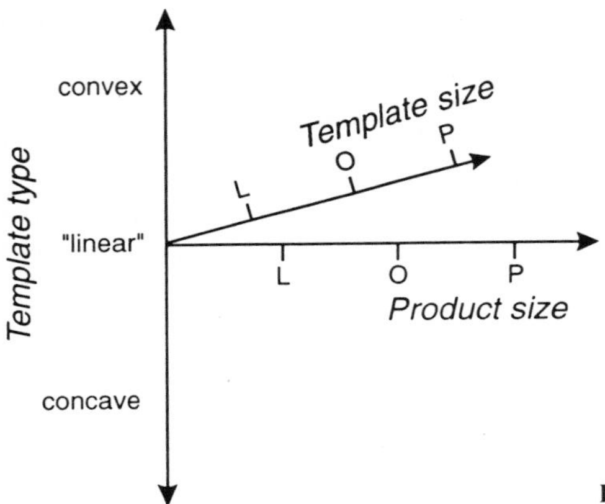

Fig. 7. Summary of templating opportunities

the binding sites, now attached to the polymer network, are in a fixed geometrical arrangement which, together with the surrounding polymer chains, form a cavity complementary in shape and electronic features to the template. Choice of the proper reaction conditions allows the formation of cavities which are selective towards re-binding of the original template molecule. The approach therefore has some features in parallel with the procedures used in preparing catalytic antibodies [10, 11]. Here a binding site or an antibody is synthesised biochemically in response to a challenge from a hapten, which is itself designed as a mimic for the transition state of the reaction which the researcher seeks to catalyse.

The basic concepts in forming a molecularly imprinted polymer are therefore rather simple. Indeed this apparent simplicity has misled some would-be users of this approach who have failed to appreciate that realising this in practice, particularly with any degree of efficiency, has proved enormously difficult. Not the least, most polymer chemists would appreciate that to produce a crosslinked polymeric network sufficiently rigid to retain some "memory" of an imprint molecule, and yet allow ready mass transfer of molecules to and from the memory cavities, is no small undertaking. The early workers in the field have made enormous efforts to bring the technique to a point where materials capable of application and exploitation are now becoming available, and this is as much a tribute to their tenacity as it is to their scientific invention.

6 Binding Sites

A binding site consists of a polymerisable group, generally a vinyl-type monomer, with a second functional group attached, capable of interacting with the template molecule. After polymerisation the vinylic residue is part of the

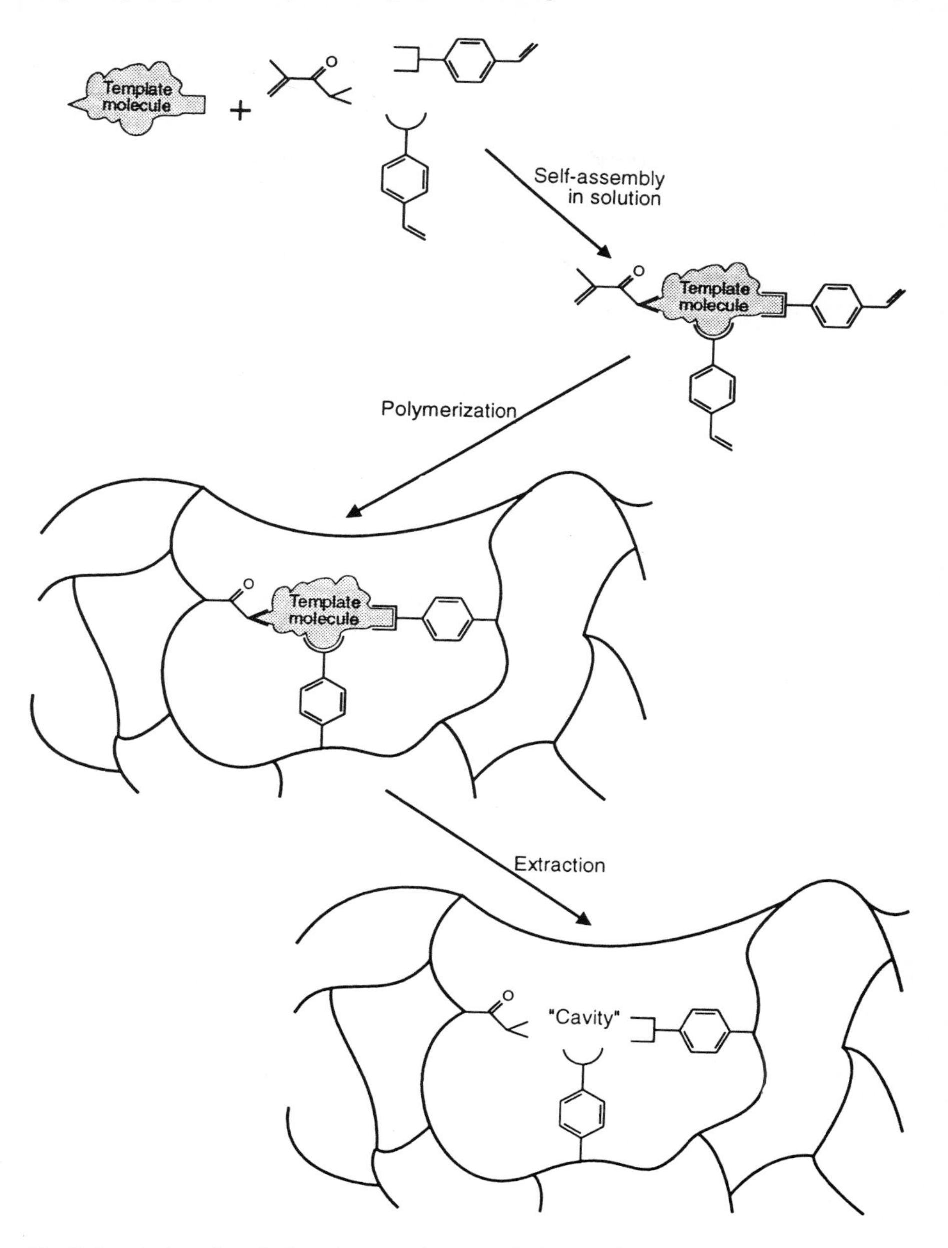

Fig. 8. Imprinting of synthetic polymers using a moelcular template

backbone of the network, and ideally the other functional group exists on the surface of the cavity left by the template, readily available for re-binding of the template or other molecules challenged to the cavity. The best results have been obtained using templates attached to more than one binding site, and interactions between the binding sites and template having some directionality to them. Interactions which have been exploited are: a) covalent bonding; b) π–π interactions; c) H-bonding; d) hydrophobic/van der Waal's interactions; e) (transition) metal/ligand binding; f) crown ether-type interactions with ions; and g) ionic bonding. These are shown in Fig. 9.

The various types of interaction of course imply different levels of specificity. Covalent bonding tends to be very limited, i.e. it is specific for particular functional groups, whereas ionic bonding and hydrophobic interactions are applicable to broad groups of compounds and are therefore less specific. Covalent bonding is generally more directional, whereas ionic and hydrophobic interactions are much less so. All this tends to suggest that an approach based on covalent bonding should yield the optimum imprinted polymers. However, the other vital factor in this is the re-binding of the template or template-like molecules. Often this needs to be rapid (e.g. in chromatographic separations) and covalent bond formation often offers the worst possible prospect from this point of view, whereas the interchange of ionic and hydrophobic interactions can be extremely rapid.

In many respects the exploitation of π–π interactions (often reasonably directional) and H-bonding (often very directional) offer a good compromise. In the case of H-bonding there remains perhaps great scope for further ex-

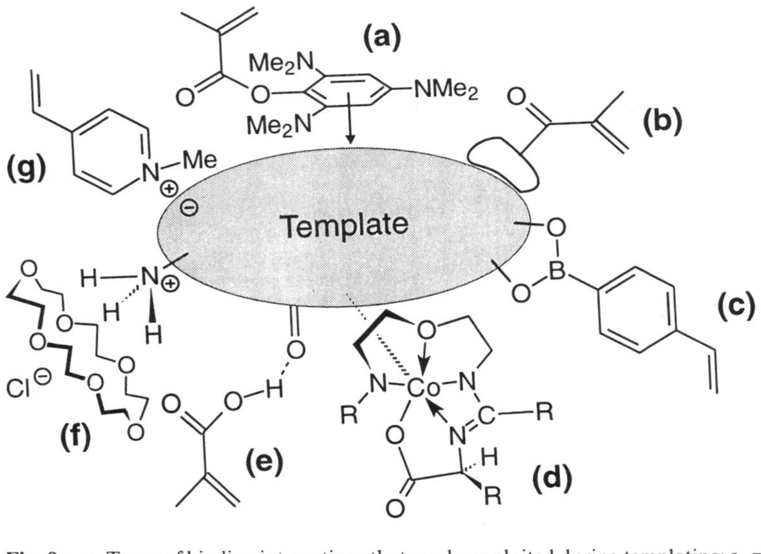

Fig. 9a–g. Types of binding interactions that can be exploited during templating: **a** π–π interaction; **b** hydrophobic or van der Waals interaction; **c** covalent bonds; **d** (transition) metal–ligand binding; **e** hydrogen bonding; **f** "crown ether"–ion interaction; **g** ionic interaction

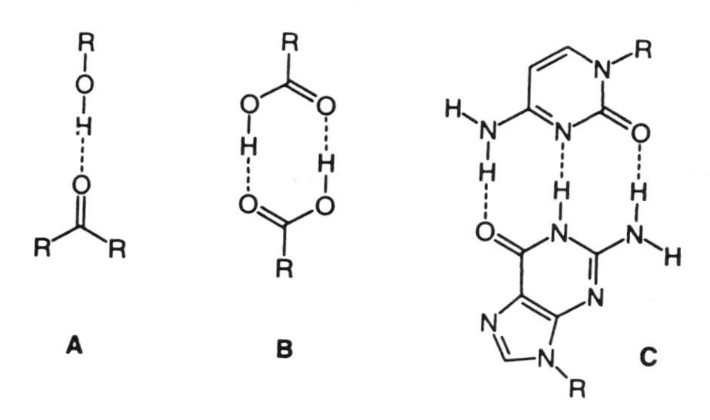

Fig. 10a–c. Hydrogen bonding possibilities for binding templates: **A** one hydrogen, two centres; **B** two hydrogens, four centres; **C** three hydrogens, six centres

ploitation, particularly via multiple interactions. The situation which has been used most to date is the "one hydrogen two centres" hydrogen bond (Fig. 10). In this case the hydrogen atom tunnels between two positions, each close to electronegative atom. A "two hydrogens four centres" bonding mode is realised, e.g., in the dimeric structure of carboxylic acids; "a three hydrogens six centres" bonding pattern is found in nucleic acid base pairs, or in the complexation of primary ammonium ions to crown ethers, and the more unusual binding mode, a "two hydrogens three centres" interaction is known between a carbonyl group and two alcohol functionalities. To date the more complex hydrogen bonding patterns, i.e. beyond "two hydrogens, four centres", have not been exploited in imprinting polymers, nor indeed has the highly directional interaction between crown ether-type molecules and, for example, primary ammonium ions. Preliminary work aimed at the latter have been carried out [29].

In developing binding sites it is important to remember that the interaction with the template must be sufficiently strong *in solution* for it to be exploitable. As we shall see later the formation of the polymer matrix must be performed in the presence of a suitable pore-forming liquid (a porogen), and the solvation properties of the polymerisation medium will be a composite of those of its constituents, primarily the comonomers and the porogen. These components must also not interfere with the template/binding site interaction being exploited, e.g. if H-bonding is to be a key feature, solutions must be dry, and use of aliphatic alcohols would not be appropriate.

7 Polymer Networks

A polymer network or infinite network arises when all the polymer chains in a system are linked to each other or are crosslinked. The network as a consequence is not molecularly soluble since even a thermodynamically "good" solvent, despite solvating each individual polymer chain very effectively, cannot induce the chains to migrate apart since they are covalently linked. A "good" solvent will, however, swell the network until it reaches its elastic limit in that solvent. In molecular imprinting inorganic, synthetic organic and biopolymers, have all been exploited but the largest body of work have involved synthetic organic polymers. In principle these might be polycondensation or addition (chain reaction) polymers (from vinyl and related monomers and cyclic monomers), and again in practice most work has involved crosslinked vinyl-type polymers. In general these are random networks, but scope exists for imprinting in polymers where there is intrinsic order in the system. The remarkable enantioselective separations achieved on modified cellulose chromatographic stationary phases are undoubtedly due in part to the occurrence of highly regular crystalline regions in the natural polymer.

In forming crosslinked vinyl polymer networks two categories of material can result: i) gel-type and ii) macroporous networks. The former are generally rather lightly crosslinked (typically 0.5–10% of backbones residues connected) and these are microporous in the dry state. They will swell on contact with a good solvent to generate a substantial solvent-filled volume within the network. The distance between crosslinks is however rather large and consequently on contact with solvents these networks are highly flexible and mechanically soft and rather fragile. These therefore are totally unsuited for molecular imprinting. The rigidity can be enhanced simply by increasing the crosslink ratio and in a vinyl polymer this is readily achieved by increasing the proportion of divinyl crosslinking monomer in the comonomer mixture. However, as the rigidity rises, the degree of swelling in a good solvent falls rapidly with the increasing crosslink ratio, and mass transfer in such networks can become totally impaired. Again such network;s are not suitable for molecular imprinting.

Macroporous (or porous) networks are generated by polymerising a vinyl and divinyl comonomer mixture with a porogenic species present at the outset. This is usually an organic solvent, but can be a mixture of a solvent and a linear polymer. Two classes of porogen can be distinguished, the first are good solvents for the comonomer mixture but are precipitants for the incipient polymer. The second are good solvents for both the comonomers and the incipient polymer. Precipitant porogens tend to encourage formation of large pores (up to $\sim 10\,000$ Å) and yield rather low surface area (internal) materials (~ 10–$50\ m^2\,g^{-1}$) (Fig. 11a). Solvating porogens tend to favour formation of small pores (down to ~ 30 Å) and yield materials with high surface areas (up to $\sim 1000\ m^2\,g^{-1}$) (Fig. 11b). For the latter species it is also necessary to have high levels of crosslinker to produce mechanically stable materials.

Macroporous networks are required for molecular imprinting because the pore system allows relatively unimpeded access to the interior of the network by solvents, templates and template-like molecules. In addition, of course, the polymer network itself must be highly crosslinked in order to retain a memory or an imprint of the original template. Generally networks formed with solvating porogens yielding a dense system of relative small pores and accompanying

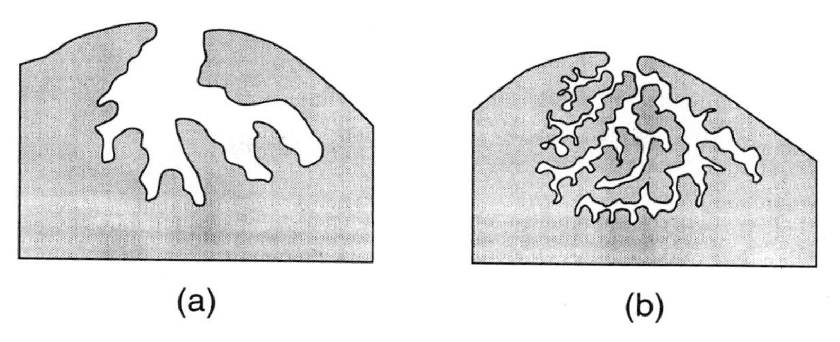

Fig. 11a, b. Macroporous polymer networks with: **a** rather low, **b** rather high surface areas

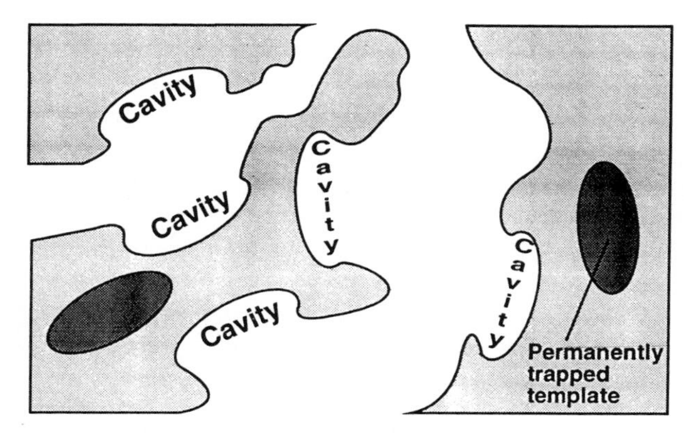

Fig. 12. Schematic representation of an imprinted polymer network, showing accessible cavities in the surface of pores, and inaccessible sites deep within the polymer structure

high surface area tend to provide the best results in imprinting. A schematic representation of the structure and location of imprinted cavities is shown in Fig. 12. In some cases the template is incorporated into the polymer matrix well removed from where pores are formed and such templates remained locked within the network.

It must be stressed that this model of macroporous networks is a rather simplistic one and far more details are available in [30].

8 The First "Crossroad" in Molecular Imprinting

Perhaps not surprisingly the development of molecular imprinting in polymers has been stimulated by the development of theories for antibody formation and the mechanism of enzymatic reactions. As early as 1932, Mudd [31] postulated not only that "...antibody synthesis occurs from amino acids or peptides at the antigen surface...", but he also reflected on the antibody specificity which "...is due to the stereochemical correspondence with the antigen". It took eight years until Pauling [32], replaced speculation with experimental evidence and also proposed a detailed mechanism for antibody formation, on which he elaborated in later years with particular emphasis on the hapten-antibody interactions [33]. It was also at this time that first of the fruitful interdisciplinary adaptations took place. Inspired by Pauling's theory in 1949, Dickey [34] prepared the first synthetic antibody analogue. By polymerisation of an acidified silicate solution in the presence of the template (hapten) methyl orange, he obtained a material which, after drying and elution of the dye with methanol, adsorbed methyl orange significantly better than ethyl orange compared to a blank silica gel. Dickey continued to study this system and found that preference for a certain molecule could be tailored through the choice of the template [35]. The highest specific adsorption was always obtained for the template and the smaller the degree of structural relationship between template and any other chosen molecule became, the less strongly these molecules were adsorbed. He also showed that low pH, low electrolyte concentration and low temperatures favour the development of specific adsorption properties. A wide range of other dyes were used as templates, but their adsorption properties were not studied to any great extent. Compounds not adsorbed on silica under the given conditions did not yield specific adsorbents, and weakly adsorbed templates only introduced a small effect. He further concluded that those template molecules which cannot be removed from the silica without destroying the matrix have a different mode of interaction than those involved in the specific adsorption phenomena. Furthermore, entrapped templates such as methyl orange have strongly modified acid-base properties and are not affected by UV radiation.

In the following years Bernhard [36], Erlenmeyer and Bartels [37], Haldeman and Emmett [38] and Curti et al. [39] repeated Dickey's experiments and confirmed his initial observations. Additional investigations were also carried out to elucidate the underlying mechanism of the imprinting process. Anionic functional groups on the template were found not to influence the selectivity [36], and more evidence was found through nitrogen adsorption measurement to support the imprint theory [38, 40] and to reject the association mechanism hypothesis. These mechanistic findings were fiercely questioned by Morrison et al. [40], but equally strongly rebuffed by Bartels and Prijs [41], based on their own studies of X-ray powder diffraction of the template, an imprinted silica and a blank silica gel [42]. The imprinted silica with the template still in place showed a genuine diffraction pattern when compared with the patterns recorded for the template and the blank.

Enantiomers were found to be partially separated when the silica was imprinted with one enantiomer. This was first shown by Curti and Colombo [43] in 1952. Camphorsulfonic acid was obtained with 30% ee and mandelic acid with 10% ee [44]. Other researchers followed suit. They introduced other chiral [44–46] and achiral [47–51] templates and studied the selectivity of the imprinted polymer by varying the structural features of the template molecule and relating them to the difference in specific adsorption by crossover tests [47,48]. Biologically active molecules seemed to offer particular attractions and were therefore studied extensively [44–51]. All investigations confirmed previous finding. The template is always adsorbed most strongly, and the selectivity towards other molecules decreased, with an increase in structural differences. [47,48]. In addition it was found that molecules which only differ in one methyl group can be distinguished [48] and that there seemed to be a lower limit for a suitable template size in these systems since, e.g., dimethylaniline and diethylaniline exhibited the same adsorption power [37]. The most recent trend in imprinted silicas uses aluminium ions as dopants to introduce more specific interactions with the template. The silica gel is only surface-modified in the presence of e.g. phenanthrene by treatment with diazomethane. After removal of the template, the silica shows a stronger retention for all polycyclic aromatic compounds, but no particular affinity towards the template [52].

In 1972, twenty-three years after Dickeys's first experiments on imprinting in inorganic polymers, Takagishi and Klotz prepared imprints in synthetic organic polymers [53]. Poly(ethylenimine) was thiolated with thiobutyrolactone and the thiol groups crosslinked in the presence and absence of a dye. Yet again methyl orange was employed as the template. The imprinted polymer exhibited more binding sites and stronger binding for the template than did a blank polymer. Takagishi et al. [54] continued to optimise the behaviour of the polymers and also studied their thermodynamic characteristics in detail. This was not undertaken until ten years later in 1982 when *N*-vinylpyrrolidone was crosslinked with methylenebisacrylamide (BIS) in the presence of methyl orange and its homologues. A template effect was observed but not for the polymer imprinted with the butyl analogue, although methyl orange showed a lower affinity to the butyl orange imprinted polymer than to a blank [54]. With increasing crosslinking, polymers revealed an increasing affinity towards the template [54–56]. Template recognition improved also at lower temperatures which, together with other thermodynamic studies, suggests that the recognition process on these imprinted polymers is due to an entropic effect [56]. In the following years studies were extended to anthraquinone and naphthoquinone sulfonates as templates [57–59]. Surprisingly 1,2-naphthoquinone-4-sulfonic acid, probably due to its size, did not produce a template effect. [58, 59]

There is only one example which makes use of a biopolymer, starch, imprinted with a dye [60]. Methylene Blue could be adsorbed on starch which was crosslinked with cyanuric chloride to almost 100% with respect to the concentration of theoretically available binding sites. The blank adsorbed a mere 17% under identical conditions. It is estimated that 7–8 glucose units form a single cavity for the template, somewhat reminiscent of α- or β-cyclodextrins, but shape selectivity has not been investigated. It seems remarkable but these investigations seem to have been initiated without any awareness of the earlier work on imprinting.

9 The Second "Crossroad" in Molecular Imprinting

It was at the time when the two possible mechanisms of substrate binding by an enzyme, "lock and key" and "induced-fit" [61], were subject to much speculation and postulation, that a second beneficial adaptation of an evolving concept from the biological sciences was advanced. This time it was taken from the theory of mechanistic aspects of enzyme action. It was believed that a substrate embedded itself into a complementary pocket in the enzyme. Functional groups which are located in as defined spatial relationship to each other in this pocket then catalysed the reaction. Apparently this scenario sparked off the idea of designing synthetic polymers with functional groups which would posses a defined three-dimensional relationship to each other. In 1973, Wulff pioneered an ingenious design strategy, very much as in an enzyme, for preparing so-called "enzyme-analogue built polymers" [1, 27, 62, 63]. This really initiated the work on molecular imprinting in polymers. A suitable template and properly chosen polymerisable binding sites are copolymerised by a free radical process in the presence of a large amount of crosslinker and solvent. Due to the rigidity of the polymer network obtained, the binding sites are fixed in space. The template can be removed from the polymer when a reversible binding interaction has been chosen, and inside the polymer remain cavities which are complementary to the template, and in which functional groups are held in defined spatial positions. Note that in contrast to the work of Dickey, Takagishi and Klotz, Wulff's aim was to achieve *well-defined interactions* between the template and specifically designed binding sites and not rely on rather less well-defined interactions with the polymer network.

From here research proceeded mainly towards detailed investigations of the parameters which influence the imprinting process complemented by scrupulous studies. The early success of Wulff was slowly improved upon with selectivity properties gradually being optimised. In the years to come the field started to grow and began to branch out in several distinct directions including: the use of covalently bound templates; the use of non-covalently bound templates; imprinting on polymer surfaces; specific adsorbents via ion templates; imprinting using mainly hydrophobic interactions; and metal-ion complexes as binding sites.

9.1 Studies Using Covalently Bound Templates

Covalently bound templates have been extensively studied from 1972 [27, 28] up to the present day by Wulff et al. [62–64] who were joined by Damen and Neckers (1980) [65,66], by Sarhan et al. (1982) [67–69] and by Shea and Thompson (1978) [70]. A variety of parameters have been investigated and their influence on the selectivity of binding and the application of these polymers as chiral stationary phases have been quantified. The parameters investigated are the degree of crosslinking [65, 66, 71–79] the structure of the crosslinker [71, 72,

74–79], additional comonomers [68, 74, 75, 80], the polymerisation temperature [71–73], a number of binding sites for different functional groups [27, 78, 79, 81–86], and a variety of template molecules. These include monosaccharides [63, 64, 72, 75], aromatic ketones [83, 84], amino acids [81, 82, 89], and hydroxy acids [27, 68, 69, 78, 79]. Other parameters investigated are the influence of pH [69, 71, 72, 75, 87], temperature [71, 72, 87], solvent [72] and the number of binding sites [63, 76, 80, 83, 84, 86, 88]. In chromatographic applications separation factors of around 5 and higher, accompanied by resolution factors of 1.5 or slightly higher under optimised conditions [71, 72, 75] have been obtained.

9.2 Studies Using Non-Covalent Bound Templates

Polymers imprinted with non-covalent interaction have been addressed, particularly by Mosbach et al. [90–94] (from 1981 onwards) and later also by Sellergren et al. [95–98] and Andersson [99]. Very similarly to the investigations on imprinted polymers prepared with covalent interactions, almost the same parameters as in the list given above have been investigated. Careful attention has been focused on the effect on selectivity of the type [81, 91, 94, 100–103] and number of binding sites [90, 96, 102–104], complex formation in solution [90, 96], different crosslinkers [99, 102], reaction conditions [96, 98, 105, 106], structural variations of the templates [95–97, 103, 104, 107, 108], and the effect of polymer morphology on selectivity [106]. Finally, the elucidation of the recognition mechanism has been the centre of some activity [90, 95–98, 102–194, 108–110]. A large number of amino acid enantiomers have been separated on chiral stationary phases based on these polymers. Separation factors of 10 and even higher have been achieved together with base line separations of the racemates [93, 98, 102–104, 108]. Recently β-adrenergic blockers [94] were efficiently separated on a chiral stationary phase using non-covalent interactions. Very recently it has even been possible to carry out drug assays on this type of polymer imprinted by theophylline and diazepam, which showed selectivities comparable to those obtained with the enzyme-multiplied immunoassay technique [4]. Imprinting with nucleotide bases also revealed remarkable selectivities when 9-ethyl adenine was used as template molecule. The retention volume for adenine was half the value found for the template molecule as shown by Sellergren et al. [111]. Benzylamine, although being a much stronger base than 9-ethyl adenine, was hardly retarded at all. These results support once again the idea that the recognition process involves cavities with defined binding sites and cannot be described as a simple ion exchange process [110]. It further stresses the potential of imprinted polymers for useful applications such as immunoassays (see above) due to their remarkable selectivities. Finally, it highlights a parallel development with that involving chiral template molecules in the use of template molecules characterised by a predominantly planar structure. Further work may confirm that imprinting with "two-dimensional" template molecules using non-covalent interaction is more demanding than using "three-dimensional" ones. A recent publication by Sherrington et al. [112] also supports this possibility.

9.3 Imprinting on Polymer Surfaces

A third theme was introduced by Sagiv in 1979 [113]. Instead of creating a spatially imprinted three-dimensional polymer network, surfaces were imprinted [113–117], usually by modification of silica layers. Using this approach it has proved possible to prepare silica layers with distance-selective functional groups [114–118]. Chromatographic studies have shown that variation in the length of template of just one methylene group is detectable [117]. Crucial for the successful application of these polymers as chromatographic stationary phases is the right choice of binding interactions [119, 120]. Schiff's bases seem best to fulfil the criteria for rapid and reversible bond formation, and hence are to be found in the most selective sorbents [114, 115, 117]. Polymers, where the same template as that used on silica is attached to styryl-based binding sites and incorporated during polymerisation, showed a lower selectivity towards the template, and the kinetics are affected unfavourably compared to the silica counterparts as well [119, 120]. Polymerisable mono- and diacetals of a variety of aromatic ketones were used in the same manner. Imprinted polymers showed distance accuracy towards the template when evaluated in a batch procedure [84]. More detailed studies, evaluating the control of mircoenvironment shape [83, 121] and modes of binding associated with it, followed [83]. Other types of covalent bonds have been employed as well [119, 120], but no examples of non-covalent interactions are known. This type of imprinted polymer has only been studied on a few occasions. So far they seem to have only academic value in contrast to their imprinted chiral counterparts where there is a clear opportunity for technological exploitation. The recent synthesis of a silica gel onto which triple-helix bundle proteins are placed in a defined manner may soon change this [116].

9.4 Specific Absorbents Exploiting Ion Templates

Two years after Dickey first published, in 1951 the first imprinted polymers based on an ion template were prepared by Kljatschenko [122]. A nickel 1,2-dimethylglyoxime complex was incorporated during the condensation step in the preparation of a phenolic resin. The selectivity towards the ion template was very low. This was also observed by Stanberg et al [123] in 1958. They studied the reaction in more detail and found that formaldehyde, the crosslinker, was detrimental to the metal complex during polymerisation and also caused large amounts of ligand to leach off when the resin was demetallated afterwards. Almost twenty years had passed by before, in 1976, Nishide et al. [124, 125] could report more successful results based on the same general idea. Poly(4-

vinylpyridine), which forms complexes with metal ions in solution, was crosslinked with 1,4-dibromobutane in the presence of the metal ion template. The polymer was washed after polymerisation to free it from the metal ion and this was followed by sorption capacity and competitive metal binding studies. The template is the most strongly absorbed ion and in this first example particularly high selectivities were obtained for the metal pairs Cu^{2+}/Fe^{3+} and Co^{2+}/Zn^{2+} [124, 125]. This system was optimised. A wider range of ion templates and the effect of the amount of crosslinker used were also studied. Higher degrees of crosslinking improved polymer specificity [126, 127].

Most importantly for a successful imprinting by an ion template is the formation of a well-defined polymerisable metal complex in solution prior to polymerisation. Hence metal specific ligands are of particular interest and attempts to improve selectivity have been focused on them. Vinylimidazoles [127], diesters of vinylphosphonic acid [128, 129], 4-vinyl-2,2′-bipyridines [130, 131] and methacrylic acid with vinylpyridine [132] are the most intensely studied ligands [133].

After selective imprinted polymers for a range of metals became available, slow sorption kinetics proved to be of considerable concern and indeed have remained so. Both sorption and desorption are presently too slow for industrial application. To try to overcome this a modified 2-step imprinting procedure has been developed to prepare porous polymer beads where the metal ligands occupy only the bead surfaces [134]. In another attempt the original imprinting strategy by Nishide has been used to synthesise the polymer on a silica gel [135]. In both cases sorption kinetics have improved significantly.

Although they can be considered as mere variations on the theme, three very different and exciting approaches have appeared recently. Choi used chitin derivatives as complex forming agents for Cu^{2+} and UO_2^{2+} [136]. The biopolymer was crosslinked with glutaraldehyde. Selectivity for both ions improved. Polymerisable ionophores were used for the first time by Mosbach et al. to synthesise sorbents selective for Ca^{2+} [137]. The imprinted polymer showed a six-fold increase in selectivity for the ion template. Based on their "molecular knot" strategy, Sauvage et al. synthesised polymerisable 2,9-diphenyl-1,10-phenanthroline ligands which form a pseudotetrahedral metal complex with metal ions such as Cu^{1+}, Co^{2+}, Zn^{2+}, Ag^{1+} [138]. Pyrroles employed as substituents on these complex-forming phenanthroline ligands have been polymerised electrochemically. After polymerisation the polymers are electroactive, they re-complex metal ions and show a template effect. These final examples certainly provide a promising outlook for future developments in ion imprinted polymers.

9.5 Metal Ion Complexes as Binding Sites

As pointed out earlier, ionic, covalent and hydrogen-bonding interactions are particularly useful for introducing functional groups in a defined spatial arrangement. Surprisingly nobody has yet exploited crown ethers as possible binding sites nor tried to use more complex hydrogen bonding receptors [29]. Metal complexes which can interact with a template via one or more of their vacant coordination sites represent an example of a binding site. This is despite the fact that chiral metal complexes in particular would add a new dimension to the possible interactions within the template concept. The first of the few attempts was carried out by Belokon et al. [139]. The crosslinked acrylamide-based polymeric matrix exhibited a stabilising effect on a copolymerised, chiral, octahedral salicylidene Co^{3+}-complex. Deuterium exchange studies led to the conclusion that the matrix stabilised the initial conformation of the complex. The highest enantioselectivity of all imprinted polymers so far reported is that by Fuji et al [140, 141]. A specific coordination site for chiral amino acids, a Schiff's base Co^{3+} complex, was copolymerised with styrene and divinylbenzene. The amino acid bound to the metal complex was removed. Equilibration of an amino acid racemate in the presence of the imprinted polymer showed that the enantiomer used for the imprint was incorporated preferentially with an optical purity of up to 99.5%. Polymer blanks prepared in the absence of an amino acid resolved racemates with $\sim$ 50% ee. This is certainly attibutable to the chiral complex itself, and also highlights the crucial steric effect of the chiral cavity in this example. Surprisingly enough no further papers have been presented to date, although evaluation of these polymers as chiral stationary phases had been envisaged [141]. Nevertheless a chromatographic phase designed with metal complex binding sites has been synthesised very recently by Dhal and Arnold [142]. An achiral bisimidazole functioned as the template which was bound to a pentadentate Cu^{2+} complex. Copolymerisation with ethylene glycol dimethacrylate in the usual manner, followed by removal of template and metal ion, yielded a selective polymer network. Copper ions were reintroduced to the polymer and competitive binding studies between the template and a slightly shorter bisimidazole revealed the preference for the template. A column packed with this polymer separated the isomers, which were inseparable on a conventional reversed-phase material. This confirms the general validity of this new approach which by selecting the metal ion and/or complex offers a wide variety of possibilities to fine tune binding site/template interaction. It also suggests applicability for metal ion catalysed reaction mimicking and approaching enzyme-like behaviour.

9.6 Imprinting Using Mainly Hydrophobic Interactions

Hydrophobic and/or van der Waals interaction has not been well studied. Only as late as 1988 were Yamamura et al. [143] venturesome enough to choose octadecylsilyl monolayers as a polymeric material for hydrophobic interactions. Templates such as adamantane, cholesterol, vitamin K_1, pyrene derivatives, and chlorophyll, amongst others, formed ester bonds to a long alkyl- or oligooxyethylene chain and these were present when octadecylsilylchloride was covalently bound to a SnO_2 electrode surface. The electrode surface was freed from the template and potentiometric measurements followed. The electrochemical signal was almost proportional to the amount of template present in solution. The selectivity of the monolayer towards the template in the presence of hydrophobic competitors revealed that the recognition process reflects the shape and size of the template employed. Very recently Bystrom et al. [144] have employed steroid templates with hydroxy functionalities in either the 3- or 17-position with divinylbenzene monomer as the source of the network. The interactions here again are largely hydrophobic ones, and selective imprints have been achieved. More details are given in the next section because this work is essentially beyond the third "crossroad" as we shall see.

10 The Third "Crossroad" in Molecular Imprinting

When Wulff first introduced his idea of "enzyme-analogue built polymers" [1, 27, 62, 63] he was in fact introducing two new concepts to molecular imprinting. The first was the idea of designing well-defined interactions into the imprinted cavity, mimicking enzymes, and the progress in achieving this has been described. Implicit in his approach, however, was the second concept of producing selective cavities by molecular imprinting which could then go on to function as selective reagents, or better, selective catalysts, just like enzymes. The development of imprinted cavities *with selective reactivity* is therefore the third "crossroad" in this area. This particular step, at least in terms of *catalytic* cavities, is a very difficult and demanding one. Selective binding and *separation* can be achieved by equilibration, and with chromatographic processes, by *multiple* interaction with the selective binding sites. However, selective *catalysts* essentially exploits kinetic differences and, in general must arise from *single* selective interactions at an imprinted site. In other words to achieve selective catalysis the differentiation at the imprinted site must be very high. Because the site will be visited only once by reactants (or substrates), appropriate configurational and conformational factors must be exploited during that visit if selectivity is to be achieved in the reaction.

1978 saw the advent of a reaction carried out inside a polymer cavity. Shea and Thompson [70] copolymerised the distyrylester of *trans*-1,2-cyclobutane dicarboxylic acid with divinylbenzene. The diester was hydrolysed after polymerisation. Fumaric acid dichloride was reacted with the pendant alcohol groups inside the cavities. Then a carbene source was introduced and made to react with the double bond of the fumaric diesters. The ester bonds were hydrolysed again to free the products from the polymer. *trans*-1,2-Cyclopropane dicarboxylic acid was the preferred product from an enantiomeric excess of 0.05%, a small value, but in itself extremely significant, because it showed that an asymmetric reaction was indeed possible.

Two years later Damen and Neckers reported a different stereoselective synthesis, the photochemical dimerisation of *trans*-cinnamic esters [65, 66, 82]. Three templates were used in the preparation of polymers; α-truxillic, β-truxinic and γ-truxinic acid. Their styrylesters were copolymerised with styrene and divinylbenzene and afterwards hydrolysed. Competitive rebinding studies confirmed that each imprinted polymer predominantly incorporated its own template [66]. This encouraging first step led to the more adventurous idea of allowing *trans*-cinnamoyl chloride to react with the phenolic hydroxyl groups inside the cavities and subsequently irradiating to form the dimer (cyclobutanes). Hydrolysis would then allow the collection of the products formed. If the reaction takes place inside a preformed cavity, with specific location of binding group, evidence for it would be found in the product distribution. Indeed this was the case. The product distribution was very dependent on the template used in the preparation of the imprinted polymer. Random dimerisation of *trans*-cinnamic ester in solution yielded α-truxillate and this was also true for the α-truxillic acid imprinted polymer. However, when β-truxinic acid was the template, 53% of β-truxinic was formed accompanied by 47% α-truxillic acid, and the polymer imprinted with γ-truxinic acid produced 53% of the template with the rest α-truxillic acid.

The stereospecific conversion of D-mandelic acid to its enantiomer reported by Sarhan et al. [145] is thought to arise from formation of a transition state induced by base complexation of the template inside the cavity. An analogy with enzymes is clear. This report was two years before Wulff and Vietmeier presented a highly sophisticated strategy, in 1989, for the enantioselective synthesis of amino acids using polymers possessing imprinted chiral cavities [88, 89]. The basic idea was to form an enolate held by the binding sites inside the chiral cavity, so that one side of it was sterically inaccessible and reaction with an electrophile would create a single enantiomer. For various amino acids enantomeric excesses of up to 33% was achieved. This is only moderate compared to conventional methods, but unique for an enzyme-mimicking approach. In the same year Mosbach investigated the possibility of using a coenzyme-substrate analogue as a template for pyridoxal 5'-phosphate, with the same aim of synthesising amino acids enantioselectively on an imprinted chiral polymer. Together with Andersson [146], he prepared polymers imprinted with *N*-pyridoxyl-L-phenylalanineanilide. The polymers showed significant preference towards the template enantiomer ($\alpha = 2.5$) and the α-proton exchange during Schiff's base formation was facilitated. Synthesis of amino acids with this approach have not yet been reported.

The move towards catalytic reactions was initiated in 1983 by Hopkins and William. Phthalimide attached to a methacrylate residue acted as template. Hydrolysis after microgel formation with methyl methacrylate, 2-ethoxyethylmethacrylate and ethylene glycol dimethacrylate left behind cavities possessing primary amine groups. Shape selectivity was confirmed by determination of hydrolysis rates between different 4-nitrophenyl esters. An acetate was hydrolysed considerably faster than a caproate [147]. Four years later Leonhardt and

Mosbach prepared imprinted polymers with not only a more specific substrate binding but with a true catalytic turnover [148]. Certain N-protected amino acid derivatives formed pseudotetrahedral complexes in solution in the presence of Co^{2+} and two molecules of 5-vinylimidazole. These were polymerised in the presence of divinylbenzene and the metal-ion and amino acids eluted together afterwards. Hydrolysis of the corresponding 4-nitrophenylester of the amino acid template showed an increase in catalytic activity (factor 2–4) compared with a blank prepared without the template. The polymer also exhibits a clear preference for the activated ester of the template, which suggests that hydrolysis indeed takes place inside the cavities. Two years later the scope was further broadened by the preparation of the first polymer imprinted with a transition state analogue.

Robinson and Mosbach [149] used *p*-nitrophenylphosphate to model the transition state of a carboxyester hydrolysis. The phosphonate approximates closely to the pseudotetrahedral geometry and charge distribution of the assumed transition state of the reaction. Crosslinking of poly[4,5-vinylimidazole] in the presence of the analogue and Co^{2+} using 1,4-dibromobutane yielded a rigid polymer matrix. The template was removed. The nitrophenylester of acetic acid was hydrolysed eight times faster on the imprinted polymer than on the blank. That the reaction took place inside the cavities was confirmed by the fact that the transition state analogue acted as an inhibitor.

Morihara et al. coincidentally pursued the same idea of utilising transition state analogues as templates [150] In a sense they went back to the roots of imprinting and used silica as matrix. Here, however, the similarities end. Imprints were prepared with aluminium ion doped silica. In the presence of the transition state analogue further polymerisation took place and the template could be extracted afterwards. The butanolysis of benzoic anhydride (transesterification) on dibenzamide imprinted silica was studied. Only dried imprinted polymers show catalytic activity. Aluminium ions were important for rate enhancement but imprinted polymers obtained without doping were also catalytically active, but to a lesser extent. Active sites were marked with phenolphthalein. The purple colour disappeared after addition of the template. The template was also able to inhibit the reaction. Both observations suggest that the catalytic transformation occurred within the cavities or in this case within the "footprints" [150]. Further investigations followed closely. The use of 9 different transition state analogues related to dibenzamide and 16 different but structurally related inhibitors shed a lot of light on the complementarity of template and foot print [151, 152]. Competitive inhibition was measured. Each imprinted silica performed best for its own template. Partial affinities could be estimated for defined structural features of the inhibitors. The results suggest that structural variations, such as the substitution of a methyl group for a hydrogen, have considerable influence on the catalytic activity and selectivity of the footprints [151, 153]. Thermodynamic data on substrate binding led to the proposal that, depending on the template, the "footprints" sometimes bind only the selected substrate, but can also induce stress or strain in other molecules

[152]. The latest results on imprinted chiral footprints [154] have shown that enantioselective catalysis (hydrolysis) does occur, and based on kinetic measurement the authors believe that this is due to an enantioselective mechanism. Kaiser and Andersson also chose aluminium doped silica as a polymeric material to obtain phenanthrene imprints and their work has been discussed earlier [52]. No selectivity towards the template was observed when imprinted silica was used as stationary phase. Only relative retention and capacity factors increased. Furthermore, even after careful extraction in a Soxhlet, the polymer still leaked phenanthrene. They also found that diazomethane yields a side reaction forming long alkyl chains. Finally they attempted to repeat the work of Morihara et al. [150–155]. but were not able to detect any selectivity using dibenzamide as the template and instead found that the template decomposes into at least five different products when adsorbed on the silica. Clearly further work is required on these systems.

Very recently Mosbach et al. [156] have reported their latest results on β-elimination reactions catalysed by polymers molecularly imprinted with transition state analogues similar in structure to the haptens used in closely related work on catalytic antibodies [157]. There seems little doubt that this approach is likely to gather momentum, and run in parallel with the work on "abzymes".

A somewhat different approach to reactions in polymer cavities has been reported by Berstrom et al. [144] and emphasises the importance for looking at avenues other than the major ones being tackled by the leading groups. As mentioned earlier this group have exploited hydrophobic forces to generate a chiral imprinted polymer. To achieve this, the group converted steroid templates with hydroxy functionalities in either the 3- or 17-positions into their acrylates. Copolymerisation in the presence of divinylbenzene gave the imprinted polymer. The template was hydrolysed from the polymer, and $LiAlH_4$ was allowed to complex with the residual pendant hydroxy groups inside the cavities. To test the selectivity of the polymer in reduction reactions a solution of a steroid with carbonyl groups in the 3- and 17-position was stirred in the presence of the activated imprinted polymer. In the case of the template having the hydroxy group in position 17, complete preference for one product was observed (99% ee). Reaction with $LiAlH_4$ in solution in the absence of imprinted polymer preferentially reduced at the 3-position of the steroid. The polymer prepared with the other template, with the hydroxy group in the 3-position, afforded complete reduction of the carbonyl group in the 3-position. In contrast to the reduction in solution, where formation of the 3β-OH (90:10) is favoured, the 3α-OH was the predominant product in the solid phase reaction. A blank with methylmethacrylate as apparent template gave selectivities as found in solution. Not only are these results very encouraging, suggesting that one day imprinted polymers will be able to perform reactions with similar specificity to enzymes, they also mean that the scope of potentially suitable templates has been significantly enlarged. It is possible to choose molecules which can be used as templates, involving hardly any further modifications and leading to imprinted polymers which show distinctly different selectivities compared to the same reaction carried out in solution.

11 Molecular Imprinting in Biopolymers

Although the emphasis in this review has been on the manipulation of synthetic polymers it would not be appropriate to omit entirely the complementary work on biopolymers. Two important strategies are evolving here, both of which involve molecular imprinting. The first concerns specifically chemically modified proteins, and the second the generation of catalytic antibodies with functions similar to enzymes, but potentially with much greater scope in terms of the reactions catalysed.

It is well known now that natural enzymes can have their patterns of substrate and stereoselectivity influenced by changing the nature of the solvent. Perhaps the most dramatic effect is the reversal of reactivity when a hydrolytic enzyme in water becomes an esterification catalyst in hydrocarbon solvents [158]. Imprinting procedures however have also been exploited to modify enzyme activity. Klibanov and Russell [159] were the first to go down this road. A competitive inhibitor for subtilisin (a protease) was allowed to bind to the reactive site of the enzyme in aqueous solution. In this case the inhibitor acts as a template to keep in place the defined tertiary structure of the protease. After lyophilisation, the inhibitor was removed by washing with a dry solvent. At this stage the active conformation of the enzyme is "frozen in" and the active centre made available for substrate binding. In the absence of water, or with only trace levels present, the enzyme is unable to reach its stable native conformation. During the drying process new, unnatural hydrogen bonding patterns are formed and are "locked in" as long as no hydrogen bond donor, especially water, is present to break these links and allow attainment of an equilibrium conformation. Measurements in octane revealed a hundred times more active enzyme compared with the same enzyme, lyophilised in the absence of the template. Unexpectedly, up to a water concentration of 0.030%, inversion of enantioselectivity was observed. With increasing water content, the stability and affinity of the enzyme were increased as well. Above a concentration of 0.030% the enzyme resumed normal enantioselectivity.

A year later the same research group presented a second rather different approach [160] in which they described the production of abiotic receptors by imprinting of proteins. The basic idea was once again to induce a certain conformation into the tertiary structure of a protein using a template and then to "freeze it in". Bovine serum albumin (BSA) was dissolved in water, containing either *p*-hydroxybenzoic acid or tartaric acid. After lyophilisation the protein was washed with anhydrous solvent to remove the template. The resultant imprinted proteins bound up to 30-fold more of the template in anhydrous solvents then the non-imprinted blank in either anhydrous or aqueous solution. The process of binding the template to the protein was complete after 24 hours. When, for example, tartaric acid was used as template, malic acid was able to bind almost as efficiently to the abiotic receptor. Lactic and succinic acid were bound half as strongly and dimethyl tartrate or threitol were incorporated hardly at all. This approach is different to the first one in the respect that in this case potentially the *whole* protein structure is involved in providing suitable spaces for the template inside the reversibly crosslinked biopolymer network, and not just the active site. While the template is diffusing into the network and trying to establish thermodynamically stable sites within the protein structure, the hydrogen bonding network is de-crosslinked and either at the same position or elsewhere crosslinked again until equilibrium is attained. Lyophilisation causes the formation of even more hydrogen bond crosslinks, which at this stage results in lower protein flexibility, leading to a new protein conformation which is "frozen in" through interlocking hydrogen bonds.

In subsequent studies [161] the nature of the imprinting process has become clearer and for the first time abiotic protein receptors have been used as selective adsorbents for a chromatographic separation of maleic and acrylic acids. It was found that the higher the solvent's propensity to form hydrogen bonds, the lower the protein-template binding in it, thus pointing to hydrogen bonds as the driving force in this interaction as proposed earlier [160]. The critical requirement for the imprintability is the presence of a sufficiently long polymer chain, capable of forming hydrogen bonds with the template. In addition to BSA, other macromolecules, i.e. partially cleaved BSA, glutathione, poly(L-aspartic acid), dextrans and their derivatives, partially hydrolysed starch and polymethacrylic acid also could be imprinted. It is deduced from these experiments that the mechanism of imprinting and binding in water involves a multipoint hydrogen bonding of each template with two or more sites on the polymeric chain. This is supported by numerous binding studies of L-malic acid analogues and related ligands to BSA imprinted with L-malic acid. Somewhat contradictory are the results obtained from imprinted maltohexaose and poly(ethylenimine) (PEI), where imprinting does not result in any selectivity towards the template at all. This may still be rationalised since the PEI was only of low molecular weight ($\approx$ 1.000 daltons). Much higher PEI (say 60 kd) should provide more opportunity for the template to form multiple hydrogen bonds. Curiously, imprinted poly(ethyleneglycol) did demonstrate a small amount of net binding [160]. These minor inconsistencies question the assumption that multiple hydrogen bonding is exclusively responsible for the selectivity, and point to a somewhat more complex imprinting mechanism than currently proposed.

Shortly after the first imprinted protein was reported, Mosbach et al [162]. equilibrated α-chymotrypsin with *N*-Ac-D-tryptophane in a buffer, rapidly precipitated the product with 1-propanol, and dried the precipitate. After careful washing with anhydrous solvent, the "bio-imprinted" enzyme could be used to synthesise the ethylester of the template in cyclohexane. The enzyme became inactive when the enantiomer of the template was used for the imprinting procedure. Not only could the presence of water affect the reaction rate, but too much water was also found to destroy the observed effect. The enzyme stayed inactive in solvents like ethylacetate, diethylether, or acetonitrile, but showed activity not only in 1- but also in 2-propanol and acetone. Results with other *N*-protected D-amino acids support the general applicability of "bio-imprinting". The modified enzyme exhibited high degrees of substrate and unnatural (D)-enantioselectivity. It was further confirmed that α-chymotrypsin cannot be imprinted with L-enantiomers of the templates. Serin rendered the imprinted enzyme inactive, hence pointing to an active site related effect [163]. Similar observations were made by the same research group with L-amino acid oxidase from crotalus adamanteus "bio-imprinted" in the presence of D-phenylanaline or D-tryptophan. The change of enantioselectivity was attributed to the organic solvent, affecting the deamination rates more for the L-amino acids than the D-amino acids (compared to water) and the "bio-imprinting" procedure increasing the rates for the reactions with the D-aminoacids [164].

A different way to reduce enzyme flexibility and hence to introduce modified selectivity has been achieved through intramolecular crosslinking. This was applied to octameric yeast phosphofructokinase to study the effect of substrate-imprinted conformational changes on the regulatory properties. Crosslinking performed in the presence of fructose-6-phosphate yielded an imprinted enzyme which exhibited significantly higher affinity to this substrate than the native conformation. Hoffmann et al. [165] who conducted this study, tried to shed more light on the complex allosteric kinetics and oligomeric structure of this enzyme. Amongst other observations and the one mentioned above, they found that the imprinted enzyme does not show any cooperativity with respect to the template, but is still activated by adenosine monophosphate and fructose-6-phosphate. This activation increases the affinity towards the template. Interaction of phosphofructokinase with adenosine triphosphate at the substrate-binding site is not affected by the crosslinking. Although few in number [159–165], these examples demonstrate very clearly the obvious potential for modification of proteins, and particularly enzymes, through a rational choice of template.

12 Catalytic Antibodies

The development of catalytic antibodies or "abzymes" is now a major area of research in its own right [10, 11, 166, 167]. While it is not appropriate to provide an in-depth review of the topic here, because of the conceptual similarity with molecular imprinting in polymers, equally it would be inappropriate to ignore the area completely.

The function of antibodies is very different to that of enzymes, but they are closely related in terms of their molecular make-up. Antibodies have a functional region which binds other molecules with high affinity and specificity. Antibodies stabilise the thermodynamically most stable structure to which they bind. Their ultimate function is to destroy antigens. In contrast enzymes, although they bind to their substrates as well, stabilise the transition state of a chemical transformation, hence securing their unsurpassed catalytic activity. Both have in common that they are composed of the same amino acid building blocks and therefore, at least in theory, an antibody binding site could satisfy the requirements for the active site of an enzyme [168]. By virtue of their essentially unlimited range of recognition sites, antibodies may provide an enormous repertoire of new tailor-made, enzyme-like catalysts. With the advances in genetic engineering in the mid-eighties (hybridoma technology) [169, 170] it has become possible not only to synthesise monoclonal antibodies for the first time but also to synthesise them in comparatively large quantities. In the past antibodies were produced from immunised animals. since then methods for raising antibodies have further advanced and more recently allow their production without immunisation [171, 172]. This has inevitably led to an even greater diversity in available amino acid sequences as potential binding sites, and to simplification of the genetically engineered preparation process. Recent results support the remarkable possibility that there might be virtually no limit to catalytic antibody design [167, 172].

The techniques for generating catalytic antibodies are basically the same as those used for raising conventional non-catalytic ones. Antibodies are raised against a hapten. A hapten can be looked upon as being an ordinary template. A hapten is brought in contact with a solution containing antibodies with an enormous diversity of functional regions and recognition sites. Through a very sensitive immunoassay-based screening procedure only those antibodies which bind the hapten to any significant extent can be identified. These can be separated from the others and cloned. If the hapten contains a structural element which is a transition-state analogue for a particular reaction, there is a possibility that a site on the antibody might bind this structural element. When bound antibodies are separated and purified, then introduced, not to the original transition state analogue (template), but to the corresponding reactants for the reaction of interest, the antibody encourages or catalyses the reaction by favouring combination of the reactants within the transition state binding site. In practice the hapten is generally a conjugate of a small protein and the molecular sized transition state analogue, and so the process is more complicated than portrayed. Also, the challenge of selecting those few antibodies which are catalytic, out of the large number which merely bind the hapten, is a formidable task, to which biochemists have responded marvelously [173]. Nevertheless the similarity with molecular imprinting in polymers is remarkable.

The first reaction ever catalysed in this way was the hydrolysis of an ester bond [166,170]. Phosphonate haptens represent the majority of transition state analogues designed for this purpose, because their tetrahedral, negatively charged structures closely mimic the transition state of esterases and peptidases and as a result they can be powerful inhibitors [174]. For a variety of other types of bonds, i.e. glycosidic [175], trityl-protected alcohols [176], sequence specific peptide bonds [177], phosphate monoesters [178], epoxides 179], imides [1890] etc., hydrolytic antibodies are also available. The potential of catalytic antibodies becomes particularly evident when the reaction being catalysed in one for which enzymes are not available. Schultz and Braisted reported the first example of a bimolecular antibody-catalysed Diels-Alder reaction in 1990 [181]. In cases where the transition state and the product are very similar in their geometric and electronic requirements, product inhibition can become a serious problem. Ingenious design of transition-state analogues (templates) has therefore been required to minimise these effects, and this has increased the level of importance of synthetic organic chemistry in this work.

Metalloporphyrins have also been used as haptens to mimic cytochrome P450-like activity and the corresponding antibodies were able to catalyse an epoxidation reaction [182]. Catalytic antibodies can also be used to carry out enantioselective reactions, as was first shown by Danishefsky et al. [183] in 1991, but with optical purities of about 84% there is still a long way to go before enzyme-like performance is achieved. Antibodies have also been produced capable of catalysing a Claisen rearrangement [184], an aminoacylation reaction [185], a dehydration step [186], transesterification of enol esters [187], and a conformational rearrangement of a peptide bond [188]. Even incorporation of cofactors such as metal ions or complexes in specific sites within a cooperative site is possible. Here the antibody primarily exert its traditional role as a selective and tight binder, and an appropriately positioned metal ion complex functions as the catalyst [182]. Finally, site-directed mutagenesis of catalytic antibodies [189] has been reported and this offers considerable further scope for the development of these imprinted systems. While to date the activities achieved generally do not compare with the activities of enzymes, the potential for producing catalytic antibodies for non-enzymatically catalysed reactions still provides a major stimulus for this area to expand,.

13 Practical Aspects of Molecular Imprinting

13.1 Templates Used

Although there has been some activity using metal ions as templates the overwhelming majority of studies have used low molecular weight organic molecules as templates. Metal ions which has been exploited are those which form a variety of well-defined coordination complexes, i.e. the transition metal ions, notably Zn, Co, Cu, Ni, Ag, Fe and Cd in different oxidation states, although there is one paper where Ca^{2+} is the templating ion [137]. The main organic templates that have been used as are follows:

Methyl orange [34–40, 46, 53–56] and its ethyl-[34–37, 46, 54, 56], propyl-[34, 35, 38, 39, 46, 54, 56] and butyl-[34, 35, 38, 39, 46, 54–56] analogues, *p*-diamino-p' sulphonamidoazobenzene [36, 40], aniline [37, 47], other dyes like methylene blue [35, 60], triphenylmethane dyes [35], phenolphthalein and related structures [35], rhodanile blue [91], safranine [91], surface-active dyes [113], camphorsulfonic acid [41, 43], mandelic acid [41, 43, 68, 69, 145], glyceric acid [27, 28, 67, 78, 79, 86], mono-, di- and triols [160, 161], mannitol [86], tartaric acid [161], malic acid [160, 161] and other multifunctional dicarboxylic acids [159–161], *p*-hydroxybenzoic acid [160, 161], quinine [44], quinidine [44], cinchonine [44], cinchonidine [44], hydroxy- and methoxy-morphinane derivatives [45, 50], nicotine [50], 2,2′-bipyridine [47, 48], 1,10-phenanthroline [47, 48], polycondensed aromatic amines [112], benzochinoline isomers [47], 2,2′-pyridylthiazole and isomers [48], antipyrine derivatives [48], coranine-based analeptics [48], astrapholoxine [50], various methylquinolines [51], 9-ethylaniline [111], phenanthrene [52], mono- [151] and dibenzamides [150] and their sulphonamides, ring-substituted mono- and dibenzamides [150–155], phthalimide [147], anthraquinone and naphthoquinone sulfonates [57–59] 1,2-cyclobutane dicarboxylic acid [70, 82, 191], dimers of cinnamic acid ester [65, 66], aromatic and aliphatic mono- and diketones [83, 84] and aldehydes [114–117, 119, 120], steroids [143, 144], adamantane [143], vitamins K_1 [143] and B_{12} [35], NAD and aldehydes [114–117, 119, 120], steroids [143, 144], adamantane [143] vitamins K_1[143] and B_{12}[35], NAD and bisimidazoles [142], chiral metal complexes [139, 141], *N*-pyridoxyl-L-phenyl alanineanilide [146], various phosphonate, esters in particular *p*-nitrophenylmethyl phosponate [149, 178], substituted *N*-methylpyridinium ions [174], *N*-protected amino acids [81, 82, 100, 107, 148, 159, 162, 163] and their derivatives (especially involving variation of size and functionality of the protection group and substituents at the chiral carbon atom and the neighbouring amino group [89–90, 92–94, 96–99, 101, 104, 105, 107, 108], conformational amino acid isomers [95], free [63] and derivatised monosaccharides [64], especially mannosides [71–75, 78, 80], fructose-6-phosphate [165], β-adrenergic blockers [94], structures analogous in shape and substitution pattern to those found in Claisen rearrangements [184], intra- and bimolecular Diels-Alder transition-state analogous [181], metalloporphyrins [182] and β-hydroxyketones [186].

The only use of an oligomer as a template appears to be the application of different oligopeptides in successive imprinting of receptor memory in tetrahymena [192]. Apparently the micellular tetrahymena did develop a receptor-level memory. In the generation of catalytic antibodies as well, of course, polymeric templates are employed in the sense that low molecular weight transition state analogues are generally conjugated with a suitable protein for raising antibodies.

13.2 Polymer Networks Employed

The only inorganic polymer which has had serious investigation is silica [34–41, 42–51]. Some modified aluminium-doped [150–155] and methylated silicas [52] have also been used for imprinting procedures. Surface modified silicas, including encapped versions, have also been tried [115–117, 119, 120, 143]. In general however networks based on synthetic vinyl polymers have been exploited most. These include species derived from: ethyleneglycol dimethacrylate (EGDMA) [4, 63, 64, 68, 49, 71, 72, 87, 90, 92–98, 102, 104, 105, 114, 119, 120, 142, 145], EGDMA-co-methylmethacrylate (MMA) [63, 76, 87, 108, 136, 137, 147]; divinylbenzene (DVB) [75, 101, 144, 199] DVB-co-styrene (ST) [27, 28, 65–70, 75, 78, 79, 82–84, 86, 87, 89, 107, 118, 130, 131, 134 ,140, 141, 145, 148] trimethylolpropane trimethacrylate (TRIM) [73]; *N*-methylenebisacrylamide (MBA)-co-*N*-vinylpyrrolidone (NVP) [54]; MBA-co-acrylamide (AM) [128, 129, 139]; DVB-co-NVP [56–59]; MBA-co-AM [190] (also methacrylamide analogues); chiral [74, 99] and non-chiral [91] (meth)acrylate and/or acrylamide-based crosslinkers; poly(methacrylic acid) [160, 161]; poly(ethylene glycol) [160, 161]; poly(L-aspartic acid) [160, 161] poly(*N*-vinylimidazole) crosslinked by irradiation [127]; interpenetrating polymer networks of DVB-co-St and EGDMA-co-MMA [75, 87]; poly(ethyleneimine) crosslinked by disulphide linkages [53]; poly(4-vinylpyridine) crosslinked with 1,4-dibromobutane [125, 126, 135] and electrochemically polymerised pyrrole networks [138].

By far the most used systems are matrices based on methacrylate, methacrylamide and styrene; and the most common crosslinkers are ethyleneglycol dimethacrylate and divinylbenzene. Methacrylamide based species are the most hydrophilic and styrene ones the most hydrophobic, with the methacrylate systems falling in between. This therefore provides quite a wide choice and imprinted polymers with high mechanical stability and chemical inertness, suitable for example in HPLC applications, are readily achievable. To some extent the chemical nature of the matrix is of less importance than its morphology, and in this respect the type and level of crosslinker used, together with the nature and proportion of the porogen are more crucial (see later). All of these experimental parameters are of course inter-dependent.

In a typical imprinting procedure the template, polymerisable binding site(s), crosslinker, additional monomer (if appropriate), porogen and free radical initiator are mixed to form an isotropic solution. The mixture is then polymerised, generally by moderate heating (see later) to form a monolithic block. This is then crushed into fine particles and exhaustively extracted in a Soxhlet with an appropriate solvent or solvents to remove the template, initiator fragments and any non-polymerised monomer and crosslinker. After vacuum drying the particles may be sieved if, for example, they are to be tested in a chromatography column.

In a modified procedure the polymerisation can be carried out in suspension in an immiscible liquid to form spherical polymer particles [30].

Natural polymer-based networks have also been investigated. The proteins etc comprising antibodies represent the largest group [164, 166, 169, 189] but this is of course a specialised area. Poly(saccharides), in particular starch [60], dextran [161], dextrin [161] and maltohexose [161], and also natural polypeptides, mainly enzymes [162–165], embody the more accessible biopolymers. In some instances imprinting is achieved through formation of covalent bonds, with crosslinkers like cyanuric chloride or glutaraldehyde. Likewise chitin derivatives similarly crosslinked have been exploited [136].

13.3 Influence of Binding Site

Binding sites offering various types of interaction with the template have been used. The interactions exploited are: covalent bonds [66, 70, 75, 76, 78, 79, 82–84, 114, 115, 120, 140, 145, 158] electrostatic interactions [35, 95, 99, 101, 126, 128, 144, 190] hydrogen bonding [53–59, 78, 81, 91, 94, 95, 98–101, 103, 146, 150–155, 148, 149], hydrophobic [53–59, 91, 144, 147], and charge transfer interactions [87], ligand binding to metal complexes [124, 125, 127, 130–135, 139–142] and ionophore-type complexation interactions [137]. The use of covalent bonds, hydrogen bonds and ionic interactions must be emphasised, because each of these alone or in combination with the other two has provided the most impressive results in templating. [71, 73, 90, 100] Their use has lead to the most enantioselective polymers, which under optimised conditions in an HPLC mode have given rise to base-line separations with resolution factors of around 2 and separation factors of 5 [71], and even up to 10 under the most favourable conditions. [100, 104] In the context of chromatography it is valuable to point out how very different the properties of binding sites can be once they are inside a polymer network. Stimulated by some unexpectedly strong peak broadening under HPLC conditions using stationary phases with boronic acids as binding sites [71, 74, 80], binding groups exhibiting faster exchange have had to be synthesised. [80] Rate enhancements in solution for ester formation with diols of up to a factor of 10^8 have been achieved with modified boronic acids, but the chromatograms still showed the same peak broadening, especially for the template, though less so far its enantiomer. This phenomenon is also observed when covalent interactions are replaced by hydrogen bonding and ionic interactions but, as one would expect, to a lesser extent. [96, 104, 105] It is assumed that the rate limiting step mainly involves the embedding and residence of the template in the cavity, although other factors may contribute to a lesser degree [71, 108].

Arylboronic acids [63, 72, 80, 85, 194] have been most extensively studied as binding sites for derivatised monosaccharides. By using an amino group in the position ortho to the boronic acid the rate constant for the esterification with a diol has been increased by a factor of 10^8, as mentioned above [79]. It has also been shown that the flexibility of the boronic acid group is crucial. If the binding interaction is too rigid or too flexible, it is detrimental to the selectivity of the chiral cavities [79]. This is also the case regarding the distance between the aryl boronic acid moiety and the polymerisable group attached to it. Selective binding goes through a maximum as this distance is increased [79]. A similar optimum has been found in the length/flexibility of the crosslinker (see later) and it is clear that a compromise is required to achieve maximum exploitation in templating. On the one hand it is necessary to have good binding between the template and the matrix site to allow efficient formation of the cavity; likewise the matrix must be a tight one (high crosslink ratio, short crosslinker) to create a permanent cavity. On the other hand, however, it must be possible to remove the template, and to allow fast and facile access of substrate molecules to the cavity. In some respects therefore it is necessary to create a cavity which is very well defined structurally and topologically, and yet which retains sufficient flexibility, along with the surrounding matrix, to allow molecules to come and go. This has been described as a "breathing" [193] action, not unlike the flexibility associated with the binding and reactive site of enzymes.

Carboxylic acids have commonly been used as binding sites for non-covalent interactions. Although not explicitly mentioned, the change from 4-vinylbenzoic acid [101] to the sterically less demanding methacrylic acid [96] (MAA) significantly improved the selectivity of otherwise identical polymers. Hence in the majority of cases MAA has been the binding site of choice. The question remains to be answered whether the selectivity increase was due to the fact that the acid group in MAA is closer to the polymer backbone inside the cavity and therefore allows a more significant interaction of the template molecule with the cavity itself, or to the fact that the smaller MAA molecule gives less rise to steric repulsion during complex formation with the template in solution leading to the formation of a more stable complex. Only recently itaconic acid has been found to be superior to MAA. This may be due to the numerous functional groups provided by the particular template, in which case the structure of itaconic acid would suggest a higher capability for forming a higher density of multiple hydrogen bonding interactions with the template [94].

Where the template has been an aromatic amine of low basicity, interaction with carboxylic acid groups has proved too weak to allow selective imprinting. However, Sherrington et al. [112] have shown that the sulphonic acid residue in 2-acrylamido-2-methylpropane sulphonic acid is very useful under these circumstances; the much higher acidity of this acid allows protonation of the weak base template, and hence sufficiently strong electrostatic interaction for imprinting to be achieved.

13.4 Influence of Crosslinker

The "length" of a crosslinker is the distance between the polymerisable units and this defines the maximum distance between the two crosslink points attaching two polymer chains. Varying the length of an alkyl or oxyethylene chain between two methacrylate units showed that the shortest crosslinker ethylene glycol dimethacrylate (EGDMA) led to the highest selectivity in the imprinted polymer [74, 75, 87]. Divinylbenzene (DVB) represents an even shorter and more rigid crosslinker, but selectivity was lower in this case [74, 75, 87]. This points clearly towards the importance of an optimum length, and probably also optimum flexibility, of the crosslinking agent as discussed earlier. Investigations into crosslikers, which on the one hand are shorter but on the other are less rigid than DVB, have failed to improve matters so far due to their tendency to cyclopolymerise [73].

Flexibility of the polymer network is crucial to allow the template to enter the cavity and rebind to the binding sites. However, flexibility is also crucial to maximise the degree of complementarity towards the template that the cross-linker is able to provide. Even in solution, prior to polymerisation, the cross-linker associates with the template. Hence flexibility of the crosslinker allowing it to model the shape of the template is essential for selective imprinting. EGDMA has so far proven to be the optimum choice, the best compromise between flexibility/rigidity and length [1, 74, 75, 87, 98, 104].

Trimethylolpropanetrimethacrylate (TRIM) is a trifunctional crosslinker and is one carbon bond longer than EGDMA. It shows a slight decrease in the selectivity of imprinting when used as direct replacement for EGDMA keeping conditions as similar as possible [72]. Unfortunately it has not been studied as the basis of a chiral stationary phase, and thus nothing can be said about the influence of its increased flexibility. The use of the shorter and more rigid structure of DVB under similar conditions led to a decrease in selectivity. The design of novel crosslinker structures have to accommodate these observations. Dimethacrylates based on Bisphenol A resemble extended divinylbenzene analogues [74]. A strong decrease in the selectivity achieved illustrates the inability of the rigid structure to adapt to the shape of the template. The same arguments probably apply to two more rigid derivatives of EGDMA, 3,4-dihydroxytetrahydrofuran dimethacrylate and 1,4:3,6-dianhydro-D-sorbitol dimethacrylate [74]. The former is the same length as EGDMA while the latter is two carbon bonds longer. Both showed a decrease in imprinting selectivity, which was more pronounced in the case of the more rigid sorbitol-derived crosslinker structure.

The proportion (vol %) of crosslinker used in preparing the imprinted matrix, together with the level and type of porogen control the detailed morphology of the matrix. Pore size, pore distribution, pore volume and surface area are all adjustable, although they are of course closely inter-related. The crosslink ratio or "nominal" crosslink ratio is defined as the volume % of crosslinker (usually divinylic) employed.

The whole range of crosslink ratio from a few percent up to $\sim$ 100% has been studied, with the template acting as an additional crosslinker in its own right. The smaller the ratio of template to crosslinker the higher is the selectivity of imprinting found of the polymer [71, 74, 87]. Varying the amount of crosslinker and keeping the amount of template constant shows a surprising effect. At low degrees of crosslinking selectivity is poor, because the polymer matrix is not crosslinked enough to retain the shape of the cavity. Then, with increasing amount of crosslinker, a sudden remarkable increase in selectivity is observed, which rises more slowly when approaching the 100% level [66, 74, 75]. Every crosslinker studied in imprinted systems shows qualitatively the same effect. EGDMA shows this sharp increase in selectivity at bout 40% crosslinker and the effect starts to level off at about 60%. The values for butanedioldimethacrylate are 45% and 90%, for DVB 40% and 70% [74, 75] DVB has also been investigated in this context with entirely different templates. Selectivity still rises sharply with increasing amount of crosslinker [66]. This time the increase is observed between 25% and 60% crosslinking. Two observations are true for all studies. Below ~ 15% no selective cavities survive, and the highest selectivity always coincides with the highest degree of nominal crosslinking. The porogen however is also important in this context.

Chiral crosslinkers have the possibility of forming diasteriomeric associates with a chiral template, and of thus producing a more selective imprint. Introduction of chirality between the polymerisable units though, means a loss of flexibility compared to crosslinkers such as EGDMA. So far the reported attempts to improve enantioselectivity of chiral imprints have failed to produce results better than the ones obtained with EGDMA under basically identical conditions [74, 99]. *N*,*O*-bisacryloyl-L-phenylalaninol is similar in length to EGDMA and more hydrophilic Polymers imprinted with L-phenylanaline ethyl ester showed a fall in selectivity. Dianhydro-D-sorbitol was converted into a chiral crosslinker by methacrylation of its hydroxyl groups. Another derivatised sugar, α-phenyl-D-mannoside, was employed as template, but the selectivity of the imprinted polymer was smaller, probably due to the more rigid bicyclic spacer replacing the ethylene glycol linkage in EGDMA. Overall therefore it seems that the effort expended in synthesising chiral crosslinkers is not worthwhile unless some compensating flexibility can be introduced along with the chiral environment.

Relatively hydrophobic polymer networks derived from rather hydrophobic monomers and crosslinkers are readily available, but this is not so far more hydrophilic networks. Although crosslinked acrylamide resins are very well established in gel electrophoresis [195], nothing is known about their pore structure and accessibility when applied to imprinting. Only very recently has attention been drawn to more hydrophilic, hydrolytically stable crosslinkers, derived from (meth)acrylamides, suitable for imprinting. One reason for examining their properties is their potential for exhibiting much stronger and more defined interaction with templates. Another reason for resorting to more hydrophilic polymer networks is a solubility problem often encountered. The template and binding sites are often not soluble enough in the solution of porogen and monomer and this causes unwelcomed deviation from optimum composition and conditions for imprinting. Shea et al. [190] characterised a variety of (meth)acrylamide derived hydrophilic macroporous polymer networks. They showed that organic salts used as templates could be recovered from the polymer and rebinding was possible. Competitive binding studies have not been carried out so far, so how well the polymer retains the cavity shape cannot be commented on. One drawback they encountered was the need to use a larger amount of solvent (porogen) to dissolve the (meth)acrylamide monomers.

EGDMA and DVB-derived sequential interpenetrating polymer networks have been studied in imprinting to find the better compromise using EGDMA on its own [74, 75]. Different ratios of both crosslinkers and different ratios of porogen to crosslinker have been tried to optimise imprintability. At the end of the day, however, the EGDMA polymers demonstrated the highest selectivity for the resolution of a racemate in this case. From these results it seems that the contribution of a more rigid network has an adverse effect on the selectivity and to date therefore EGDMA offers the best properties of a crosslinker employed for imprinting, at least in the context of the templates used so far to probe its applicability.

13.5 Influence of Polymerisation Temperature

A lowering of polymerisation temperature leads to polymers with higher selectivity. This has been found in many systems which differ in their choice of template, binding site and crosslinker [64, 73, 96, 98, 102, 105], Temperature drops of 20 °C are sufficient to improve polymer selectivity [73, 96]. If not hampered with solubility problems, room temperature or 0 °C are the chosen polymerisation temperatures today [64, 73, 102, 105], compared to 60 or even 80 °C which were common years ago [73, 76, 99, 105]. There are two reasons why a lower temperature should produce more selective imprints. If non-covalent interactions are considered, complex formation should be favoured and therefore should be stronger at lower temperatures, yielding a more defined interaction pattern [96, 105, 109]. Regardless of the type of interaction, however, weak association forces between crosslinker and template, enhancing shape correlation between them should benefit from lower temperatures as well. Secondly, a slower polymerisation process will increase the likelihood that the chain formation does not interfere with the existing interactions. It will also lead to a network in which polymer chains are entangled with less strain, hence removal of the template will lead to less reorganisation and restructuring of the polymer network, preserving the original complementarity of the cavity.

13.6 Initiator and Initiation Method

Moving towards more reactive radical initiators and at the same time lowering the polymerisation temperature has proven to be a viable route for increasing selectivity, and experiments have been carried out at 40 °C [96, 105], 20 °C [64, 73], 3 °C [73] and 0 °C [105]. Photochemical generation of radicals has led to comparable results [102, 105]. Handling of samples is more convenient using radiation induced radical formation and the rate of initiator decay can easily be adjusted. However non-transparent polymer precipitates, and hence radiation scattering may prevent a uniform polymerisation process and so adversely affect the polymer structure.

13.7 Porogen/Solvent Selection

The influence of the polymerisation solvent on the performance of imprinted polymers arises from two separate effects. The choice of solvent defines the complexation strength of non-covalent binding sites and crosslinker with a template, and it has been shown that this affects the resulting selectivity of imprinting in the polymer [96, 105]. However, the selectivity of imprinting covalently bound templates is almost unaffected by a change of solvent [72]. The other influence is concerned with the pore structure of polymers. In this respect the solvent acts as a porogen and together with other parameters, e.g. temperature [74, 76, 105, 106, 190] determines the size, shape and size distribution of pores as opposed to cavities [198, 199]. A highly (macro)porous polymer is desirable for chromatographic applications, minimising retardation of the template as it migrates to the cavity. An optimum porogen or porogen mixture exists for each crosslinker which yields the most beneficial porosity for a given crosslinker. The process of arriving at the optimum solvent/porogen involves trial and error and usually results in a compromise between solvent and porogen effects [1, 71, 74–76, 87].

14 Future Prospects

A recent review [200] on "template syntheses" which appeared in the course of finalising this text provides a fine critical summary of the advances made in the area of low molecular weight template reactions and reacted self-assembly processes. It also highlights how the use of templates has expanded, and is likely to continue to do so in the future. Likewise there is little doubt now that the progress made in the molecular imprinting of polymers is sufficiently impressive for the area to be further developed. However, some important changes are likely to occur. There exists already very sophisticated HPLC methodology for the separation of many substrates including enantiomeric separations, and further *generic* research into imprinted polymers for separations is probably not worthwhile. However, if *particular* separations are identifiable as problematical then the molecular imprinting approach is certainly worthy of pursuing. In general this might be expected to be in collaboration with research groups which have such specific problems.

Separation applications are also worthy of investigation if some entirely novel principle is involved. For example, Dunkin et al. have recently demonstrated [29] the ability to form an anisotropic imprinted polymer. This was achieved by selective blocking or filling of cavities employing polarised photochemsitry and a photochemically active template. An anisotropic polymer of this type might offer an opportunity to use polarised spectroscopy in an optical sensor device.

In contrast to the separations the *generic* investigation of selective reactions (stoichiometric or catalytic) within imprinted cavities remains very worthwhile, since progress here has been slow and difficult. With the considerable research effort being made in the area of catalytic antibodies, it is important that parallel studies are continued with imprinted polymers, because the latter offer, in principal, an experimentally simpler and almost certainly more cost effective technology, and it would seem to be irresponsible to allow such an opportunity to pass by. Novel ideas are required here as well, and one possibility is to exploit imprinted surfaces as heterogeneous catalysts. High surface areas will be required to give good turnover frequencies, and so imprinted thin films on existing high surface area supports seems to be a potential way forward, not unlike some of the approaches already made in separations using imprinted systems.

Closely related to this is the idea of combining or interfacing an imprinted film with some other reactive surface, such as an electrode. Sauvage et al. [138] have already made a start here and the idea has particular merit if a sensor or catalytic device can be evolved. If this is to be successful, however, it is important to understand the limitations associated with molecular imprinting to date, and to have a detailed and realistic view of how the imprinting effect can be optimised. The naive picture that any molecule can be imprinted in any polymer network simply by crosslinking around the template has unfortunately been given some credence by those who have not researched the literature with sufficient rigour. Imprinting can be extremely effective but novel systems are almost certain to require a great deal of experimental optimisation, and this should not be overlooked.

The introduction of bioimprinting [162] is an exciting development and opens up new possibilities. This will be particularly important if biochemists can be stimulated to become involved. Recent pioneering work by Wu and Orgel [196, 197] has involved mimicking the polymerisation step from a single- to a double-stranded DNA. For this purpose an oligonucleotide of a single nucleic base was synthesised first and afterwards added to a solution containing phosphorylated (activated) nucleotides. Polymerisation of the activated base proceeded only in the presence of an oligonucleotide. The complementary oligonucleotide exhibited the highest catalytic activity. This example certainly highlights the wide applicability of the template concept. However it needs to be pointed out that in this particular case cavities were not formed, but rather a template polymerisation [25, 26] was carried out.

The future of molecular imprinting in polymers therefore seems assured. It is hoped that this review will provide a useful critical summary of the achievements to date. Perhaps it will stimulate the introduction of new scientists to the field, or at least provide a convenient and rapid route for potential new participants to assess realistically what this approach has to offer them.

15 References

1. Wulff G (1986) ACS Symp Ser 308: 186–130 In: Ford WT (ed) Polymeric reagnets and catalysts. Molecular recognition in polymers prepared by imprinting with templates. Washington DC
2. Watson JD, Crick FHC (1953) Nature (Lon) 171: 737
3. Todd AR (1956) In: Todd AR Perspectives in organic chemistry. Interscience, London, p 263
4. Vlatakis G, Andersson LI, Mueller R, Mosbach K (1993). Nature 361: 645–7
5. Mutter M (1989) Angew Chim Int Edn 28: 535
6. Mutter M, Tuchscherver GG, Miller GG, Altmann KH, Carey RI, Wyss DS, Labhard AM and Rivier TE (1992) J Amer Chem Soc 114: 1463
7. Mutter M and Tuchscherver GG (1988) Die Makromol. Chem Rapid 9: 437
8. Schultz PG (1988) Science 240: 426
9. Atherton E and Shepherd RC (1989) Solid Phase Peptide Synthesis, Oxford University Press
10. Lerner R, Benkovics SJ and Schultz PG (1991) Science 252: 659
11. Schultz DG and Lerner RA (1993) Acc Chem Res 26: 391
12. Lehn JM, Rigault A, Siegel J, Harrowfield J, Chevrier B and Mobras D (1987) Proc Natal Acad Sci USA 84: 2565
13. Lehn JM and Rigault A (1988) Angew Chem Intl. Edn 27: 1095.
14. Dietrich-Buchecker C and Sauvage JP (1992) New J Chem 16: 277
15. Dietrich-Buchecker C and Sauvage JP (1992) Bull Soc Chim Fr 129: 113
16. Anelli PL, Delgado M, Gandolfi MT, Goodnov TT, Kaifer AE, Philp D, Pietrazkiewiz M, Prodi L, Reddington MV, Slawin AMZ, Spencer N, Stoddart JF, Vincent C and Williams DJ (1992) J Amer Chem Soc 114: 193
17. Sauwage JP (1993) Ed New J Chem 17: pp 619–763
18. Gokel GW and Korzeniowski SH (1982) Macrocylic Polyether Syntheses, Springer Verlag, Berlin
19. Anderson S, Anderson HL and Saunders JKM (1993) Acc Chem Res 26: 469
20. Busch DH (1992) J Incl Phenom Mol Recog 12: 389–395
21. Busch DH and Stevenson NA (1990) Coord Chem Rev 100: 119–54
22. Lindsey JS (1991) New J Chem 15: 153–180
23. Broer DJ and Heynderickx I (1990) Macromols 23: 2474–77
24. Percec V, Heck J, Johansson G and Ungar G (1993) Polym Prepr March, 116–7
25. Challa G and Tan YY (1981) Pure Appl Chem 53: 627
26. Shavit N and Cohen J (1977) Polymerisation in Organised Systems, ed Elias HG, Gordon and Breach, London p 213
27. Wulff G and Sarhan A (1972) Angew Chem Int Ed Engl 84: 364–5
28. Wulff G, Sarhan A and Sabrocki K (1973) Tetrahedron Lett 37: 4329–32
29. Dunkin IR, Sherrington DC and Steinke J (1994) unpublished results. Steinke J Ph.D. Thesis University of Strathclyde, Glasgow, U.K.
30. Guyot A (1988) In: Syntheses and Separations Using Functional Polymers, Sherrington DC and Hodge P Eds J Wiley and Sons, Chichester, U.K. Chap. 1. p1.
31. Mudd S (1932) J Immunol 23: 423–7
32. Pauling L (1940) J Amer Chem Soc 62: 2643–57
33. Pressman D and Pauling L (1949) J Amer Chem Soc 71: 2893–2899

34. Dickey FH (1949) Proc Natl Acad Sci 35: 277–9
35. Dickey FH (1955) J Phys Chem 59: 695–707
36. Bernhard SA (1952) J Amer Chem Soc 74: 4946–7
37. Erlenmeyer H and Bartels H (1964) Helv Chim Acta 47: 46–51
38. Haldeman RG and Emmett PH (1955) J Phys Chem 59: 1039–43
39. Curti R, Colombo U and Clerici F (1952) Gazz Chim Ital 82: 491–502
40. Morrison JL, Worsley M, Shaw DR and Hodgson GW (1959) Can J Chem 37: 1986–95
41. Bartels H and Prijs B (1974) Adv Chrom 11: 115–43
42. Bartels H (1967) J Chrom 30: 113–6
43. Curti R and Colombo U (1952) J Amer Chem Soc 74: 3961
44. Beckett AH (1957) Nature 179: 1074
45. Beckett AH and Anderson P (1960) J Pharm Pharmacol 12: 228T–36T
46. Beckett AH and Youssef HZ (1963) J Pharm Pharmacol 15: 253T–66T
47. Erlenmeyer H and Bartels H (1964) Helv Chem Acta 47: 1285–88
48. Bartels H and Erlenmeyer H (1965) Helv Chim Acta 48: 285–90
49. Erlenmeyer H and Bartels H (1965)Helv Chim Acta 48: 301–3
50. Bartels H, Prijs B and Erlenmeyer H (1966) Helv Chim Acta 49: 621–25
51. Bartels H (1967) Z Anorg Allg Chem 350: 143–7
52. Kaiser GG and Andersson JT (1992) Fres J Analyt Chem 342: 834–9
53. Takagishi T and Klotz IM (1972) Biopolymers 11: 483–91
54. Takagishi T, Hayashi A and Kuroki N. (1982) J Polym Sci Polym Chem Ed 20: 1533–47
55. Takagishi T, Sugimoto T, Hamano H, Lim Y.-J., Kuroki N and Kuzoka H (1984) J Polym Sci Polym Lett Ed 22: 283–9
56. Takagishi T, Sugimoto T. Hamano H, Lim Y.-J. and Kuroki N (1984) J Polym Sci Polym Chem Ed (1984) 22: 4035–9
57. Takagishi T, Hamana H, Shimokado T and Kuroki N (1985) J Polym Sci Polym Lett Ed (1985) 23: 545–8
58. Kozuka H, Takagishi T, Yoshikawa K, Kuroki N and Mitsuishi M (1985) J Polym Sci Polym Chem Ed (1985) 24: 2695–700
59. Takagishi T and Okada M (1986) Chem Express 1: 359–62
60. Shinkai S, Yamada M, Sone T and Manabe O (1983) Tetrahedron Lett 24: 3501–4
61. Cleland WW (1975) Acc Chem Res 8: 145–51
62. Wulff G (1993) Makromol Chem Macromol Symp 70: 285–8
63. Wulff G, Minarik M and Schauhoff S (1991) GIT Fachz Lab 35: 10–12, 15–17
64. Wulff G and Haarer J (1991) Makromol Chem 192: 1329–38
65. Damen J and Neckers DC (1980) J Amer Chem Soc 102: 3265–7
66. Damen J and Neckers DC (1980) J Org Chem 45: 1382–7
67. Sarhan A (1982) Makromol Chem Rapid Commun 3: 489–94
68. Sarhan A, Ali MM and Abdelaal MY (1989) React Polym 11: 57–70
69. Sarhan A (1989) Makromol Chem 190: 2031–9
70. Shea KJ and Thompson EA (1978) J Org Chem 43: 4253–5
71. Wulff G and Minarik M. (1990) J Liq Chrom 13: 2987–3001
72. Wulff G, Poll H.-G, Minarik M (1986) J Liq Chrom 9: 385–405
73. Steinke JHG (1990) Diplomarbeit, Heinrich-Heine-Universität Düsseldorf
74. Wulff G, Vietmeier J and Poll H-G (1987) Makromol Chem 188: 731–40
75. Wulff G and Vesper W (1978) J Chrom 167: 171–86
76. Wulff G, Oberkobusch D and Minarik M (1985) React Polym 3: 161–75
77. Wulff G, Vesper W, Grobe-Einsler R and Sarhan A (1977) Makromol Chem 178: 2817–25
78. Sarhan A and Wulff G (1982) Macromol Chem 183: 85–92.
79. Wulff G and Sarhan A (1982) Makromol Chem 183: 1603–1614
80. Wulff G and Sarhan A (1982) Makromol Chem 188: 741–48
81. Wulff G, Best W and Akelah A (1984) React Polym Exch Sorb 2: 167–74
82. Damen J and Neckers DC (1980) Tetrahedron Lett 21: 1913–6
83. Shea KJ and Sasaki DY (1989) J Amer Chem Soc 111: 3442–4
84. Shea KJ and Dougherty TK (1986) J Amer Chem Soc 108: 1091.3
85. Wulff G (1982) Pure Appl Chem 54: 2039–102
86. Wulff G, Schultze I, Zabrocki K and Vesper W (1980) Makromol Chem 181: 531–44
87. Wulff G, Vesper W, Grobe-Einsler R and Sarham A (1977) Makromol Chem 178: 2799–816
88. Wulff G and Vietmeier J (1989) Makromol Chem 190: 1717–26

89. Wulff G and Vietmeier J (1989) Makromol Chem 190: 1727–35
90. Sellergren B, Lepistö M and Mosbach K (1988) J Amer Chem Soc 110: 5853–60
91. Arshady R and Mosbach K (1981) Makromol Chem 182: 687–92
92. Andersson LI, Miyabayashi A, O'Shannessy DJ and Mosbach K (1990) J Chrom 516: 323–33
93. Ekberg B and Mosbach K (1989) TIBTECH 7: 92–6
94. Fischer L, Mueller R, Ekberg B and Mosbach K (1991) J Amer Chem Soc 113: 9358–60
95. Lepistö M and Sellergren B (1989) J Org Chem 54: 6010–2
96. Sellergren B (1989) Makromol Chem 190: 2703–11
97. Sellergren B and Andersson LI (1990) J Org Chem 55: 3381–2
98. Sellergen B (1989) Chirality 1: 63–8
99. Andersson LI (1988) React Polym 9: 29–41
100. Andersson LI and Mosbach K (1990) J Chrom 516: 313–23
101. Andersson LI, Sellergren B and Mosbach K (1984) Tetrahedron Lett 25: 5211–14
102. Andersson LI and Mosbach K (1990) J Chrom 516: 313–23
103. Sellergren B and Nilsson KGI (1989) Meth Mol Cell Biol 1: 59–62
104. Anderson LI, O'Shannessy DJ and Mosbach K (1990) J Chrom 513: 167–81
105. O'Shannessy DJ, Ekberg B and Mosbach K (1989) Anal Biochem 177: 144–51
106. Sellergren B and Shea KJ (1993) J Chrom 635: 31–49
107. Sellergren B, Ekberg B and Mosbach K (1985) J Chrom 347: 1–10
108. O'Shannessy DJ, Andersson LI and Mosbach K (1989) J Mol Recog 2: 1–5
109. Kempe M and Mosbach K (1991) Analyt Lett 24: 1137–45
110. Sellergren B and Shea KJ (1993) J Chrom submitted
111. Shea KJ, Spivak DA and Sellergren B (1993) J Amer Chem Soc 115: 3368–9
112. Dunkin IR, Lenfeld J and Sherrington DC (1993) Polymer 34: 77–84
113. Sagiv J (1979) Isr J Chem 18: 346–53
114. Wulff G, Heide B and Helfmeier G (1986) J Amer Chem Soc 108: 1089–91
115. Toa Y-T and Ho Y-H (1988) J Chem Soc Chem Commun 417–8
116. Tahmassebi DC and Sasaki T (1992) Abstr Amer Chem Soc 204: Aug, 314 (ORGN)
117. Wulff G and Görlich T, private communication
118. Norrloew O, Månsson M-O and Mosbach K (1987) J Chrom 396: 374–77
119. Wulff G, Lauer M and Disse B (1979) Disse Chem Ber 112: 2854–65
120. Wulff G, Heide B and Helfmeier G (1987) React Polym Ion Exch Sorb 6: 299–310
121. Shea KJ and Sasaki DY (1991) J Amer Chem Soc 113: 4109–21
122. Kljatschenko WA (1951) Dokl Acad Nauk S.S.S.R. 81: 235ff
123. Stanberg J, Seidl J and Rahn J (1958) J Polym Sci (1958) B 31: 15–24
124. Nishide H, Deguchi J and Tsuchida E (1976) Chem Lett 2: 169–74
125. Nishide H, Deguchi J and Tsuchida E (1977) J Polym Sci Polym Chem 15: 3023–9
126. Nishide H and Tsuchida E (1976) Makromol Chem 177: 2295–310
127. Kato M , Nishide N, Tsuchida E and Sasaki T (1981) J Polym Sci Polym Chem 19: 1803–9
128. Efendiev AA and Kabanov VA (1982) Pure Appl Chem 54: 2077–92
129. Kabanov VA, Efendiev AA and Orujev DD (1979) J Appl Polym Sci 24: 259–67
130. Gupta SN and Neckers DC (1982) J Polym Sci Polym Chem 20: 1609–22
131. Neckers DC (1985) Polymeric bipyridines as chelating agents and catalysts, in Metal-containing polymer systems, Seats JE, Carraher CE, Pittman CU, Jr (eds), Plenum Pres NY, London
132. Kuchen W and Schram J (1988) Angew Chem Int Ed Engl 27: 1695–7
133. Harkins DA and Schweitzer GK (1991) Sep Sci Techn 26: 345–55
134. Tsukagashi K, Yu KY, Maeda M and Takagi M (1993) Bull Chem Soc Jpn 66: 114–20
135. Chanda M and Rempel GL (1992) React Polym 16: 149–58
136. Choi KS (1990) Makromol Chem Makromol Symp 33: 55–63
137. Rosatzin T, Andersson LI, Simon W and Mosbach K (1991) J Chem Soc Perkin Trans II 8: 1261–7
138. Bidan G, Divisia-Blohorn B, Lapkowski M, Kern J-M and Sauvage J-P (1992) J Amer Chem Soc 114: 5986–94
139. Belokon YN, Tararov I, Savel'eva TF, Vorob'ev MM, Vitt SV, Sizoy VF, Sukhacheva NA, Vasil'ev GV and Belekov VM (1983) Makromol Chem 184: 2213–23
140. Fuji Y, Matsutani K and Kilkuchi K (1985) J Chem Soc Chem Commun 415–7
141. Fuji Y, Matsutani K, Ota M, Adachi M, Savoji I, Haneishi and Kuwana Y (1984) Chem Lett 1487–90

142. Dhal PK and Arnold FH (1991) J Amer Chem Soc 113: 7417–8
143. Yamamura K, Hatakeyama H, Naha K, Tabushi I (the late) and Kurihara K (1988) J Chem Soc Chem Commun 79–81
144. Byström SE, Börje A and Akermark B (1993) J Amer Chem Soc 115: 2081–3
145. Sarhan A, Abou M and El-Zabah (1987) Makromol Chem Rapid Commun 8: 555
146. Andersson LI and Mosbach K (1989) Makromol Chem Rapid Commun 10: 491–5
147. Hopkins A and Williams A (1983) J Chem Soc Perkin Trans II 891–6
148. Leonhardt A and Mosbach K (1987) React Polym Ion Exch Sorbents 6: 285–6
149. Robinson DK and Mosbach K (1989) J Chem Soc Chem Commun 969–70
150. Morihara K, Kurihara S and Suzuki J (1988) Bull Chem Soc Jpn 61: 3991–8
151. Morihara K, Nishihata E, Kojima M and Miyake S (1988) Bull Chem Soc Jpn 61: 3999–4003
152. Shimada T, Nakamishi K and Morihara K (1992) Bull Chem Soc Jpn 65: 954–8
153. Morihara K, Tanaka E, Takeuchi Y, Miyazaki K, Yamamoto N, Sagawa Y, Kawamoto E and Shimida T (1989) Bull Chem Soc Jpn 62: 499–505
154. Morihara K, Kurokawa M, Kamata Y and Shimada T (1992) J Chem Soc Chem Commun 358–60
155. Morihara K, Kawasaki S and Kofuji M and Shimada T (1993) Bull Chem Soc Jpn 66: 906–13
156. Muller R, Andersson LI and Mosbach K (1993) Makromol Chem Rapid Comm 14: 637
157. Shokat KM, Leumann CJ, Sugasaware R and Schultz PG (1989) Nature 338: 169
158. Klibanov AM (1989) Trends Biochem Sci 14: 141
159. Russell AJ and Klibanov AM (1988) J Biol Chem 263: 1624–6
160. Braco L, Dabulis K and Klibanov AM (1990) Proc Natl Acad Sci 87: 274–7
161. Dabulis K and Klibanov AM (1992) Biotech Bioeng 39: 176–85
162. Ståhl M, Månsson M-O and Mosbach K (1990) Biotech Lett 12: 161–6
163. Ståhl M, Jeppssonwistrand U, Månsson M-O and Mosbach K (1991) J Amer Chem Soc 113: 9366–8
164. Ståhl M, Månsson M-O and Mosbach K (1993) Protein Eng 6: 51–51
165. Kriegel T, Schellenberger W, Kopfschläger G and Hoffmann E (1991) Biomed Biochim Acta 50: 1159–65
166. Schultz PG (1989) Angew Chem Int Ed Engl 28: 1283–95
167. Green BS (1991) Curr Opin Biotech 2: 395–400
168. Rini JM, Schulzegahmen U and Wilson IA (1992) Science 255: 959–66
169. Tramontano A, Janda KD and Lerner RA (1986) Proc Natl Acad Sci 83: 6736–40
170. Pollack SJ, Jacobs JW and Schultz PG (1986) Science 234: 1570–3
171. Winteer G and Milstein C (1991) Nature 349: 293–9
172. Lerner RA, Kang AS, Bain JD, Burton DR and Barbas CF (1992) Science 285: 1313
173. Tawfik DS, Zemel RR, Arrad-Yellin R, Green BS and Eshhar Z (1990) Biochemistry 29: 9916–21
174. Shokat KM, Ko MK, Scanlin TS, Kochersperger L, Yonkovich L, Thaisrivongs S and Schultz PG (1990) Angew Chem Int. Ed Engl 29: 1296–1303.
175. Reymond JL, Janda KD and Lerner RA (1991) Angew Chem Int Ed Engl 30: 1711–3
176. Iverson BI, Cameron KE, Jahangiri GK and Pasternak DS (1990) J Amer Chem Soc 112: 5320–3
177. Iverson BL and Lerner RA (1989) Science 243: 1184–8
178. Scanlan TS, Prudent JR and Schultz PG (1991) J Amer Chem Soc 113: 9397–8
179. Sinha SC, Keiman E and Reymond J-L (1993) J Amer Chem Soc 115: 4893–4
180. Liotta LJ, Benkovic PA and Miller GP (1993) J Amer Chem Soc 115: 350–1
181. Braisted AC and Schultz PlG (1990) J Amer Chem Soc 112: 7430–1
182. Keinan E, Sinha SC, Sinha-Bagchi A, Benory E, Ghozi MC, Eshhar Z and Green BS (1990) Pure Appl Chem 62: 2013–9
183. IKeda S, Weinhouse MI, Janda KD, Lerner RA and Danishefsky SJ (1991) J Amer Chem Soc 113: 7763–4
184. Jackson DY, Liang MN, Bartlett PA and Schultz PG (1992) Angew Chem Int Ed Engl (1992) 3: 182–3
185. Jackson JR, Prudent JR, Kochersperger L, Yonkovich S and Schultz PG (1992) Science 256: 365–7
186. Uno T and Schultz PG (1992) J Amer Chem Soc 114: 6573–4
187. Fernholz E, Schloeder D, Liu KKC, Bradshaw CW, Huang HM, Janda KD, Lerner RA and Wong CH (1992) J Org Chem 57: 4756–61
188. Gibbs RA, Taylor S and Benkovic SJ (1992) Science 258: 803–6

189. Jackson DY, Prudent JR, Baldwin EP and Schultz PG (1991) Proc Natl. Acad Sci 58: 58–62
190. Shea KJ, Stoddard GJ, Shavelle DM, Walmi F and Choate MM (1990) Macromols 23: 4497–507
191. Shea KJ, Thompson EA, Pandy SD and Beauchamp PS (1980) J Amer Chem Soc 102: 3149–3155
192. Csaba G, Kovács P, László V (1989) Acta Protozoologica 28: 175–82
193. Wulff G, Private Communication
194. Wulff G and Stellbrink H (1990) Rec Trav Chim Payes-Bas (J Roy Soc Neth Chem Soc) 109: 216–21
195. Gelfi C and Righetti PG (1981) Electrophoresis 2: 220–28
196. Wu TF and Orgel LE (1992) J Amer Chem Soc 114: 5496–500
197. Wu TF and Orgel LE (1992) J Amer Chem Soc 114: 7963–8
198. Rosenberg J-E and Flodin P (1987) Macromols 20: 1518–22
199. Rosenberg J-E and Flodin P (1986) Macromols 19: 1543–6
200. Hoss R and Vogtle F (1994) Angew Chem Int Edn Engl 33: 375
201. Mullis KB, Scientific American, April 1990, p 56
202. Erlich HA, Gelfand D and Sninsky JJ, Science (1991) 252: 1643
203. Brock TD and Freeze H J. Bacteriol (1969) 98: 289
204. Saiki RK, Science (1988) 239: 487

Editor: Prof. Ledwith
Received: June 1994

Polymers as Free Radical Photoinitiators

C. Carlini and L. Angiolini
Dipartimento di Chimica Industriale e dei Materiali, University of Bologna, Viale Risorgimento 4, 40136 Bologna, Italy

The present review classifies, discusses and critically summarizes the results in the field of free radical polymerizations as well as of photocrosslinking and photografting reactions promoted by polymeric photoinitiators. Particular emphasis is given to the synthesis and structure-properties relations of the cited polymeric systems in order to optimize their molecular design for obtaining the desired performances. The advantages of polymeric photoinitiators with respect to the corresponding low-molecular-weight analogues are also examined. Attention is given to special applications from both the technological and fundamental point of view. Future developments based on the peculiarity of the polymeric systems to give synergistic effects are finally evidenced.

Advances in Polymer Science, Vol. 123

1 Introduction

Polymers acting as free radical photoinitiators can be defined as macromolecular systems containing side-chain or main-chain photosensitive moieties which, through the action of light absorption at the appropriate wavelength, produce primary radicals able to initiate polymerization and crosslinking reactions of mono- and multi-functional monomers and oligomers. Polymeric systems bearing either photodissociable groups along the backbone or photomodifiable end-groups to be utilized, in the presence of a proper monomer, for obtaining block copolymers, also fit well with the above definition. On the other hand, photocrosslinkable polymers, containing pendant photoreactive moieties such as azidoaryl, cynnamoyl, chalchonyl, maleimido etc. groups, largely applied as photoresists for imaging systems, will not be considered here, as their application technology does not provide photoinitiated polymerization and crosslinking of any added monomer, but gives rise to a network by connecting linear polymer chains through a cure process in which each reaction steps require the absorption of at least one photon.

In the last few decades UV curing has found extensive application in the area of surface coatings, printing inks and more recently printed circuits as well as optical discs. The need for advanced materials in the above specific fields can be satisfied not only by improving the industrial technology in terms of capability of controlling drying and crosslinking processes, but also with a better knowledge of the relationship between physical properties of the finished materials and molecular structure of formulation components. Taking into account that the photoinitiator can appreciably affect the overall performance of the polymerizable systems, much effort has been spent in order to optimize important aspects of this component, such as absorption in the appropriate wavelength region, quantum yield for generation of active species and their reactivity towards monomers, solubility, stability in the dark as well as yellowing, odouring and toxicological properties.

Indeed, with the aim not only to increase cure speed, process productivity and energy saving, but also to improve the ultimate properties of the final product, new tailor-made low-molecular-weight photoinitiators have been designed for specific fields of application such as pigmented lacquers, clear coatings, printing inks, thick layered coatings, adhesives, optical discs, printing plates and electronic circuits. More recently, polymeric photoinitiators have also been subjected to an increasing research interest as they are expected to offer several advantages compared with their corresponding low-molecular-weight analogues.

Some of the most important benefits connected with the macromolecular architecture of polymeric photoinitiators can be summarized as follows.

1) Improvement of photoinitiation activity resulting from either energy migration between excited and ground state photosensitive moieties along the polymer chain or intramolecular reactions responsible for the formation of more active species. Higher activity can also be achieved by the photogeneration of active species which, being protected, like in a cage, by the microenvironment of the

polymer chain, reduce their tendency to coupling processes, thus favouring their reaction with monomers.

2) The copolymerization of photosensitive monomers with conventional comonomers enables synthesis of polymeric systems bearing the photoreactive moieties at different distance from each other, thus allowing modulation of their initiation activity for specific applications.

3) The design of polymeric systems with pendant photosensitive moieties at different distance from the backbone allowing modification of their photoreactivity as a consequence of the microenvironment variation.

4) The synthesis, through copolymerization routes, of polymeric systems bearing on the same macromolecule different photosensitive groups may provide potential synergistic effects of activity.

5) Light stability improvement of the coating in terms of non-yellowing properties, as most of the residual photosensitive groups remaining attached to the polymer network strongly reduce their migration capability onto the film surface.

6) Possibility of manufacturing low-odouring and non-toxic coatings due to the substantial absence of volatility of the polymeric systems as well as the reduced release of photofragments, most of which remain linked to the polymer matrix.

The investigation of the relationships between structural requirements and photoinitiation activity of the above polymeric systems, as far as points 1–4 are concerned, is very important. In fact, this may produce not only a better comprehension of the real working mechanism of the photosensitive moiety when bound to a macromolecular chain, but also the ability to design more advanced photoinitiators with a gain of industrial productivity and a reduction of irradiation exposure. Points 5 and 6, being intrinsically connected with the polymeric nature of the photoinitiator, assume a relevant importance from the technological point of view, in terms of enchancement of ultimate properties of the finished product.

The present overview is divided in three main sections where the polymeric photoinitiators are presented according to their mechanism of action, i.e. photoinduced hydrogen abstraction, electron transfer and cleavage reactions. In each section, particular attention is devoted to the synthesis and relationship between molecular structure and photoinitiation properties of the polymeric systems. A further section deals with some special applications having particular interest from the point of view of fundamental research and technological development.

2 Photoinitiation by a Hydrogen Abstraction Mechanism

It is well known [1–4] from extensive research on low-molecular-weight systems that, under UV light exposure, aromatic ketones, the most representative being benzophenone, anthraquinone and thioxanthone, give rise to free radical species with a mechanism which provides the formation of a ketone singlet excited state, its evolution by an intersystem crossing process (ISC) to a triplet state and finally its photoreduction by a hydrogen donor (RH). In Scheme 1 the above process is represented for benzophenone.

$$Ar_2C{=}O \xrightarrow{h\nu} [Ar_2C{=}O]_{S1} \xrightarrow{ISC} [Ar_2C{=}O]_{T1} \xrightarrow{RH} Ar_2\dot{C}{-}OH + R\cdot$$
$$\qquad\qquad n \rightarrow \pi^* \qquad\qquad n \rightarrow \pi^*$$

Scheme 1

The most effective hydrogen donors are reported [1, 2, 5] to be alcohols and ethers, hydrocarbons behaving similarly although to a lesser extent.

It is also well established [2, 6–8] that in the photoinduced vinyl polymerization promoted by benzophenone, the free radical deriving from the hydrogen donor is active in the initiation step, whereas the semipinacol radical mainly undergoes self-coupling and combination with the growing polymer chains. It is pointed out, however, that this termination reaction may involve both hydrogen transfer and direct combination (Scheme 2).

Initiation:

$$R\cdot + M \longrightarrow P_1\cdot$$

Propagation:

$$P_1\cdot + (n-1)M \longrightarrow P_n\cdot$$

Termination:

$$P_m\cdot + P_n\cdot \longrightarrow \text{Polymer}$$

$$2Ar_2\dot{C}{-}OH \longrightarrow Ar_2C(OH){-}C(OH)Ar_2 \quad \text{(self-coupling)}$$

$$Ar_2\dot{C}{-}OH + P_n\cdot \longrightarrow Ar_2C(OH){-}P_n \quad \text{(direct combination)}$$

$$Ar_2\dot{C}{-}OH + P_n\cdot \longrightarrow Ar_2C{=}O + P_nH \quad \text{(hydrogen transfer)}$$

Scheme 2

Preliminary work dealing with UV irradiation of long-chain *n*-alkyl esters of benzophenone-4-carboxylic acid [9, 10] allowed cyclic products to be obtained, thus suggesting the formation of intermediate di-radicals by intramolecular hydrogen abstraction (Scheme 3).

The possibility of applying similar intramolecular hydrogen abstraction reactions to the photoinduced polymerization of methyl methacrylate (MMA), has been tested [1] by using long-chain *n*-alkyl N-substituted imides of 3,3′,4,4′-benzophenone tetracarboxylic dianhydride (BTDA) (Scheme 4).

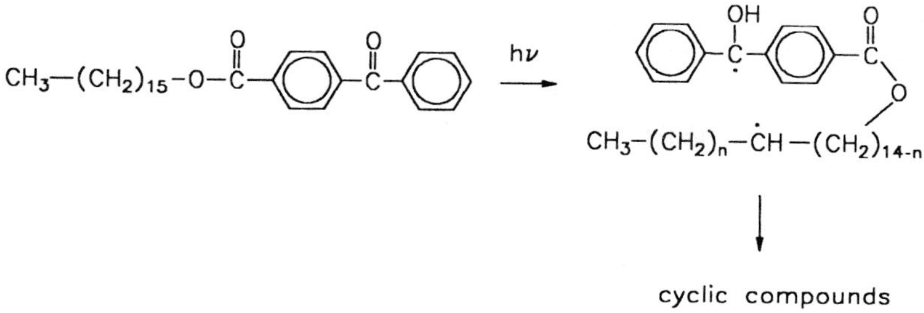

$$CH_3-(CH_2)_{15}-O-\overset{O}{\overset{\|}{C}}-C_6H_4-\overset{O}{\overset{\|}{C}}-C_6H_5 \xrightarrow{h\nu} C_6H_5-\overset{OH}{\overset{|}{\dot{C}}}-C_6H_4-\overset{O}{\overset{\|}{C}}-O-(CH_2)_{14-n}-\dot{C}H-(CH_2)_n-CH_3 \longrightarrow \text{cyclic compounds}$$

Scheme 3

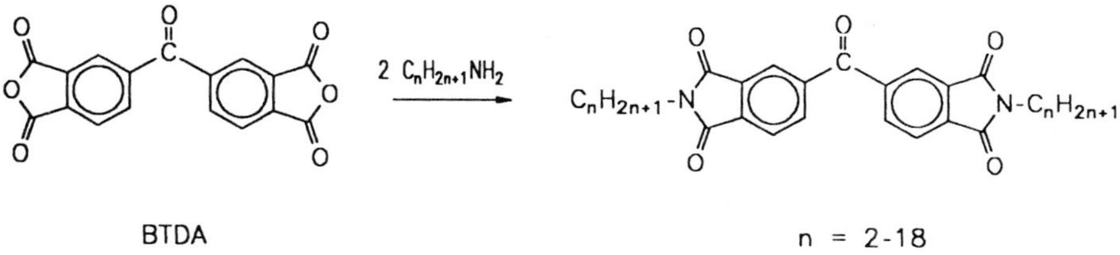

$$\text{BTDA} \xrightarrow{2\ C_nH_{2n+1}NH_2} C_nH_{2n+1}\text{-N(imide)}-C_6H_3-\overset{O}{\overset{\|}{C}}-C_6H_3-\text{(imide)N-}C_nH_{2n+1}$$

BTDA

n = 2-18

Scheme 4

It has been found that the C_{18}-imide derivative exhibits, as compared with BTDA, the highest efficiency, according to an intramolecular hydrogen abstraction activation mechanism.

In this context, polymeric systems containing side-chain benzophenone groups appeared very interesting regarding crosslinking and grafting to photoinitiating properties. Accordingly, the photograft polymerization of MMA onto poly(4-vinylbenzophenone-*co*-styrene) [poly(VBP-*co*-St)], performed in benzene solution and in the absence of any hydrogen donor, has been successfully carried out [11], thus confirming that intramolecular hydrogen abstraction occurs from the polymer backbone.

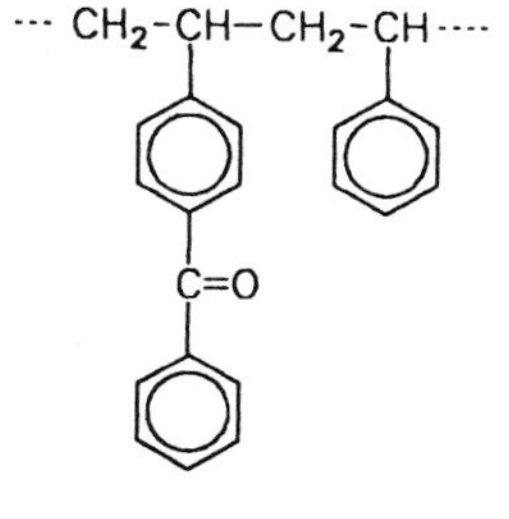

$$\cdots CH_2-\underset{C_6H_4-C(=O)-C_6H_5}{\underset{|}{C}H}-CH_2-\underset{C_6H_5}{\underset{|}{C}H}\cdots$$

poly(VBP-*co*-St)

The solubility of the graft copolymers obtained at room temperature strongly suggests that the semipinacol radicals are mainly involved in intramolecular coupling reactions. On the other hand, when the reaction is carried out at higher temperature [11], the formation of a graft copolymer which swells in benzene is observed, thus indicating that crosslinking may also occur.

UV irradiation of poly(4-vinylbenzophenone) [poly(VBP)] in benzene solution, and in the presence of isopropanol as hydrogen donor, gives rise to a more complex picture [12, 13]. Indeed, intra- and inter-molecular coupling reactions by the side-chain benzophenone ketyl radicals (K·) markedly change the macromolecular morphology with the occurrence of cyclic and network structures as well as chain scission processes (Scheme 5).

According to a similar mechanism, the photolysis of poly[bis(4-benzoylphenoxy)phosphazene] [poly(BPP)] in air equilibrated CH_2Cl_2 solution induces chain scission and extensive degradation of the macromolecules [14].

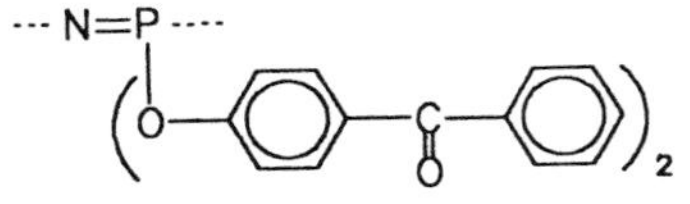

poly(BPP)

As intramolecular hydrogen abstraction from the backbone is not possible in this case, the above process is believed to originate from the cleavage of peroxy radicals formed by reaction of oxygen with phosphorus macroradicals derived from residual unreacted P-Cl groups in the phosphazene main chain (Scheme 6).

On the contrary, in the absence of molecular oxygen, or in the presence of hydrogen donors, bimolecular processes involving benzophenone ketyl radicals analogous to those depicted in Scheme 5 are reported to lead to extensive photocrosslinking and gel formation [14]. Similarly, thin films of poly(4-benzoylphenoxymethyl styrene) [poly(BPMS)], upon UV exposure, also give rise to crosslinking reactions [15].

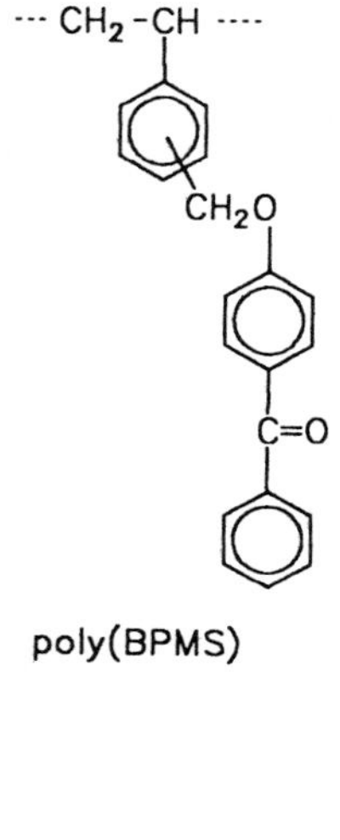

poly(BPMS)

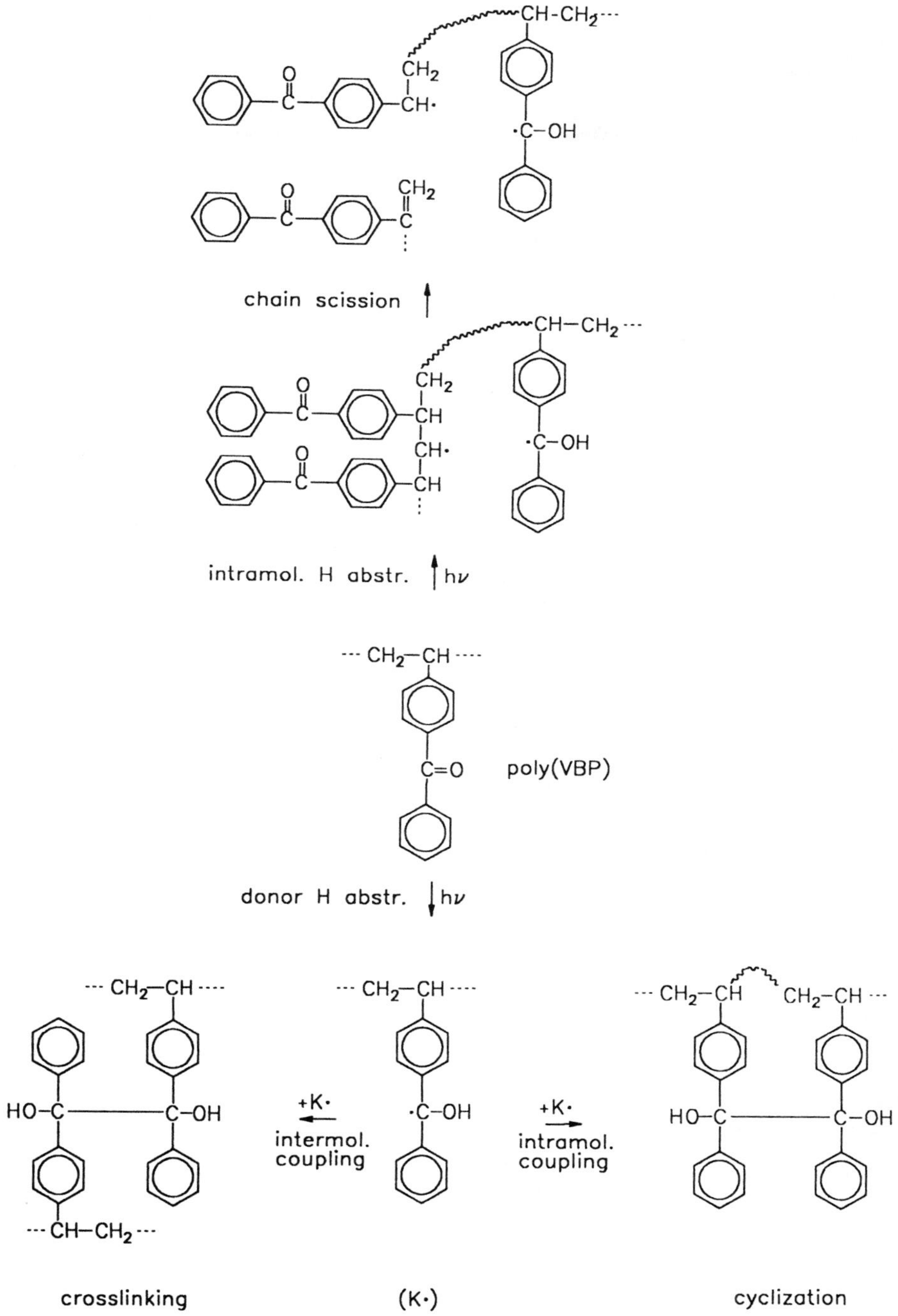

chain scission

intramol. H abstr. $h\nu$

poly(VBP)

donor H abstr. $h\nu$

$+K\cdot$ intermol. coupling

$+K\cdot$ intramol. coupling

crosslinking

(K·)

cyclization

Scheme 5

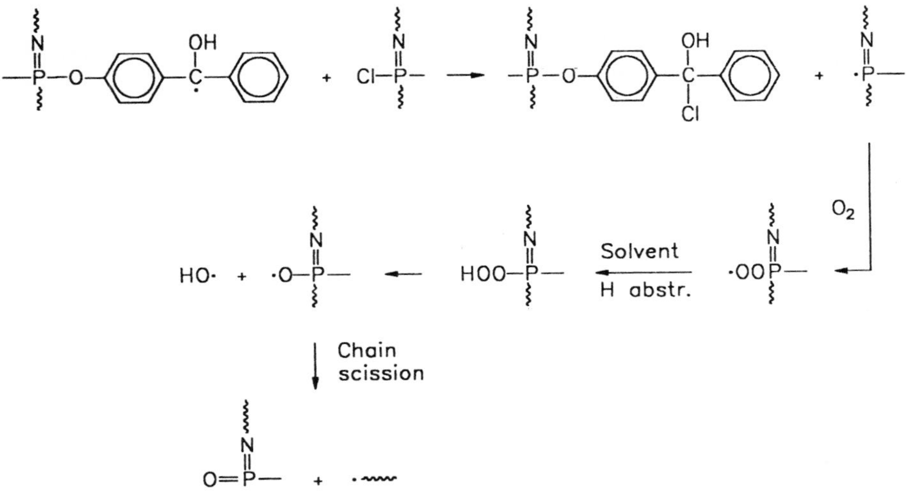

Scheme 6

Benzoylated polystyrenes [poly(VBP-*co*-St)] having different contents of VBP co-units have been checked [16] as initiators for the photoinduced polymerization of several unsaturated monomers using solvents with different hydrogen donating capability, such as dimethylformamide (DMF), tetrahydrofuran (THF), cyclohexane and aromatic hydrocarbons.

The rate of polymerization in DMF, upon UV irradiation at 365 nm, decreases in the sequence vinyl acetate (VAc) $>$ acrylonitrile (AN) $>$ MMA $\gg$ St ≈ 0. The lack of polymerization in the case of styrene, under the above conditions, is explained by the authors on the assumption that the energy of the triplet state of the pendant benzophenone moieties, analogous to free benzophenone [17], is higher than that of styrene which therefore behaves as a quencher. The increase of hydrogen donating properties of the solvent is accompanied, at least in the AN polymerization, by an improvement of poly(VBP-*co*-St) photoinitiation activity. These results indicate that the hydrogen donor efficiency increases in the sequence benzene $<$ cyclohexane $<$ toluene $\approx$ DMF $<$ THF $\ll$ p-xylene. Poly(VBP-*co*-St) and benzophenone show comparable efficiency in the UV initiated polymerization of the above unsaturated monomers [16]; however, no detailed investigation concerning the relations between photoinitiation activity and structure of the polymeric system, in terms of content of benzophenone units, has been performed. More recently, the influence of chemical structure of polymers bearing side-chain benzophenone moieties on the photoinitiated polymerization and crosslinking reactions has been studied [18–22] for mono- and bifunctional acrylic monomers as models of more complex commercial formulations, usually applied for UV curable clear coatings.

Firstly, poly(4-acryloxybenzophenone) [poly(ABP)] has been compared with the corresponding low-molecular-weight analogues, such as the monomeric

4-(2-methylpropionyloxy)benzophenone (IBP) and the dimeric bis(4-hydroxybenzophenone) glutarate (GBP), in the UV curing in film matrix of 1,6-hexanediol-diacrylate (HDDA)/2-acryloxy-2′-propionyloxy-diethylether (APDG) or HDDA/*n*-butyl acrylate (BA) equimolar mixtures.

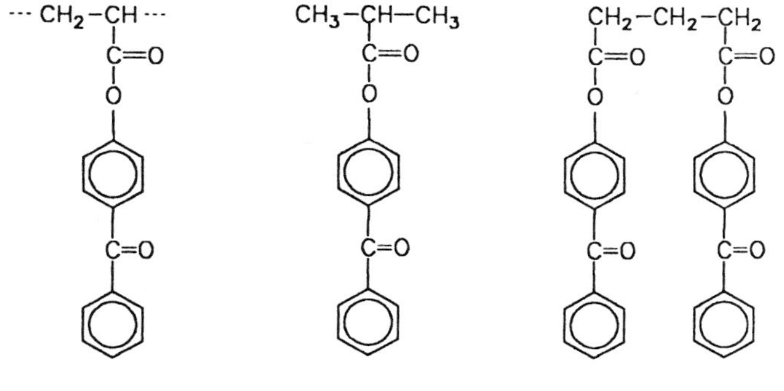

poly(ABP) IBP GBP

$$CH_2{=}CH{-}\overset{O}{\overset{\|}{C}}{-}O{-}(CH_2)_6{-}O{-}\overset{O}{\overset{\|}{C}}{-}CH{=}CH_2$$

HDDA

$$CH_3CH_2{-}\overset{O}{\overset{\|}{C}}{-}O{-}(CH_2)_2{-}O{-}(CH_2)_2{-}O{-}\overset{O}{\overset{\|}{C}}{-}CH{=}CH_2$$

APDG

$$CH_2{=}CH{-}\overset{O}{\overset{\|}{C}}{-}O{-}(CH_2)_3{-}CH_3$$

BA

Under the same experimental conditions, poly(ABP) is found [18, 20, 22] to display much higher photoinitiation activity than IBP and GBP, as revealed by a remarkable shortening of the half-time of the curing processes (Tables 1 and 2). These results were firstly explained assuming the presence of energy migration between excited and ground state neighbouring benzophenone moieties along the polymer chain. This occurence would in fact increase the effective collision distance between triplet-state benzophenone groups and hydrogen donors, thus enhancing the probability of hydrogen abstraction, with an improvement of the initiation efficiency.

Analogous considerations have been invoked [23] in order to explain the shorter lifetime of the triplet state in poly(VBP) against free benzophenone in the presence of THF and the higher efficiency of the polymeric system in the photoinduced hydrogen abstraction from the above solvent (Table 3) as well as in

Table 1. UV curing in film matrix of the HDDA/APDG equimolar mixture, under nitrogen, by low- and high-molecular-weight photoinitiators based on benzophenone (BP) moieties[a] [18–22]

Photoinitiating system			Polymerization kinetics	
Type	ABP units (mol%)	Mean sequence length of ABP co-units	$t_{1/2}$[b] (s)	R_c[c] (s^{-1})
IBP	–	–	10.6	7
Poly(ABP)	100.0	–	6.0	24
Poly(ABP-*co*-MA)	54.3	2.26	6.3	17
Poly(ABP-*co*-MA)	33.4	1.42	6.3	17
Poly(ABP-*co*-MA)	10.2	1.10	8.3	10
Poly(ABP-*co*-MtA)	80.0	5.07	5.0	29
Poly(ABP-*co*-MtA)	57.6	2.35	3.4	50
Poly(ABP-*co*-MtA)	32.5	1.44	3.4	50
Poly(ABP-*co*-AEE)	55.0	n.d.	5.0	28
Poly(ABP-*co*-HEA)	90.0	n.d.	4.2	40

[a] Photoinitiator concentration: 1 mol% of benzophenone moieties in the HDDA/APDG mixture. Irradiation performed by a medium-pressure linear (15 cm) Hg lamp (500 W) on a 10 μm thick liquid film between two NaCl discs located at a distance of 20 cm
[b] Half-time of the process as determined by IR spectroscopy
[c] Polymerization rate evaluated at $t_{1/2}$ and expressed as % conversion over time

Table 2. UV curing in film matrix of the HDDA/BA equimolar mixture, under nitrogen, by polymeric systems bearing the benzophenone (BP) moiety at different distance from the backbone as compared with low-molecular-weight analogues[a] [20–22]

Photoinitiating system		Polymerization kinetics	
Type	BP units (mol%)	$t_{1/2}$[b] (s)	R_c[c] (s^{-1})
Poly(ABP)	100.0	72	0.9
GBP	–	112	0.6
IBP	–	143	0.5
Poly(VBP)	100.0	78	0.8
Poly(VBP-*co*-MtA)	81.9	68	1.0
Poly(VBP-*co*-MtA)	55.4	52	1.4
Poly(VBP-*co*-MtA)	21.7	38	1.7
Poly(VBP-*co*-MtA)	5.0	31	2.3
PBP	–	165	0.4
Poly(UP36)	100.0	44	1.5
Poly(UP36-*co*-MtA)	72.2	41	1.6
Poly(UP36-*co*-MtA)	52.5	41	1.6
Poly(UP36-*co*-MtA)	30.1	37	1.7

[a] Photoinitiator concentration: 1 mol% of benzophenone moieties in the HDDA/BA mixture. Irradiation at 330 nm by a high-pressure Hg lamp (53 W/m^2) on a 300 μm thick liquid film loading a quartz cuvette
[b] Half-time of the process, as determined by microwave dielectrometry at 9.5 GHz in terms of ε'' (loss factor)
[c] Polymerization rate evaluated at $t_{1/2}$ and expressed as % conversion over time

Table 3. Triplet state lifetime (τ) of poly(VBP) and benzophenone (BP) in benzene as a function of tetrahydrofuran concentration [THF] and rate constants (k_a) of hydrogen abstraction from THF at room temperature in the same solvent[a] [23]

τ (μs)					$10^6 k_a$ ($l\ mol^{-1} s^{-1}$)[b]
[THF] ($mol\ l^{-1}$)	0	0.1	0.3	0.6	
Poly(VBP)	2.27	0.92	0.33	0.28	7±2
BP	3.57	1.12	0.82	0.46	3±1

[a] [poly (VBP)] = [BP] = $5 \cdot 10^{-3}$ $mol\ l^{-1}$

[b] Evaluated according to the usual Stern-Volmer treatment: $1/\tau = 1/\tau_0 + k_a[THF]$, where τ_0 is the triplet state lifetime at [THF] = 0

the photoinitiated polymerization of MMA. Investigation of the phosphorescence yield in the presence of naphthalene as quencher for poly(VBP) and copolymers of VBP either with MMA [poly(VBP-*co*-MMA)s] [23] or with styrene [24] at different compositions confirms this picture. Indeed, the results clearly show (Table 4) that the efficiency of energy transfer from the excited side-chain benzophenone chromophores to naphthalene, measured in terms of critical transfer distance (R_0) according to Hirayama's theory [25], decreases on increasing the average distance between the benzophenone moieties along the polymer chain, that is decreasing the content of VBP units in the macromolecules.

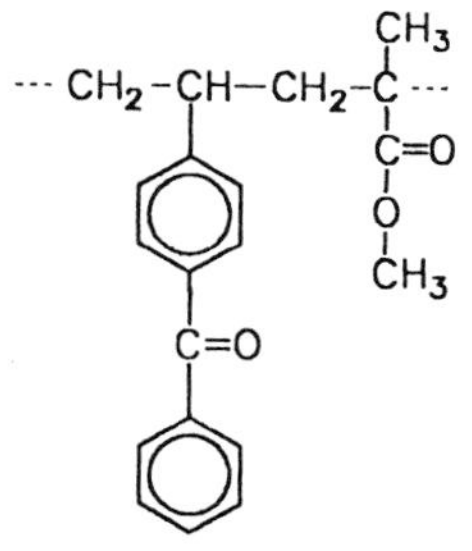

poly(VBP-*co*-MMA)

Finally, phosphorescence polarization measurements, performed on poly(VBP-*co*-St)s and poly(VBP) in glassy solution at 77 K, clearly show that the emission is subjected to a progressive depolarization on increasing the content of VBP co-units in the polymeric systems, becoming complete in poly(VBP) [26]. These data have again been explained in terms of energy transfer between side-chain excited- and ground-state benzophenone moieties, the efficiency of the above process increasing on decreasing the distance between VBP co-units in the macromolecules.

However, the improvement of photoinitiation activity under nitrogen atmosphere, found with poly(ABP), as compared with IBP and GBP, could also be

Table 4. Critical transfer distance (R_0) for poly(VBP), poly(VBP-*co*-St)s and poly(VBP-*co*-MMA)s against benzophenone (BP) [23, 24]

Sample	VBP co-units (mol%)	R_0[a] (Å)
Poly(VBP)	100	19.7
Poly(VBP-*co*-St)	77	19.0
Poly(VBP-*co*-MMA)	30	15.1
Poly(VBP-*co*-St)	50	14.5
Poly(VBP-*co*-MMA)	9	12.3
Poly(VBP-*co*-St)	9	11.9
BP	–	12.4

[a] Determined by phosphorescence measurements in glassy solution at 77 K, as a function of naphthalene concentration used as a quencher

explained in principle by assuming that intramolecular hydrogen abstraction occurs within the polymeric system. Therefore, with the aim of clarifying this point, copolymers with variable content of benzophenone units, as well as of co-units having different hydrogen donating properties, have been prepared [18, 20, 22] from 4-acryloxybenzophenone (ABP) and several acrylic comonomers, such as methyl acrylate (MA), *n*-decyl acrylate (DA), (−)-menthyl acrylate (MtA), 2-hydroxyethyl acrylate (HEA) and 1-acryloxy-2-ethoxyethane (AEE). Indeed, poly(ABP-*co*-MA)s show a lower activity with respect to poly(ABP) in the photoinitiated polymerization of the HDDA/APDG equimolar mixture on decreasing the mean sequence length of ABP units in the polymeric systems, that is on increasing the distance between benzophenone chromophores, as clearly evidenced in Table 1 by the kinetic parameters of the UV curing processes. The above results [18], taking into account the low hydrogen donating properties of the methyl group in MA co-units, strongly support the view that activity is appreciably affected by energy migration between neighbouring excited- and ground-state benzophenone chromophores along the polymer chain, an enhancement of this migration being responsible for an improvement of photoinitiation activity.

poly(ABP-*co*-MA) (R = methyl)
poly(ABP-*co*-DA) (R = *n*-decyl)
poly(ABP-*co*-MtA) (R = (-)-3-menthyl)
poly(ABP-*co*-AEE) (R = 2-ethoxyethyl)
poly(ABP-*co*-HEA) (R = 2-hydroxyethyl)

However, as shown in Table 1, poly(ABP-*co*-MtA)s display [18, 20] an appreciably higher photoinitiation activity as compared with poly(ABP), despite the mean sequence length of ABP units in the copolymers being practically the same as found for poly(ABP-*co*-MA)s having comparable compositions. These findings clearly confirm that the photoexcitation energy migration along the polymer chain is not the main parameter affecting the activity of the benzophenone containing polymers, at least when co-units with hydrogen donating properties are present in the macromolecules. Indeed, the photoinitiation activity of poly(ABP-*co*-MtA)s increases [18, 20] on decreasing the mean sequence length of ABP co-units, that is increasing the local concentration of hydrogen donating groups around the benzophenone chromophores (Table 1). This behaviour strongly indicates that intramolecular hydrogen abstraction from the donors present as side-chain in the co-units, is even more important than energy migration.

In order to give a more general confirmation of this hypothesis, photoinitiated polymerization experiments on HDDA/APDG equimolar mixtures have been extended [18, 20] to poly(ABP-*co*-AEE) and poly(ABP-*co*-HEA), AEE and HEA co-units containing, respectively, ethereal and alcoholic functions which are well known to be the most reactive in hydrogen abstraction processes by aromatic ketones [1, 17]. Indeed, in the above copolymers a small amount of hydrogen donating units (less than 10 mol%) is enough to induce an appreciably higher photoinitiation activity against poly(ABP) (Table 1). Considering that the ethereal functionality of APDG in the curing formulation could compete with AEE units in hydrogen abstraction reaction and hence reduce the difference of activity of the copolymer with respect to poly(ABP), HDDA/APDG has been replaced by the HDDA/BA equimolar mixture under the same irradiation conditions [18]. Thus, the photoinitiation activity of poly(ABP-*co*-AEE), containing 55 mol% of AEE units, has been compared with that of poly(ABP) alone or in combination with the corresponding amount of 1-(2-methylpropionyloxy)-2-ethoxy-ethane (IEE), the low-molecular-weight analogue of AEE co-units.

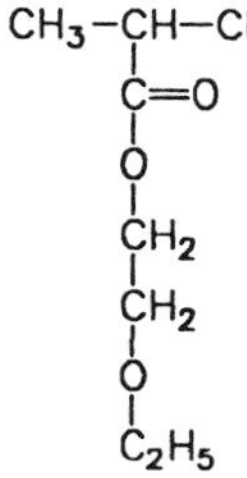

$$
\begin{array}{l}
CH_3-\underset{|}{C}H-CH_3 \\
\quad\ C{=}O \\
\quad\ \ | \\
\quad\ \ O \\
\quad\ \ | \\
\quad\ CH_2 \\
\quad\ \ | \\
\quad\ CH_2 \\
\quad\ \ | \\
\quad\ \ O \\
\quad\ \ | \\
\quad\ C_2H_5
\end{array}
$$

IEE

Indeed, as shown in Table 5, the difference of activity of poly(ABP-*co*-AEE) with respect to poly(ABP) in the photocuring of the HDDA/BA equimolar mixture is remarkably higher than in the corresponding HDDA/APDG formulation (Table 1).

Table 5. UV curing in film matrix of the HDDA/BA equimolar mixture, under nitrogen, by poly(ABP-*co*-AEE)s and by the corresponding combination of poly(ABP) with the ethereal low-molecular-weight structural model IEE[a] [18]

Photoinitiating system			Polymerization kinetics	
Type	Benzophenone units (mol%)	Ethereal moiety (mol%)	$t_{1/2}$ [b] (s)	R_c [c] (s^{-1})
Poly(ABP-*co*-AEE)	77	23	14	7
Poly(ABP-*co*-AEE)	55	45	10	9
Poly(ABP)/IEE	55	45	17	6
Poly(ABP)	100	0	17	6

[a] Polymerization conditions: see Table 1
[b] Half-time of the process, as determined by IR spectroscopy
[c] Polymerization rate evaluated at $t_{1/2}$ and expressed as % conversion over time

It is noteworthy that the addition of IEE to poly(ABP) (Table 5) does not appreciably change the activity of poly(ABP), thus suggesting that the photoreduction of excited benzophenone chromophores is much more efficient when both ethereal groups and aromatic ketone moieties are attached to the same macromolecule. This behaviour can easily be explained in terms of a higher local concentration of ethereal functions around the benzophenone chromophores in poly(ABP-*co*-AEE) than in the corresponding poly(ABP)/IEE system.

In order to have a better insight on the parameters affecting the activation mechanism of benzophenone containing polymeric photoinitiators, a sample of poly(VBP-*co*-MtA), having quite a low amount of VBP co-units (21.7 mol%), has been checked [18] in the UV curing of the HDDA/BA equimolar mixture against various combinations of different homopolymers and low-molecular-weight models at the same overall molar content of benzophenone moieties: poly(VBP)/menthyl acetate (MtAc), poly(VBP)/poly (menthyl acrylate) [poly(MtA)], 4-isopropyl benzophenone (PBP)/poly(MtA) and PBP/MtAc:

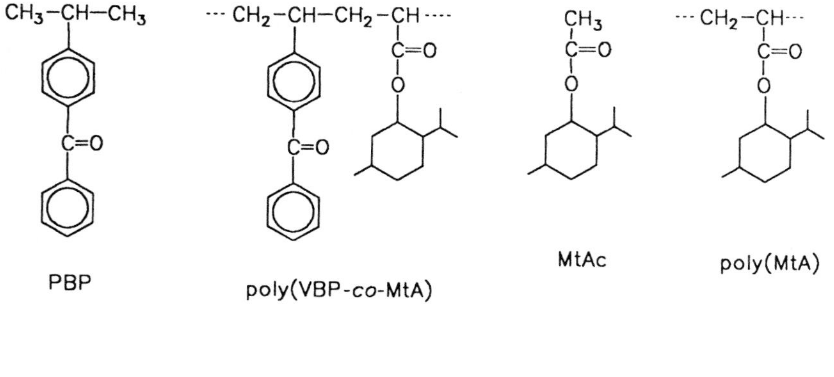

PBP

poly(VBP-*co*-MtA)

MtAc

poly(MtA)

Table 6. UV curing in film matrix of the HDDA/BA equimolar mixture, under nitrogen, by poly-(VBP-*co*-MtA) containing 21.7 mol% of VBP co-units and by mixtures of homopolymers and low-molecular-weight structural models having the corresponding amount of benzophenone (BP) and menthyl moieties[a] [22]

Photoinitiating system			Polymerization kinetics	
Type	BP moiety (mol%)	Menthyl moiety (mol%)	$t_{1/2}$[b] (s)	R_c[c] (s^{-1})
Poly(VBP-*co*-MtA)	21.7	78.3	38	1.7
Poly(VBP)/MtAc	21.7	78.3	41	1.4
Poly(VBP)/poly(MtA)	21.7	78.3	50	1.2
Poly(VBP)	100.0	0.0	78	0.8
PBP/MtAc	21.7	78.3	114	0.6
PBP/poly (MtA)	21.7	78.3	127	0.5
PBP	100.0	0.0	165	0.4

[a] Polymerization conditions: see Table 2
[b] Half-time of the process, as determined by microwave dielectrometry at 9.5 GHz in terms of ε'' (loss factor)
[c] Polymerization rate evaluated at $t_{1/2}$ and expressed as % conversion over time

The relative photoinitiation activities of the above systems (Table 6) decrease as follows: poly(VBP-*co*-MtA) $\geq$ poly(VBP)/MtAc $>$ poly(VBP)/poly(MtA) $>$ poly(VBP) $\gg$ PBP/MtAc $>$ PBP/poly(MtA) $>$ PBP.

The highest activity found for poly(VBP-*co*-MtA) nicely confirms that when both benzophenone and menthyl moieties are pendant to the same macromolecule, intramolecular hydrogen abstraction occurs with the largest efficacy, energy migration playing a minor role due to low content of VBP co-units. The efficiency improvement caused by the addition of MtAc to poly(VBP) furtherly proves that hydrogen abstraction takes place more likely from the menthyl group than from the main chain of poly(VBP). The same occurrence can therefore easily explain the higher activity of poly(VBP-*co*-MtA) as compared with poly(VBP). The similar behaviour of poly(VBP)/MtAc and poly(VBP-*co*-MtA) suggests that poly(VBP) may generate domains where a higher local concentration of MtAc along the polymer chain is present, probably caused by an intercalation process. Finally, the much lower photoinitiation efficiency of all the systems based on PBP with respect to the corresponding polymeric counterparts can be attributed to the absence of intramolecular hydrogen abstraction.

On the basis of the above described results, tailor-made benzophenone-containing polymeric photoinitiators have been studied in order to establish the effect of the relative distance from the polymer chain of the side-chain benzophenone groups. Thus, the homopolymer of Uvecryl P36 [poly(UP36)] and its copolymers with MtA [poly(UP36-*co*-MtA)s], bearing the benzophenone moiety at quite a large distance from the backbone, have been checked as photoinitiators in the polymerization of the HDDA/BA equimolar mixture against poly(ABP), poly(VBP) and poly(VBP-*co*-MtA)s [20–22].

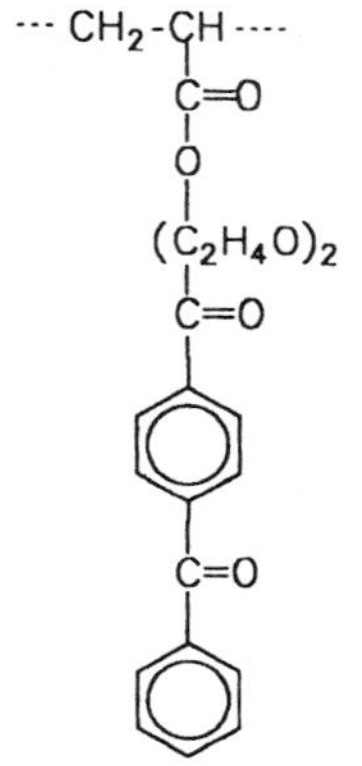

poly(UP36) 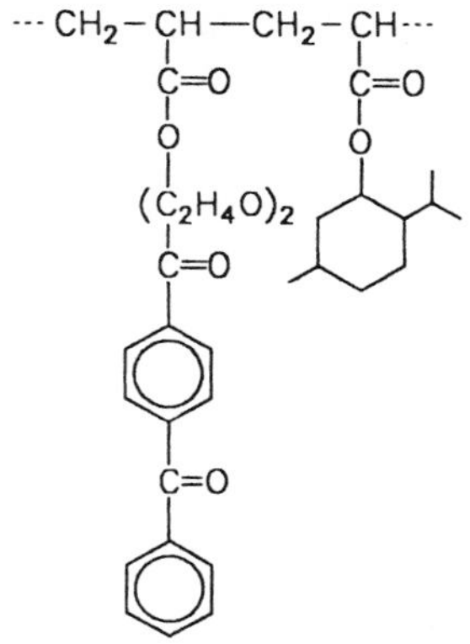 poly(UP36-*co*-MtA)

On the basis of the kinetic results (Table 2), poly(ABP) exhibits substantially the same activity as poly(VBP), despite the longer distance of benzophenone moiety from the backbone. However the homopolymers results indicate much more activity than with the corresponding low-molecular-weight monomeric analogues IBP and PBP, the dimeric model GBP showing an intermediate efficiency.

The above findings have been interpreted assuming that the conformational rigidity of the macromolecules does not appreciably affect the photoinitiation activity of the polymeric systems. The remarkably higher efficiency (Table 2) of poly(UP36) as compared with poly(ABP) and poly(VBP), is explained [22] more in terms of hydrogen donating capability of the ethereal groups inserted in the spacer between the benzophenone group and the backbone, than in terms of improvement of conformational mobility due to the longer distance of benzophenone moieties from the main chain. Indeed, this is confirmed by the observation that poly(UP36-*co*-MtA)s exhibit (Table 2) comparable or slightly higher activities with respect to poly(UP36), the replacement of UP36 by MtA units not appreciably changing the intramolecular hydrogen donating properties in both systems.

More detailed information on the activation pathway, which distinguishes the polymeric photoinitiators from the corresponding low-molecular-weight structural models, has been obtained by photophysical measurements [22, 27, 28] in terms of quantum yield and average lifetime of the triplet excited state of benzophenone moieties in the above systems. However, in order to get a better comprehension of this point, it is necessary to introduce some basic concepts about the kinetic treatment of a photoinitiated chain polymerization.

It is well established that the overall polymerization rate R_p for the above process can be expressed [29–31] by the following equation:

$$R_p = k_p/k_t^{1/2}[M]R_i^{1/2} \quad (1)$$

where k_p and k_t represent the propagation and termination rate constants, respectively. [M] stands for the instantaneous monomer concentration and R_i is the initiation rate of the polymer chain.

As previously noted [31, 32], the above equation holds for steady-state radicals concentration under low light intensity and low fractional light absorption conditions. R_i can be written as follows:

$$R_i = 2.303 \; \Phi_i \varepsilon \, c \, l \, I_0 \tag{2}$$

where c and ε are, respectively, the molar concentration and the molar extinction coefficient at the wavelength adopted, l is the light path length, I_0 the incident light intensity and Φ_i the overall quantum yield of the initiation process, that is the number of starting polymer chains per absorbed photon. The overall quantum yield Φ_i, moreover, is related to several parameters such as the individual yields of benzophenone triplet state formation (Φ_T), its photoreduction by a hydrogen donor to give rise to the primary radicals (Φ_H) and their reaction with the monomer to produce the initial growing chain radicals (Φ_{RM}):

$$\Phi_i \propto \Phi_T \cdot \Phi_H \cdot \Phi_{RM} \; . \tag{3}$$

When the photoinitiated polymerization process occurs under the same experimental conditions, in terms of c, l, I_0 and type of monomers mixture, and ε is assumed to be equal for each photoinitiator based on the benzophenone moiety, R_i results proportional to Φ_i and hence R_p to $\Phi_i^{1/2}$, according to Eqs. (1) and (2).

From transient absorption measurements in benzene solution, the triplet state of the benzophenone moiety in poly(ABP) is estimated [27] to be formed with lower Φ_T (about 2/3) than in the corresponding low-molecular-weight model IBP. Taking into account that, despite the lower value of Φ_T, poly(ABP) exhibits an appreciably higher photoinitiation activity and hence larger Φ_i value as compared with IBP, it can be concluded that the polymeric photoinitiators display markedly higher Φ_H values, Φ_{RM} being assumed to be probably of the same order of magnitude in both systems. Indeed, in accordance with a more efficient hydrogen abstraction process, a three-fold shorter lifetime of benzophenone triplet state has been observed [27] in poly(ABP) with respect to IBP (Table 7). Moreover, the evidence of ketyl radicals formation and the lack of triplet-triplet annihilation phenomena in both the systems [27], strongly support the above hypothesis.

Finally, as the above experiments have been performed in benzene, a very weak hydrogen donor, they clearly indicate that in poly(ABP) the hydrogen abstraction occurs through an intramolecular mechanism.

Photophysical studies have also been extended to different copolymers based on ABP or VBP co-units and to poly(VBP) [22, 28].

The benzophenone triplet lifetime in poly(ABP) and poly(ABP-*co*-MtA)s is found to be shorter than in GBP and poly(ABP-*co*-MA)s (Table 7). It is worth noting that in this last system the triplet lifetime of benzophenone moieties decreases on increasing the content of ABP units, that is decreasing the photoinitiation activity. Moreover, poly(VBP) exhibits a triplet lifetime equal to that of poly(ABP), according to the similar activity found in the two systems, and six

Table 7. Benzophenone (BP) triplet lifetime in high- and low-molecular-weight photoinitiators and their activity in the UV curing, under nitrogen, of HDDA/BA[a] and HDDA/APDG[b] equimolar mixtures[c] [22]

Photoinitiating system		HDDA/BA		HDDA/APDG		BP triplet lifetime[d]
Type	BP units (mol%)	$t_{1/2}$[e] (s)	R_c[f] (s^{-1})	$t_{1/2}$[e] (s)	R_c[f] (s^{-1})	(μs)
Poly(ABP)	100.0	72	0.9	6.0	24	1.6 ± 0.1
Poly(ABP-*co*-MtA)	57.6	–	–	3.4	50	1.7 ± 0.1
Poly(ABP-*co*-MtA)	32.5	41	1.4	3.4	50	1.8 ± 0.1
Poly(ABP-*co*-MA)	33.4	–	–	6.3	17	2.2 ± 0.2
Poly(ABP-*co*-MA)	10.2	–	–	8.3	10	3.5 ± 0.2
GBP	–	112	0.6	–	–	4.2 ± 0.2
IBP	–	143	0.5	10.6	7	4.8 ± 0.2
Poly(VBP)	100.0	78	0.8	–	–	1.6 ± 0.1
Poly(VBP-*co*-MtA)	55.4	52	1.4	–	–	1.6 ± 0.1
Poly(VBP-*co*-MtA)	21.7	38	1.7	–	–	1.6 ± 0.1
Poly(VBP-*co*-MtA)	5.0	31	2.3	–	–	2.3 ± 0.2
PBP	–	165	0.4	–	–	9.7 ± 0.2

[a] Determined by microwave dielectrometry. Polymerization conditions: see Table 2
[b] Determined by IR spectroscopy. Polymerization conditions: see Table 1
[c] Photoinitiator concentration: 1 mol% of benzophenone moiety in the acrylic mixture
[d] In benzene solution, in the absence of acrylic monomers
[e] Half-time of the process
[f] Polymerization rate evaluated at $t_{1/2}$ and expressed as % conversion over time

times shorter than PBP (Table 7). This nicely confirms that a shortening of the benzophenone triplet lifetime is accompanied by an improvement of photoinitiation activity, due to an intramolecular hydrogen abstraction mechanism which is particularly favoured in macromolecular systems. An apparent disagreement with the above correlation is found for poly(ABP-*co*-MtA)s and poly(VBP-*co*-MtA)s against the corresponding benzophenone-containing homopolymers, in the sense that a higher photoinitiation activity corresponds to a longer triplet lifetime in the former systems (Table 7). This has been explained [22] by assuming that, even if the primary free radicals are formed by intramolecular hydrogen abstraction from the side-chain menthyl group with a lower Φ_H, these radicals are more efficient in the reaction with the monomers which produces the starting chain radicals, thus leading to higher Φ_{RM} values. This last occurrence is interpreted [22] in terms of steric requirements hindering the recombination reaction between side-chain menthyl radicals without appreciably affecting their capability to initiate the polymerization of acrylic monomers.

In conclusion, on the basis of the collected results, polymeric photoinitiators bearing side-chain benzophenone moieties show, at least under nitrogen atmosphere, a large improvement of activity as compared with low-molecular-weight structural models. Polymeric systems having even higher activity can easily be prepared by introducing in the macromolecules, via a copolymerization route, monomeric units containing hydrogen donors able to promote intramolecular hydrogen abstraction mainly from the side-chains rather than from the backbone. In addition, the lowering of the amount of benzophenone-containing units in the copolymers usually produces an enhancement of photoinitiation activity as the microenvironment of the photoreactive moieties favours to a larger extent intramolecular hydrogen abstraction reactions, due to the increased local concentration of hydrogen donating moieties. Thus, a well established knowledge of the parameters affecting the activity may allow the design of tailor-made benzophenone-containing polymeric photoinitiators for specific applications.

3 Photoinitiation by an Electron Transfer Mechanism

It is now well recognized that aromatic carbonyl compounds may be efficiently reduced by a large variety of electron donor molecules, particularly amines [2, 30, 32–35]. Usually, tertiary amines are more effective than secondary and primary derivatives, although specific solvation phenomena may change this order of reactivity [2]. Aromatic ketones, in the presence of tertiary amines and under UV light, give rise [30, 34–36] to the formation of "exciplexes" (excited states having the highest electron-transfer character which may be preceded by a variety of collisional complexes), regardless of the excited state undergoing reaction being (n, π^*) or (π, π^*) in nature. The exciplex may be formed either by excitation of a ketone/amine ground-state charge transfer complex or by interaction of the excited ketone with the amine in the ground-state. Usually, ketone and amine behave as electron acceptor and donor, respectively. The photoinduced reduction of ketone successively occurs by a hydrogen transfer mechanism from the carbon atom in α-position to the amine nitrogen, thus forming ketyl and amino-derived radicals, the latter being the actual polymerization initiating species, when vinyl monomers are present. An oversimplified example is represented by the benzophenone/tertiary amine system in Scheme 7.

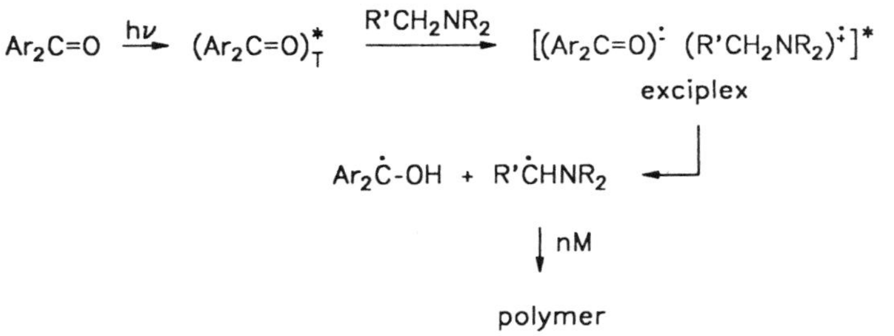

$$Ar_2C{=}O \xrightarrow{h\nu} (Ar_2C{=}O)^*_T \xrightarrow{R'CH_2NR_2} \underset{\text{exciplex}}{[(Ar_2C{=}O)^{\dot{-}}\ (R'CH_2NR_2)^{\dot{+}}]^*}$$

$$\underset{\text{exciplex}}{[(Ar_2C{=}O)^{\dot{-}}\ (R'CH_2NR_2)^{\dot{+}}]^*} \longrightarrow Ar_2\dot{C}\text{-}OH + R'\dot{C}HNR_2 \xrightarrow{nM} \text{polymer}$$

Scheme 7

Usually, photoinitiating systems based on aromatic ketone/tertiary amine combinations are more active in the polymerization of unsaturated monomers compared with ketones alone [37], particularly in the air, not only because the formation of exciplexes competes efficiently with the quenching of ketone triplet state by oxygen, but also because amines behave as oxygen scavengers through a chain process [34] (Scheme 8):

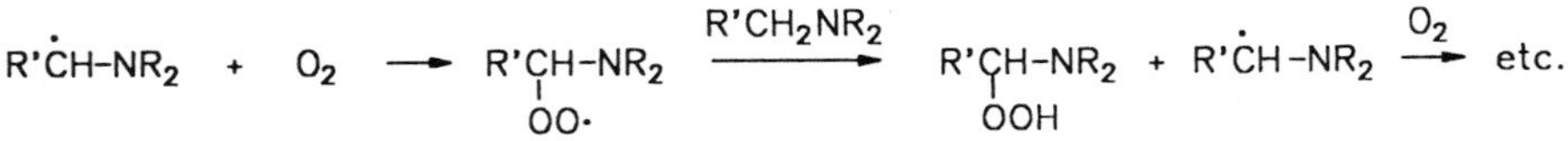

$$R'\dot{C}H{-}NR_2 + O_2 \longrightarrow R'CH(OO\cdot){-}NR_2 \xrightarrow{R'CH_2NR_2} R'CH(OOH){-}NR_2 + R'\dot{C}H{-}NR_2 \xrightarrow{O_2} \text{etc.}$$

Scheme 8

The photoreduction of aromatic ketones by tertiary amines is reported [38] to proceed at rates which are substantially faster than those observed for the corresponding photoinduced hydrogen abstraction from, e.g. alcohols. A limit case is given by fluorenone, the photoreduction of which does not occur in alcohol, ether or alkane solution, but readily takes place in the presence of amines, tertiary amines being the most effective [39, 40]. Xanthone has also been reported to be easily photoreduced by *N,N*-dimethylaniline [41], but not by 2-propanol [42]. However, the oxidation of tertiary amines photosensitized by fluorenone and xanthone is much less efficient than when sensitized by benzophenone, apparently because of lower rates of hydrogen abstraction [43]. Fluorenone/tertiary amine systems have been used successfully to photoinitiate the polymerization of MMA, St, MA and AN [30, 38, 44] and rather similar results have been obtained in the photoinitiated polymerization of MA by the benzophenone/Et_3N system [45]. Thus, the great variety of substrates participating in exciplex formation has been readily extended to polymer-based systems.

Indeed, copolymers containing 2-(*N,N*-dimethylaminoethyl) methacrylate units as electron donor moieties, when allowed to interact with photoexcited fluorenone in the presence of vinyl monomers, are effective in causing rapid crosslinking and gelation through a graft copolymerization process [30, 38] (Scheme 9).

Copolymers of 2-vinyl fluorenone with MMA [poly(2VF-*co*-MMA)s], containing 2VF units in a wide range of compositions, have been used as photoinitiating systems in combination with Et_3N, or indole-3-yl acetic acid (IAA), in the polymerization of MMA in benzene solution [30, 38] and compared with fluorenone (FLO) and 2-methyl fluorenone (2MF).

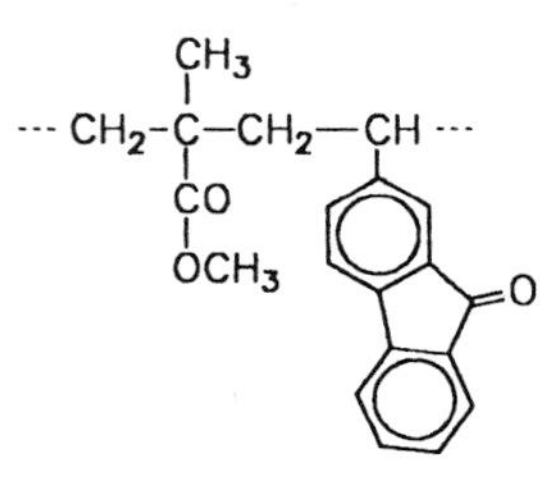

$\cdots CH_2{-}C(CH_3)(CO{-}OCH_3){-}CH_2{-}CH\cdots$ (2-fluorenonyl)

poly(2VF-*co*-MMA)

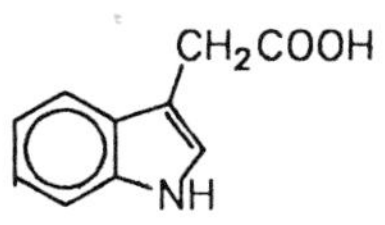

IAA (CH_2COOH, NH)

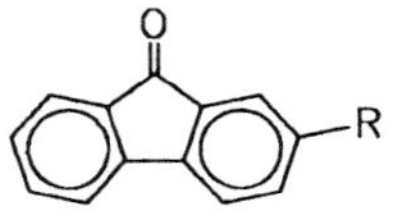

FLO (R = H)

2MF (R = CH_3)

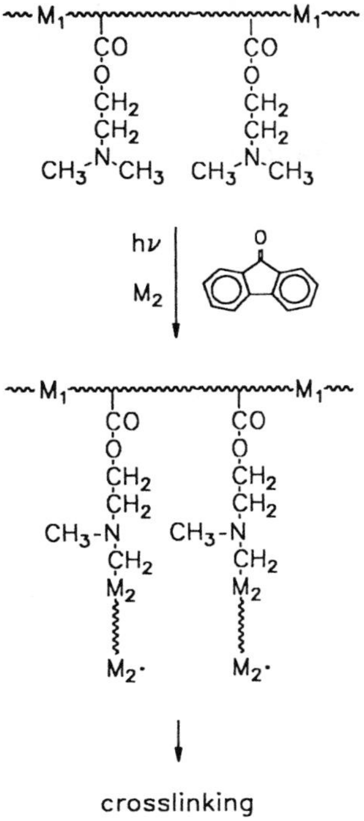

~M$_1$~~~M$_1$~ (CO–O–CH$_2$–CH$_2$–N(CH$_3$)–CH$_3$ side groups)

$h\nu$, M_2, fluorenone

~M$_1$~~~M$_1$~ (CO–O–CH$_2$–CH$_2$–N(CH$_3$)–CH$_2$–M$_2$~~M$_2$· side groups)

crosslinking

Scheme 9

The above copolymers are found to exhibit quantum yields for the photoreduction of ketone moiety and MMA polymerization rate similar to or even higher than those displayed by 2MF and FLO with the same amine (Tables 8 and 9). Polymers isolated before gelation and examined by GPC give clear evidence of formation of both poly(MMA) and graft poly(MMA) when the reaction is initiated by the poly(2VF-*co*-MMA)/IAA system. Rather similar results have been observed when IAA is replaced by Et_3N in the 0.01 M concentration range. By contrast, identical polymerizations carried out with 0.1 M Et_3N show no evidence of graft poly(MMA) formation, gelled systems being obtained only after prolonged irradiation.

The above results clearly indicate that high concentrations of Et_3N retard or prevent coupling of polymeric semipinacol radicals and their reactions with growing chains or monomer. This dependence on Et_3N concentration has been tentatively explained [30] by assuming that hydrogen bonding of semipinacol radical to the amine is important, thus providing a mechanism for either amine mediated hydrogen transfer towards growing chain radicals (Scheme 10) or a diminished capability of $Ar_2\dot{C}$-OH (amine) radicals to react by self-combination as well as to terminate growing radicals by coupling.

In any case, this result is very important for the design of systems involving polymer based photoinitiators having anticipated fast cure characteristics.

Table 8. Quantum yields for photoreduction of low- and high-molecular-weight systems based on fluorenone moiety by different amines in the presence or in the absence of MMA[a] [30]

Fluorenone system		TEA[b]		IAA
Type	2VF units (mol%)	10^{-1} M	10^{-2} M	$3 \cdot 10^{-3}$ M
2MF	–	0.27	–	0.25
2MF + MMA	–	0.13	0.20	0.28
Poly(2VF-*co*-MMA)	18.4	0.29	–	0.15
Poly(2VF-*co*-MMA) + MMA	18.4	0.15	0.21	0.21

[a] Ketone and MMA concentrations: $3 \cdot 10^{-3}$ and 5.0 M, respectively, in benzene at 30 °C
[b] TEA = Et_3N

Table 9. Photoinitiated polymerization rate (R_p) of MMA in benzene solution by low- and high-molecular-weight systems based on fluorenone moiety in combination with different amines[a] [30]

Fluorenone system			Amine		10^4 R_p ($M\,s^{-1}$)
Type	2VF units (mol%)	[Ketone] (10^{-3} M)	TEA (10^{-1} M)	IAA (10^{-3} M)	
2MF	–	3	1	–	0.86
FLO	–	3	1	–	0.98
Poly(2VF-*co*-MMA)	1.1	3	1	–	1.20
Poly(2VF-*co*-MMA)	18.4	3	1	–	1.20
Poly(2VF-*co*-MMA)	67.9	3	1	–	1.30
Poly(2VF-*co*-MMA)	18.4	3	0.1	–	1.75
2MF	–	1	–	1	0.92
FLO	–	1	–	1	1.04
Poly(2VF-*co*-MMA)	18.4	1	–	1	1.15

[a] MMA concentration: 5.0 M at 30 °C

$$Ar_2\dot{C}\text{-}OH\ (\text{Amine}) + P_n^{\cdot} \longrightarrow Ar_2C{=}O\ (\text{Amine}) + P_nH$$

Scheme 10

On the basis of the above findings, it can be concluded that grafting efficiencies and curing rates for commercial systems may be improved more easily by employing polymers having pendant amino groups rather than fluorenone moieties.

Thioxanthone compounds, having an intense absorption band around 380 nm, have been industrially developed as initiators of polymerization in UV curing of heavily pigmented (TiO_2) resins [3]. However, conventional thioxanthone photoinitiators show problems associated with their poor solubility producing migration which may result in loss of adhesion and, possibly, gloss. To overcome these problems, polymeric photoinitiators with pendant thioxanthone moieties have been designed in order to enhance the compatibility with the UV curable formulation. Indeed, a polystyrene-bound thioxanthone [poly(StX-*co*-St)] with 1.5% of photoreactive groups has been prepared and compared, in terms of photophysical properties and photoinitiation activity, with the low-molecular-weight analogue 2-benzyloxy-thioxanthone (BOTX) [46].

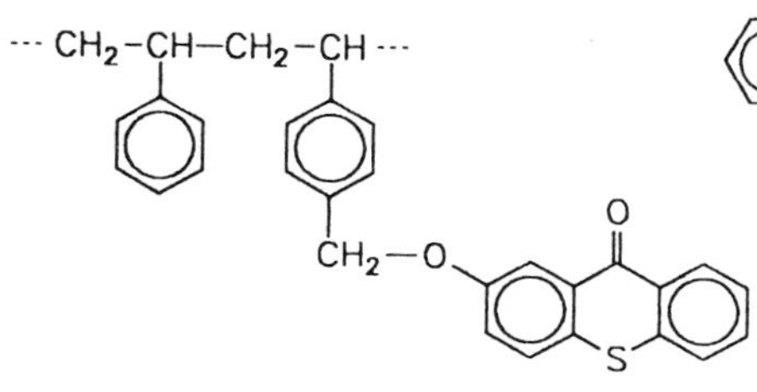

poly(StX-*co*-St) 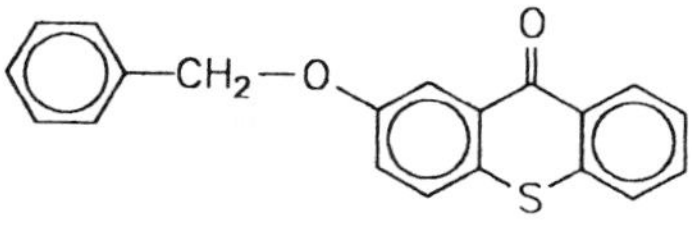 BOTX

Free and polymer-bound 2-benzyloxy-thioxanthone exhibit similar flash photolysis behaviour and the same photoreduction quantum yield in the presence of 2-(*N,N*-diethylamino) ethanol. This clearly shows that the polymeric nature does not appear to affect photophysical properties of the thioxanthone moiety. The photoinitiated polymerization of MMA in benzene solution, using BOTX and poly(StX-*co*-St) in combination with 2-(*N,N*-diethylamino) ethanol, indicates that the polymer-bound chromophore seems to operate in the same way and with similar efficiency as the free photoinitiator, at least in conditions of dilute chromophore concentration.

Copolymers of 2-acryloyl-thioxanthone (TXA) with MMA [poly(TXA-*co*-MMA)] have also been prepared [47,48] and the photophysical properties as well as the free radical photoinitiation in the polymerization of MMA studied [49,50].

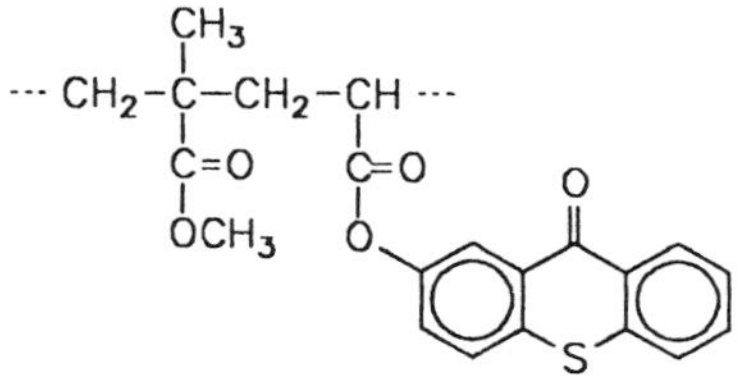

poly(TXA-*co*-MMA) 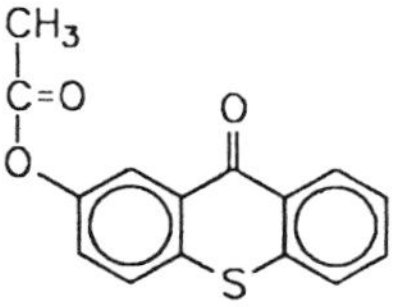 AcOTX

Microseconds flash photolysis measurements on poly(TXA-*co*-MMA) and on the corresponding low-molecular-weight model 2-acetoxy-thioxanthone (AcOTX) show that, in a hydrogen donating solvent, ketyl radical formation decreases on passing from the model to the copolymers, thus suggesting that intramolecular self-quenching is important in the copolymers due to the coiling of the polymer chains. Photoreduction of the thioxanthone moiety attached to the macromolecule, in the presence of a tertiary amine, is significantly reduced as compared with that of AcOTX. Furthermore, in the photoinitiated polymerization of MMA, the monomer quenching of the thioxanthone triplet state is also impaired by the macromolecular coiling in the bound chromophore. This favours the photoinitiation activity, the copolymers thereby exhibiting greater efficiency than AcOTX.

Polymers bearing side-chain anthraquinone moieties have also been reported. In particular, poly(2-acrylamidoanthraquinone) [poly(AAQ)] is found [51] to exhibit, in combination with a tertiary amine such as *p*-diethylaminobenzoate, higher solubility and the same photoactivity as the monomer in a mixed vinylurethane/triethyleneglycol dimethacrylate prepolymer system, when irradiated by visible light.

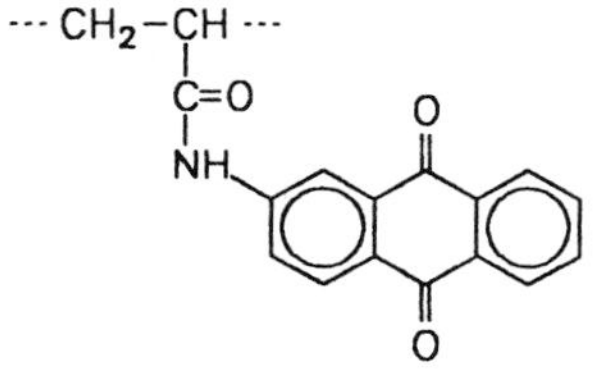

poly(AAQ)

Poly(AAQ) results even more effective than camphorquinone, the most widely established low-molecular-weight visible photoinitiator, at an equivalent concentration in the same formulation [51].

Combinations of benzophenone and tertiary amine groups, one of which was anchored to a polymer matrix, have also been checked in the photoinduced polymerization and photografting of 2-ethylhexyl methacrylate (EHMA) [52]. Indeed, poly(VBP-*co*-St)s, in the presence of *N,N*-diethylaniline (DEA), are found to display lower photoinitiation activity and photografting efficiency than the corresponding copolymers of styrene with 4-(*N,N*-diethylamino) styrene [poly(St-*co*-DEAS)] in the presence of benzophenone (BP), as revealed by conversion and fractionation data (Table 10).

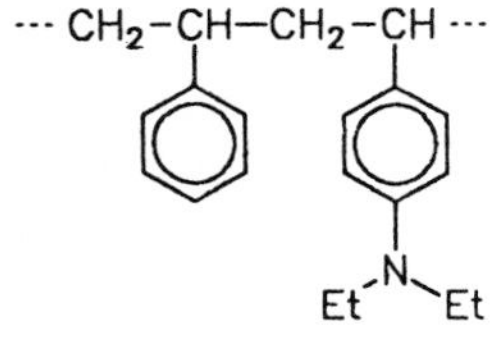

poly(St-*co*-DEAS)

These results also confirm that, when the amine function is linked to the polymer, the grafting efficiency is greatly enhanced, the amine-derived radical being a much better initiating species than the semipinacol radical.

Table 10. Photoinduced homopolymerization and graft-copolymerization of EHMA in the presence of equimolar combinations of low- and high-molecular-weight systems based on benzophenone/tertiary amine moieties[a] [52]

Polymeric component		Low-mol.-wt. component	Polymeric product		
Type	St (mol%)		Conversion (%)	Poly(EHMA) (%)	Graft poly(EHMA) (%)
Poly(VBP-*co*-ST)	67	DEA	4.1	69	31
Poly(St-*co*-DEAS)	75	BP	11.1	23	77

[a] In benzene solution at 22 °C. Subjected to 5 min UV irradiation. Ketone and EHMA concentrations: 0.02 and 0.5 M, respectively

Polymeric photoinitiators bearing both side-chain benzophenone and tertiary amine functions on the same macromolecule have also been prepared and checked in the UV curing of HDDA/APDG and HDDA/BA equimolar mixtures with the aim of obtaining systems displaying very high activity, especially in air. However, in contrast to the expected results, copolymers of 4-acryloxy benzophenone with 4-(*N,N*-dimethylamino) styrene [poly(ABP-*co*-DMAS)] are found [18] to exhibit, in the UV curing of the above acrylic formulations, much lower photoinitiation activity with respect not only to the poly(ABP)/*N,N*-dimethylaniline (DMA) system, but also to the corresponding mixture 4-isobutyroyloxy benzophenone (IBP)/DMA (Table 11).

The photoinitiation activity of poly(ABP)/DMA and IBP/DMA, as compared with that found for poly(ABP) and IBP alone, clearly confirms that the presence of amine activates both the systems. 4-[4-(*N,N*-dimethylamino)phenyl] butanoate of 4-hydroxy benzophenone (DMABP), the low-molecular-weight structural model of poly(ABP-*co*-DMAS), having benzophenone and tertiary amine functions in the same molecule, is found [18, 22, 53] to display an intermediate activity with respect to those observed for poly(ABP)/DMA and poly(ABP-*co*-DMAS) systems (Table 11).

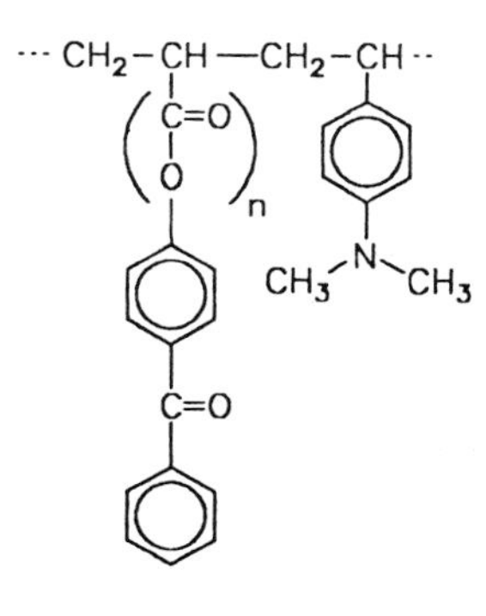

poly(VBP-*co*-DMAS) (n = 0)
poly(ABP-*co*-DMAS) (n = 1)

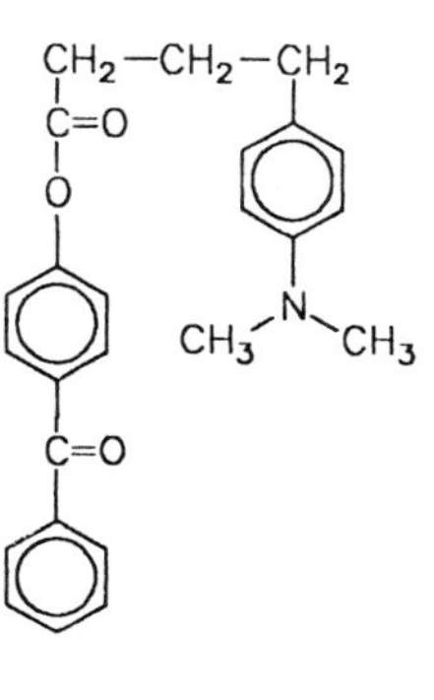

DMABP

Table 11. UV curing in film matrix of HDDA/APDG[a] and HDDA/BA[b] equimolar mixtures, under nitrogen, by photoinitiators based on both benzophenone and tertiary amine moieties[c] [18, 22, 53]

Photoinitiating system			HDDA/APDG		HDDA/BA	
Type	Benzophenone moiety (mol%)	Tertiary amine moiety (mol%)	$t_{1/2}$[d] (s)	R_c[e] (s^{-1})	$t_{1/2}$[d] (s)	R_c[e] (s^{-1})
Poly(ABP-*co*-DMAS)	43	57	12.6	3.1	290	0.08
Poly(ABP)	100	0	6.0	20.0	72	0.9
Poly(ABP)/poly(DMAS)	43	57	5.7	22.0	44	1.5
Poly(ABP)/DMA	43	57	2.6	31.8	11	4.9
DMABP	50	50	–	–	43	1.5
IBP	100	0	10.6	6.0	143	0.5
IBP/poly(DMAS)	43	57	8.2	7.1	–	–
IBP/DMA	43	57	7.3	7.7	–	–

[a] Determined by IR spectroscopy. Irradiation conditions: see Table 1
[b] Evaluated by microwave dielectrometry. Irradiation conditions: see Table 2
[c] Benzophenone moiety concentration in the HDDA/APDG and HDDA/BA mixtures: 1 mol%
[d] Half-time of the curing process
[e] Polymerization rate of the process evaluated at $t_{1/2}$ and expressed as % conversion over time

Table 12. UV curing in film matrix of HDDA/BA equimolar mixture, under nitrogen, by polymeric photoinitiators bearing both benzophenone and tertiary amine moieties in the side chain[a] [18, 22, 53]

Photoinitiating system			$t_{1/2}$ (s)	R_c (s^{-1})
Type	Benzophenone units (mol%)	Amine units (mol%)		
Poly(VBP-*co*-DMAS)	53.5	46.5	329	0.07
Poly(ABP-*co*-DMAS)	43.2	56.8	290	0.08
Poly(VBP-*co*-DEEA)	72.0	28.0	77	0.8
Poly(ABP-*co*-DEEA)	48.3	51.7	20	2.2
Poly(ABP-*co*-DAPA)	60.7	39.3	27	2.1
Poly(UP36-*co*-DEEA)	43.6	56.4	38	1.7
Poly(UP36-*co*-DAPA)	28.7	71.3	18	2.3

[a] Benzophenone moiety concentration in the HDDA/BA equimolar mixture: 1 mol%. Irradiation conditions and meaning of $t_{1/2}$ and R_c: see Table 2

In conclusion, the relative activity sequence of the above systems is the following: poly(ABP)/DMA $>$ poly(ABP)/poly(DMAS) $\approx$ DMAPB $>$ poly(ABP) $>$ IBP/DMA $>$ IBP/poly(DMAS) $>$ IBP $\gg$ poly(ABP-*co*-DMAS).

Furthermore, copolymers of 4-vinylbenzophenone with 4-(*N,N*-dimethylamino) styrene [poly(VBP-*co*-DMAS)] are found [18, 22, 53] to exhibit even lower photoinitiation activity (Table 12) than poly(ABP-*co*-DMAS).

Taking into account that for poly(VBP-*co*-DMAS) a pronounced tendency to exciplex formation is observed by phosphorescence measurements [54–57], the above results can be interpreted [20, 22] by assuming the occurrence of cage self-combination of the amine-derived radicals (formed by hydrogen transfer within the radical ion pairs) due to their high local concentration along the polymer chain. However, coupling reactions involving neighbouring amine-derived and semipinacol radicals in a cage could give the same result. Indeed, a large tendency to alternation of monomeric units in poly(VBP-*co*-DMAS) systems has been observed [56, 58]. According to the above hypothesis, the latter coupling reactions are found to occur [59] to a larger extent in copolymers of 4-(*N,N*-dimethylamino)-4′-vinyl benzophenone (AVBP) with styrene [poly(AVBP-*co*-St)] rather than in the corresponding low-molecular-weight structural model 4-(*N,N*-dimethylamino)-4′-isopropyl benzophenone (AIBP).

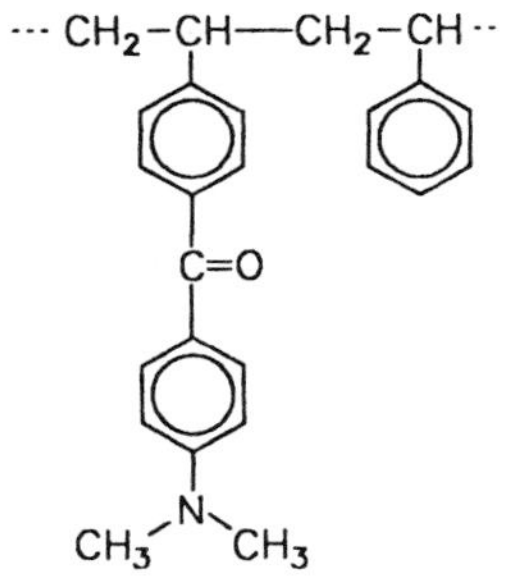

poly(AVBP-*co*-St) 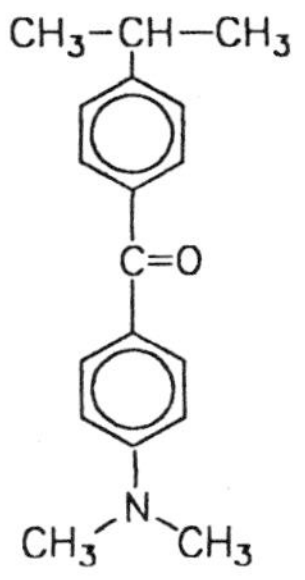 AIBP

Moreover, the polymerization of MMA, photoinitiated by poly(AVBP-*co*-St), proceeds at a lower rate than that promoted by the low-molecular-weight analogue AIBP and is enhanced in solvents where the macromolecules are more expanded, thus reducing free radical recombinations [60]. The presence of cage recombinations in this type of systems is also substantiated by the occurrence of a very efficient crosslinking when films of poly(VBP-*ca*-DMAS) are subjected to UV irradiation [54].

The remarkably higher activity of DMABP as compared with poly(ABP-*co*-DMAS), notwithstanding the probability of exciplex formation being quite similar in both systems, has been explained [22] in terms of reduced local concentration of the resulting amine-derived radicals as this parameter largely favours polymerization initiation rather than radical recombination. In this context, the maximum of activity shown by the poly(ABP)/DMA system may therefore arise from an efficient energy migration between excited- and ground-state benzophenone moieties along the backbone, thus improving the quantum yield of exciplex formation. Moreover, the amine-derived radicals thus obtained can readily diffuse and initiate the polymerization process, as they are not linked to the polymer matrix. The lower activity of poly(ABP)/poly(DMAS) against poly(ABP)/DMA substantially confirms this picture.

With the aim of improving the photoinitiation activity of benzophenone/tertiary amine containing polymers, new polymeric systems having flexible spacers of different length inserted between the above side-chain functional groups and the backbone, have been proposed [22]. Indeed, copolymers of ABP, VBP and UP36 with either 2-(*N,N*-diethylamino)ethyl acrylate (DEEA) or 3-(*N,N*-dimethylamino)propyl acrylate (DAPA) [poly(ABP-*co*-DEEA), poly(VBP-*co*-DEEA), poly(UP36-*co*-DEEA), poly(ABP-*co*-DAPA) and poly(UP36-*co*-DAPA), respectively] have been checked [22, 53, 61] in the UV initiated polymerization, under nitrogen, of the HDDA/BA equimolar mixture.

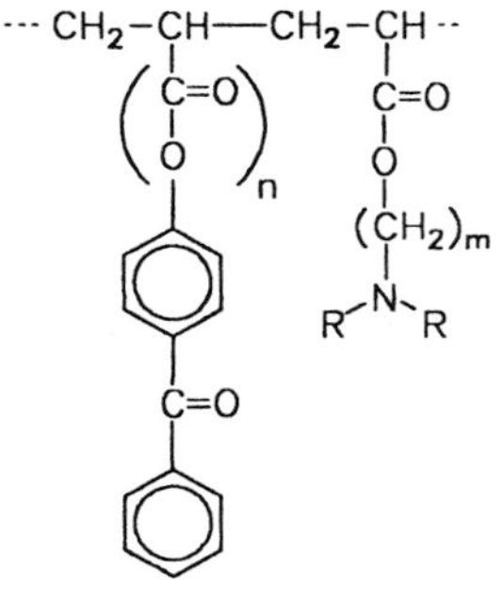

poly(VBP-*co*-DEEA) (n = 0, m = 2, R = Et)
poly(ABP-*co*-DEEA) (n = 1, m = 2, R = Et)
poly(ABP-*co*-DAPA) (n = 1, m = 3, R = Me)

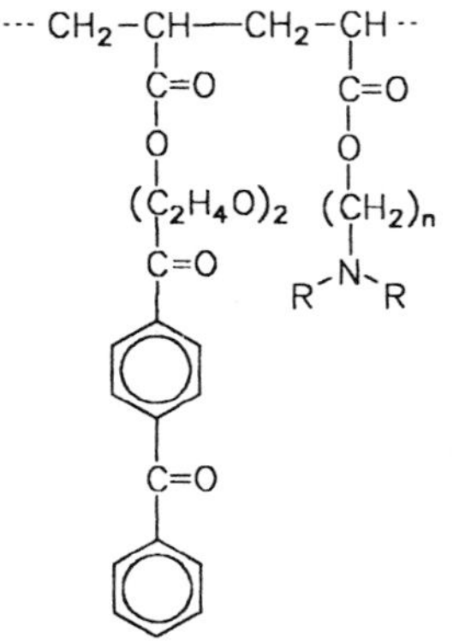

poly(UP36-*co*-DEEA) (n = 2, R = Et)
poly(UP36-*co*-DAPA) (n = 3, R = Me)

The kinetic data (Table 12), determined in the presence of the above polymeric photoinitiators, clearly indicate the following order of activity: poly(UP36-*co*-DAPA)≈poly(ABP-*co*-DEEA)>poly(ABP-*co*-DAPA)>poly(UP36-*co*-DEEA)>poly(VBP-*co*-DEEA)≫poly(ABP-*co*-DMAS)>poly(VBP-*co*-DMAS).

Such a sequence allows one to conclude that the insertion of a flexible spacer between the tertiary amine function and the main chain causes a sharp increase of photoinitiation efficiency. In addition, the relative distance of tertiary amine and benzophenone groups from the polymer chain appears to be quite an important structural requirement. Indeed, the above findings suggest that, provided the spacer is flexible, an improvement of activity is observed when both amine and ketone moieties are located at similar distances from the backbone.

The relevant increase of photoinitiation activity found in copolymers based on DEEA and DAPA co-units, with respect to poly(ABP-*co*-DMAS) and poly(VBP-*co*-DMAS), is explained [22] assuming that, in the former systems, no cage self-combination of the amine-derived radicals may occur, probably due to a lower local concentration of exciplexes along the polymer chain. Accordingly, no evidence of ground-state electron donor-acceptor (EDA) complex formation is found in UV spectra of DEEA and DAPA containing benzophenone copolymers [22, 53], contrary to what is observed in poly(ABP-*co*-DMAS) and poly(VBP-*co*-DMAS). Indeed, the existence of complexes in the ground-state of the latter systems usually favours the formation of excited EDA-complexes (exciplexes) [56].

An additional contribution to the activity improvement is also supposed [22] to be caused by the flexibility of the spacers between amine moieties and polymer chain, as the shielding of the amine-derived radicals by the polymer matrix is expected to be drastically reduced, thus markedly favouring the initiation of the polymerization process.

In conclusion, it is noteworthy that, although poly(ABP-*co*-DEEA) and poly (UP36-*co*-DAPA) display a slightly lower photoinitiation activity as compared with the poly(ABP)/DMA system, their use in UV curable surface coating formulations has to be preferred for some specific advantages. Indeed, in the copolymeric systems the unreacted amine groups remain still anchored to the polymer matrix thus preventing their diffusion onto the coating surface and hence reducing volatility. This eliminates or strongly reduces some drawbacks such as bad smell as well as photoinduced toxic and/or coloured decomposition products.

It is also well established [62, 63] that in benzil derivatives/tertiary amine systems the hydrogen transfer takes place via an intermediate exciplex formed by the excited benzil triplet state and the ground state amine. In this context, selected benzil derivatives have been advanced for preparing polymeric systems in which the benzil moiety is an integral part of the polymer backbone [64]. In particular, polyurethanes based on 4,4′-dihydroxybenzil and either 1,6-diisocyanatohexane or isophorone diisocyanate [HBIH and HBII, respectively] as well as epoxy resins based on benzil-4,4′-diglycidylether and 5,5′-dimethylhydantoin [BGEDH] have been prepared and checked in the UV curing of triethyleneglycol diacrylate (TEGDA) in the presence of *N*-methyldiethanolamine.

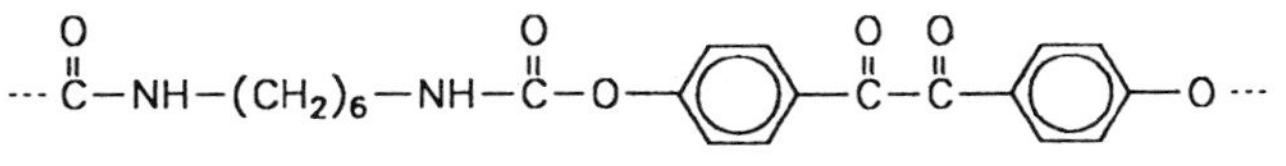

HBIH

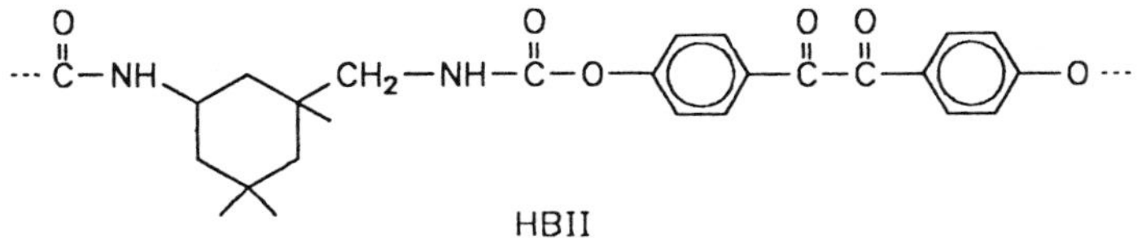

HBII

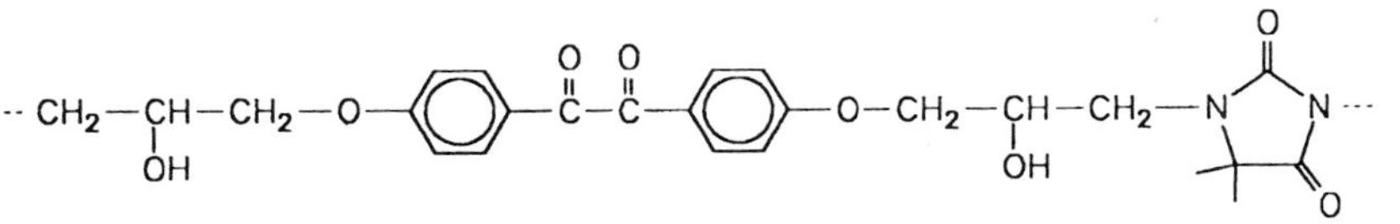

BGEDH

Incorporation of 4,4′-dihydroxybenzil into the backbone of a polyurethane produces a significant improvement of photoinitiation activity, whereas the insertion of benzil-4,4′-diglycidylether into a polymer chain fails to affect the photoinitiation efficiency to any notable degree. The photoactivity enhancement afforded in the polyurethane systems is rationalized in terms of steric and, possibly, energy migration effects.

Charge transfer (CT) complexes, different from aromatic ketone/amine systems, such as quinoline-bromine, pyridine-bromine, tetrahydrofuran-bromine etc. have also been reported to behave as initiators of vinyl polymerization, particularly under photoactivation [65–68].

In this context, the polymerization kinetic of MMA, under light irradiation, using a CT complex between a polymeric donor such as poly(*N*-vinylcarbazole) [poly(NVC)] and bromine as the acceptor has been investigated [69]. A radical mechanism for the polymerization is suggested. Moreover, the presence of bromine atoms as end groups in the obtained poly(MMA) and the lack of any evidence of poly(NVC) chemically linked to poly(MMA) have allowed the authors to propose that the radical generation process may be described as in Scheme 11:

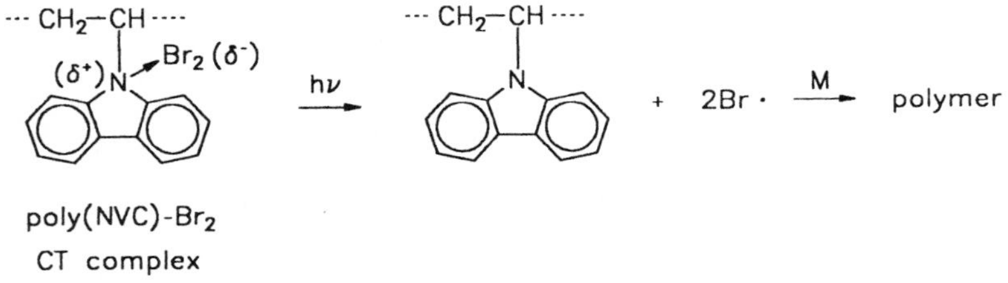

··· CH_2—CH ···· (δ^+) N → Br_2 (δ^-) —$h\nu$→ ··· CH_2—CH ···· N + 2Br · —M→ polymer

poly(NVC)-Br_2
CT complex

Scheme 11

Whereas in the MMA photoinitiated polymerization by quinoline-bromine CT complex the formation of radicals is preceded by an instantaneous complexation reaction between the CT complex initiator and the monomer [68], no evidence of this occurrence is observed in the case of the poly(NVC)-Br_2 CT complex, probably due to the steric hindrance provided by the polymeric chain. The behaviour of the above system should however be compared with that of the corresponding low-molecular-weight *N*-alkyl carbazole-Br_2 CT complex in order to clarify this point.

4 Photoinitiation by a Cleavage Mechanism

Excitation energy (71–73 Kcal/mole), usually available from commercial UV light sources, is enough to give homolytic scission of a chemical bond, provided that the molecule is able to absorb the incident light in order to be promoted to an excited state. Several types of low-molecular-weight molecules have been discovered [1, 2, 70] to possess the above requirements, thus undergoing homolytic

$$R\text{-}R' \xrightarrow{h\nu} (R\text{-}R')^{*} \longrightarrow R\cdot + R'\cdot \xrightarrow{M} \text{polymer}$$

Scheme 12

fragmentation of the photoexcited state to produce free radical species causing the polymerization initiation of vinyl monomers (Scheme 12).

All the members of this class of compounds contain an aromatic ketone group, the most representative systems being benzoin [71], benzoin ethers [72], dialkoxy-acetophenones [73], hydroxyalkylphenones [74], benzoyloxime esters [75] and, more recently, benzoylphosphine oxides [76], α-hydroxymethylbenzoin sulfonic esters [77], sulfonyl ketones [78] and morpholino ketones [70].

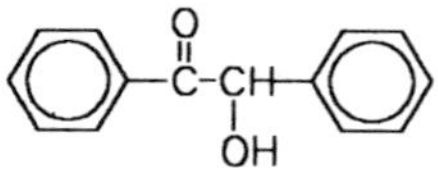

benzoin

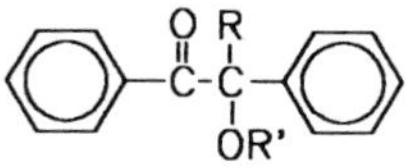

benzoin ethers (R = H, OR')

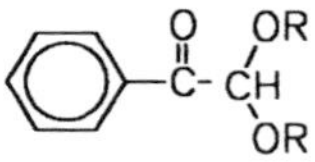

dialkoxy acetophenones

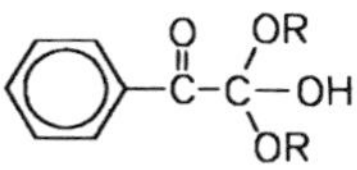

hydroxyalkylphenones

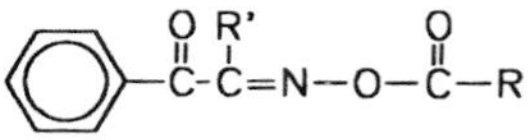

benzoyloxime esters

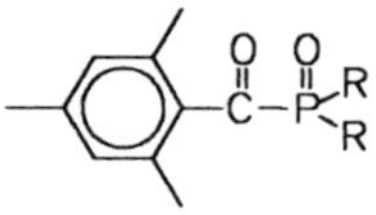

benzoyl phosphinoxides

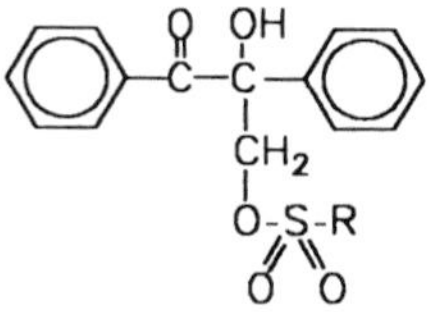

α-hydroxymethyl benzoin sulphonic esters

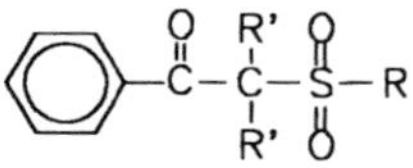

sulphonyl ketones

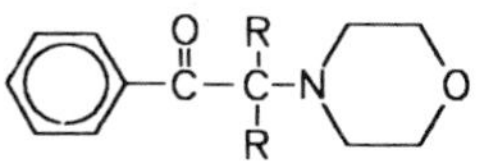

morpholino ketones

Generally speaking, this class of photoinitiator is found to be very convenient for application in the area of photocurable coatings due to peculiar advantages with respect to the previously cited initiators working with hydrogen abstraction or electron transfer mechanisms. Indeed, they usually display easy availability, high quantum yields of radical generation and high efficiency in promoting the polymerization of vinyl monomers. In addition, the excited triplet state of these

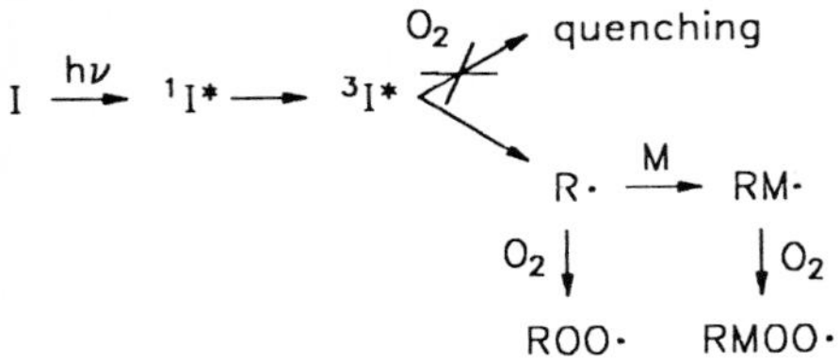

$$I \xrightarrow{h\nu} {}^1I^* \longrightarrow {}^3I^* \begin{cases} \xrightarrow{O_2} \text{quenching (blocked)} \\ \longrightarrow R\cdot \xrightarrow{M} RM\cdot \end{cases}$$

$$R\cdot \xrightarrow{O_2} ROO\cdot \qquad RM\cdot \xrightarrow{O_2} RMOO\cdot$$

Scheme 13

molecules, due to their very short lifetimes, is not appreciably quenched by oxygen, thus making photoscission substantially unaffected by the presence of air, which may however exert its usual scavenging effect on the primary and growing radicals, yielding peroxy radicals (Scheme 13).

The large variety of molecular structures capable of undergoing photofragmentation processes has readily been extended to polymer-based systems.

One of the earliest examples of photodecomposable, but thermally stable, polymeric systems contains pendant perester substituted benzophenone moieties, the photodissociation of the former group being promoted by the benzophenone chromophore which behaves as a triplet sensitizer. Indeed, the homopolymer of 4-vinylbenzoyl-4′-*tert*-butyl perbenzoate and its copolymers with styrene [poly(VBPE) and poly(VBPE-*co*-St), respectively] have been prepared and checked as photoinitiators in the polymerization of styrene and MMA by irradiation at 366 nm [79, 80].

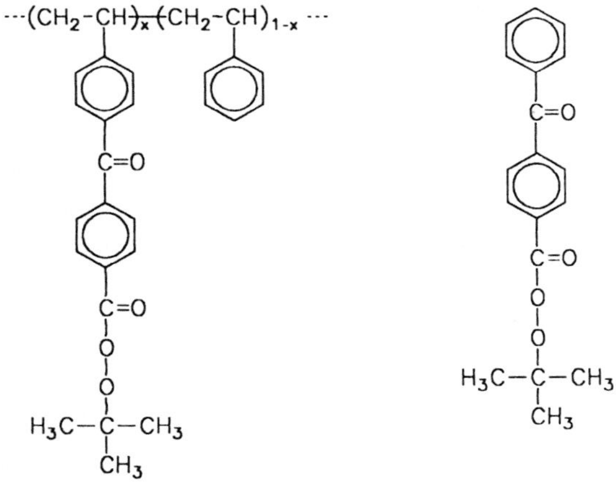

$\cdots(CH_2-CH)_x-(CH_2-CH)_{1-x}\cdots$ with pendant $-C_6H_4-C(=O)-C_6H_4-C(=O)-O-O-C(CH_3)_3$ and $-C_6H_5$ groups

poly(VBPE) (x = 1)

poly(VBPE-*co*-St) (0 < x < 1)

$C_6H_5-C(=O)-C_6H_4-C(=O)-O-O-C(CH_3)_3$

BPE

The generation mechanism of the initiating radical species is reported in Scheme 14:

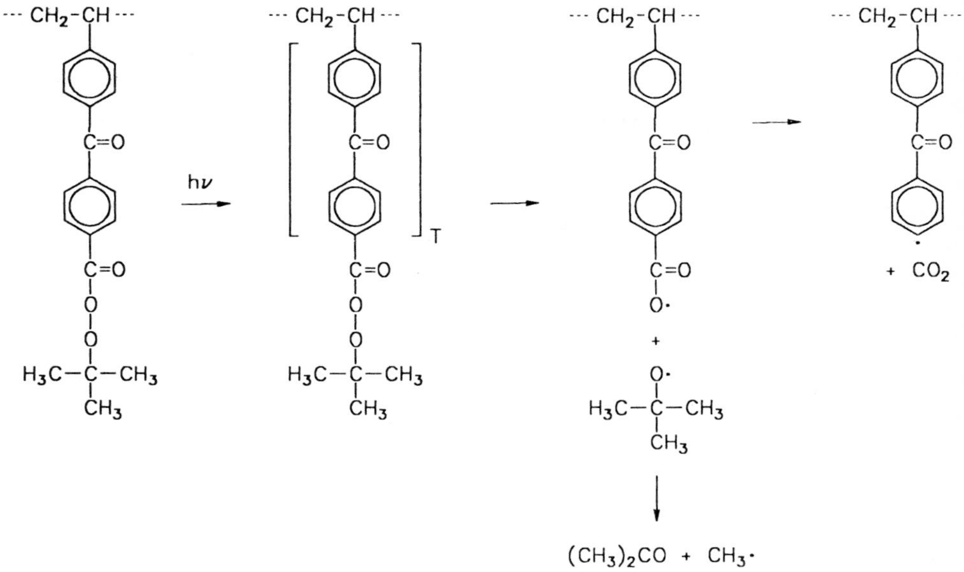

$$
\cdots CH_2-CH\cdots \text{(p-C}_6\text{H}_4\text{-CO-C}_6\text{H}_4\text{-CO-O-O-C(CH}_3)_3) \xrightarrow{h\nu} [\cdots]_T \longrightarrow \cdots CH_2-CH\cdots(\text{C}_6\text{H}_4\text{-CO-C}_6\text{H}_4\text{-C(=O)O}\cdot) + \cdot O-C(CH_3)_3 \longrightarrow \cdots CH_2-CH\cdots(\text{C}_6\text{H}_4\text{-CO-C}_6\text{H}_4\cdot) + CO_2
$$

$$
\cdot O-C(CH_3)_3 \longrightarrow (CH_3)_2CO + CH_3\cdot
$$

Scheme 14

According to Scheme 14, *tert*-butoxy and polymer carboxylate primary radicals are formed by photodissociation. However, the polymer-bound carboxylate radical may firstly lose CO_2 giving rise to a polymer-bound phenyl radical which is also able to initiate the polymerization. The *tert*-butoxy radical may evolve to give acetone and a very reactive methyl radical.

Poly(VBPE) is found to be more effective than the corresponding low-molecular-weight analogue *p*-benzoyl-*tert*-butyl perbenzoate (BPE) in the photoinitiated polymerization of styrene. The increase of cure rate is attributed to a rapid gel formation protecting the polymer-bound free radicals from termination, although the contribution by energy transfer within the polymer may also be important. Poly(VBPE-*co*-St) also promotes the photografting of MMA, the amount of poly(MMA) grafted to the photosensitive polymers increasing with the increase of VBPE co-units content in the copolymers. However, considering that low-molecular-weight radicals are also generated (Scheme 14), mixtures of graft and free poly(MMA) are finally obtained.

Taking into account [81, 82] that photodegradation of poly(4,4-dimethyl-1-penten-3-one) [poly(BVK)] and poly(3-methyl-3-buten-2-one) [poly(MIK)] proceeds predominantly through a Norrish type I mechanism via the triplet state (Scheme 15), the above homopolymers have been studied as initiators in the photoinduced polymerization of vinyl monomers such as MMA, St, AN and VAc [83].

Since α-cleavage reactions yield polymeric and fragment radicals, in all the polymerization experiments the polymer obtained is constituted by a mixture of graft copolymer and homopolymer (Scheme 16).

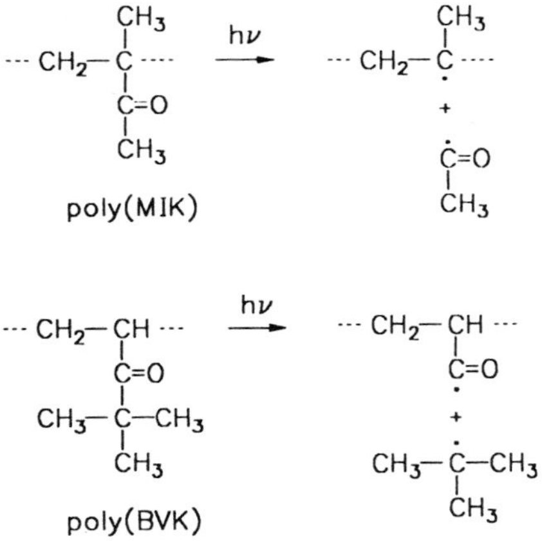

$$\cdots CH_2-\overset{CH_3}{\underset{C=O\atop CH_3}{C}}\cdots \xrightarrow{h\nu} \cdots CH_2-\overset{CH_3}{\dot{C}}\cdots + \dot{C}(=O)CH_3$$

poly(MIK)

$$\cdots CH_2-\underset{C=O\atop CH_3-C(CH_3)-CH_3}{CH}\cdots \xrightarrow{h\nu} \cdots CH_2-\underset{\dot{C}=O}{CH}\cdots + CH_3-\dot{C}(CH_3)-CH_3$$

poly(BVK)

Scheme 15

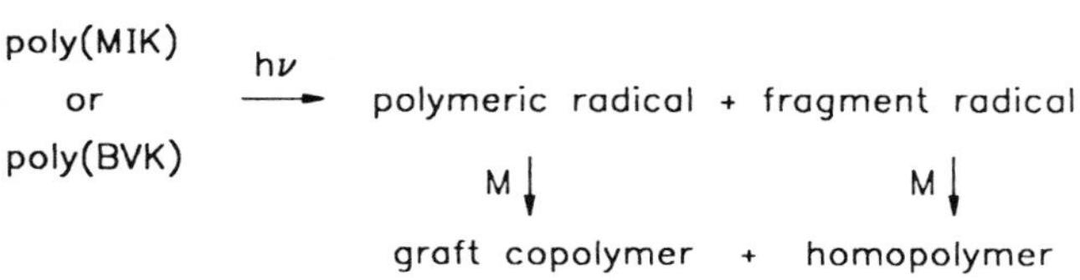

poly(MIK) or poly(BVK) $\xrightarrow{h\nu}$ polymeric radical + fragment radical

M ↓ M ↓

graft copolymer + homopolymer

Scheme 16

Fractionation data clearly indicate that the amount of graft copolymer is very close to that of the homopolymer. It can therefore be concluded that the initiation efficiency of polymeric radicals approximates that of fragment radicals.

It is moreover worth mentioning that the polymerization rate R_p is [83] appreciably higher for poly(MIK) and poly(BVK) than for the model compound 3,3-dimethyl-2-butanone (MBK) in the photoinitiated polymerization of AN and St, which are known to act as singlet and triplet quenchers, respectively (Table 13).

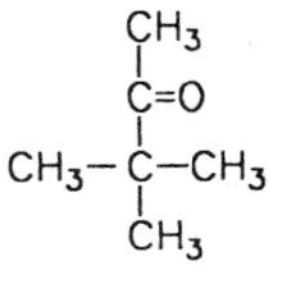

$$CH_3-C(=O)-C(CH_3)_2-CH_3$$

MBK

These data suggested to the authors that the excited states of the polymeric photoinitiators are more difficult to be quenched by the above monomers than in the low-molecular-weight model, due to the steric hindrance of the polymer chain. However, this explanation does not fully justify the higher polymerization rate displayed by MBK with respect to the polymeric photoinitiators in the polymerization of MMA and VAc (Table 13).

Table 13. Photoinitiated polymerization rate (R_P) of various monomers in the presence of high- and low-molecular-weight aliphatic ketone initiators[a] [83]

Initiator	Vinyl monomer			
	AN	St	MMA	VAc
Poly(BVK)	31.56	1.61	9.71	1.91
Poly(MIK)	6.63	0.35	7.37	1.32
MBK	1.26	0.61	12.91	2.37

[a] R_p values as 10^5 mol l^{-1} s^{-1}. Polymerization experiments carried out in benzene solution at 30 °C. Monomer concentration: 4 mol l^{-1}, irradiation wavelength: 313 nm. The photoinitiators concentration was adjusted so as to have an optical density of 0.23 at 313 nm

Copolymers of 2,3-butanedione-2-*O*-methacryloyl-oxime with MMA [poly (BOMA-*co*-MMA)], containing different amounts (7–30 mol%) of side-chain *O*-acyloxime moieties, have also been used [75] as macro-initiators for the photochemical grafting of styrene and acrylamide in benzene and dioxane solution, respectively.

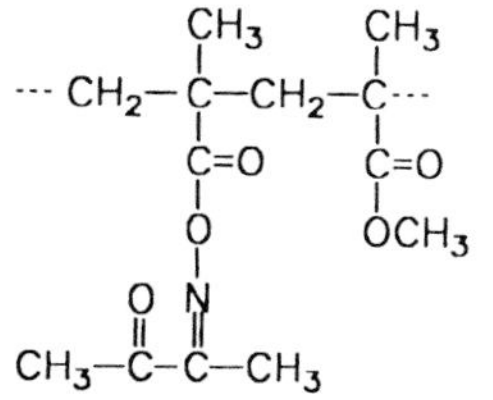

```
          CH3       CH3
          |         |
··· CH2 — C — CH2 — C ···
          |         |
          C=O       C=O
          |         |
          O         OCH3
          |
      O   N
      ||  ||
CH3 — C — C — CH3
```

poly(BOMA-*co*-MMA)

Fractionation data, IR spectroscopy and elemental analysis on the polymers obtained allow us to conclude that, due to the contemporary presence of polymer-anchored and low-molecular-weight free radicals, a mixture of graft-copolymers and homopolymers is obtained (Scheme 17).

When poly(BOMA-*co*-MMA)s are subjected to UV irradiation in benzene solution, in the absence of any monomer, extensive degradation takes place [55, 75], thus confirming the occurrence of a photocleavage mechanism. Similar results have been found [55] by irradiation in monomer-free benzene solution of the copolymers of 1-phenyl-1,2-propanedione-2-*O*-methacryloyloxime with MMA [poly(POMA-*co*-MMA)], although a much higher efficiency of main chain degradation is obtained (Scheme 18 and Table 14).

It is noteworthy that the above copolymers do not undergo any chain scission when irradiated as unplasticized casted films, provided that the temperature is below their glass transition [55]. It has been assumed that free radical recombination in cage prevails on account of lack of diffusion below T_g, due to the reduced mobility of the polymeric chains. Indeed, when the film is plasticized

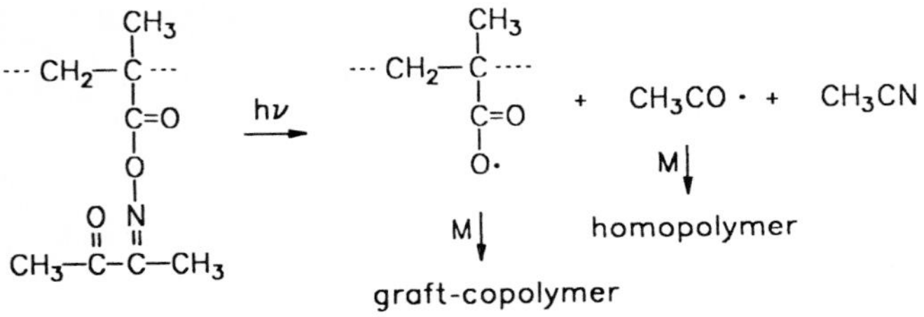

Scheme 17

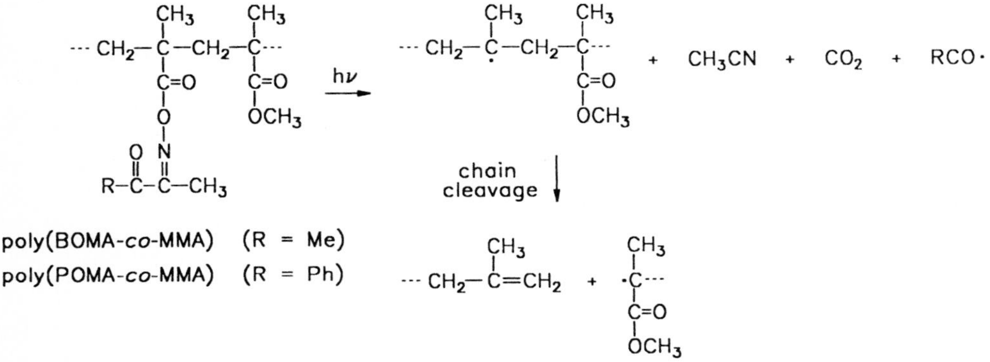

poly(BOMA-*co*-MMA) (R = Me)
poly(POMA-*co*-MMA) (R = Ph)

Scheme 18

Table 14. Photodegradation at 60 °C, in the absence of monomer, of poly(BOMA-*co*-MMA) and poly(POMA-*co*-MMA) in benzene solution or plasticized film [55, 75]

Photosensitive polymer			Irrad. time (min)	10^9 I_a ($E\ cm^{-2} s^{-1}$)	$\overline{M}_n$ [a]
Type	*O*-acyloxime moiety (mol%)	Medium			
Poly(POMA-*co*-MMA)	5.4	solution	0	4.2	29100
			15		10200
			60		4100
Poly(POMA-*co*-MMA)[b]	3.0	film	0	1.7	20900
			15		17500
			30		6700
Poly(BOMA-*co*-MMA)[c]	2.0	solution	0	1.6	22500
			30		19700
			60		15400
Poly(BOMA-*co*-MMA)[b]	2.0	film	0	0.8	28000
			15		22000
			30		21300

[a] Determined by GPC measurements
[b] Plasticized with 30 wt% of di-*n*-butyl phthalate
[c] At 25 °C

in order to lower its T_g below the irradiation temperature, extensive degradation again occurs [55] (Table 14).

Polymeric systems based on side-chain acyloxime moieties and prepared by copolymerization of 1,2-diphenyl-1,2-ethanedione-2-*O*-acryloyloxime with menthyl acrylate [poly(BMOA-*co*-MtA)], have been used as photoinitiators in the UV curing, under nitrogen, of the HDDA/BA equimolar mixture and their activity compared with that of the corresponding low-molecular-weight structural model compound 1,2-diphenyl-1,2-ethanedione-2-*O*-acetyloxime (BMOAc) [61, 84].

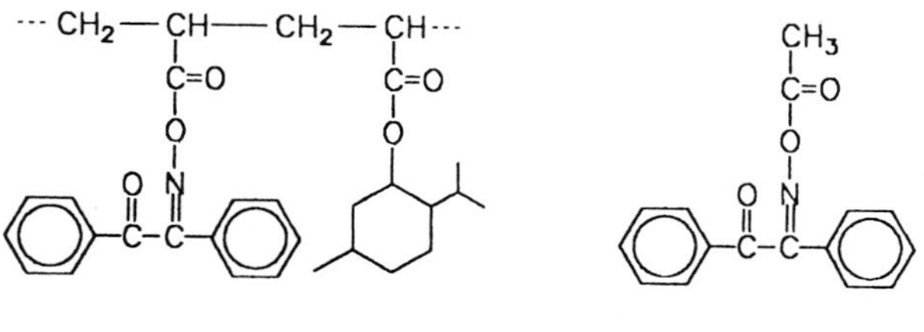

poly(BMOA-*co*-MtA) BMOAc

As reported in Table 15, the kinetic data clearly indicate that the photoinitiation activity of poly(BMOA-*co*-MtA) is not substantially affected by the content of BMOA co-units along the polymer chain and is of the same order of magnitude as that found for the model compound BMOAc. The absence of a "polymer effect" in the above photoinitiators has been interpreted [84] in terms of a photodegradation mechanism of the macromolecules involving the free radical species anchored to the main chain, even in the presence of acrylic monomers, analogous to what is reported in Scheme 18. Moreover, the induction period of the HDDA/BA photoinduced polymerization increases, on decreasing the content of

Table 15. UV curing in film matrix of the HDDA/BA equimolar mixture, under nitrogen, by poly(BMOA-*co*-MTA)s and the corresponding low-molecular-weight analogue BMOAc[a] [61, 84]

Photoinitiating system		Polymerization kinetics[b]		
Type	*O*-acyloxime moiety (mol%)	t_0[c] (s)	$t_{1/2}$[d] (s)	R_c[e] (s^{-1})
Poly(BMOA-*co*-MtA)	75.0	18.4	29	4.8
Poly(BMOA-*co*-MtA)	53.0	20.2	29	5.7
Poly(BMOA-*co*-MtA)	25.2	26.2	36	5.2
Poly(BMOA-*co*-MtA)	13.7	27.9	37	5.2
BMOAc	100.0	27.7	37	5.0

[a] Photoinitiator concentration: 0.5 mol% of *O*-acyloxime moiety in the HDDA/BA mixture. Irradiation conditions: see Table 2
[b] Determined by microwave dielectrometry
[c] Induction period
[d] Half-time of the process
[e] Polymerization rate evaluated at $t_{1/2}$ and expressed as % conversion over time

BMOA units in the copolymers, up to the value observed for BMOAc [61, 84] (Table 15). Photophysical measurements as well as photografting experiments on the above systems, combined with average molecular weight determinations could probably give a deeper insight on these aspects.

The relationship between structure and photoinitiation activity has been examined for polymeric systems bearing side-chain 1-substituted cyclohexyl-phenyl ketone moieties in the UV curing of the HDDA/BA equimolar mixture [19, 20]. Indeed, the activity of poly[(1-acryloxycyclohexyl)phenyl ketone] [poly (APK)] and styrene/4-chloromethyl-styrene/1-(4-styrylmethyloxy)cyclohexyl phenyl ketone copolymers (PABOK) has been compared with that of the corresponding low-molecular-weight structural models such as 1-hydroxy-cyclohexyl phenyl ketone (HPK), 1-acetoxy-cyclohexyl phenyl ketone (ACPK) and 1-(4-isopropyl-benzyloxy) cyclohexyl phenyl ketone (PIBOK).

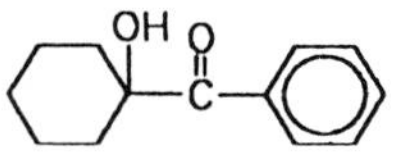

HPK 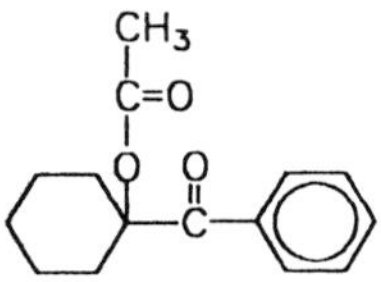 ACPK 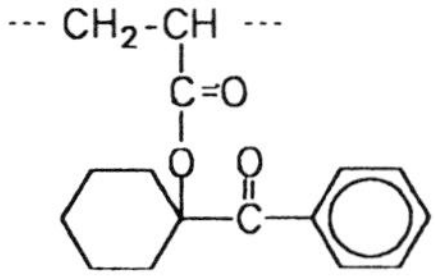 poly(APK)

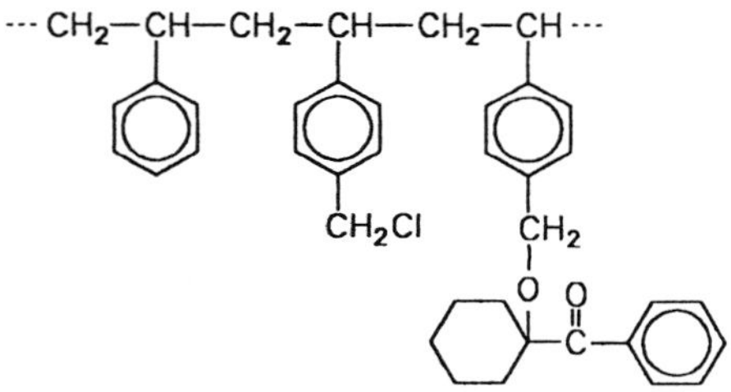

PABOK 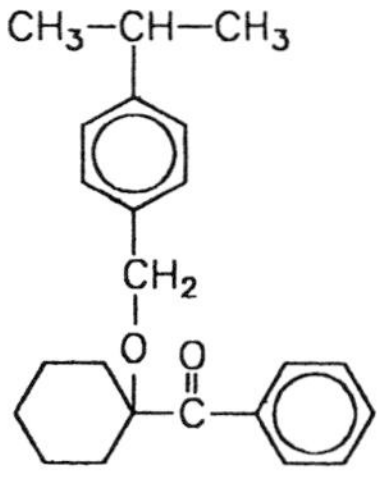 PIBOK

Analogous to what has been established [70] for HPK, UV irradiation should also induce in the abovementioned cyclohexyl phenyl ketone derivatives a Norrish I type fragmentation, i.e. α-cleavage with respect to the benzoyl moiety (Scheme 19):

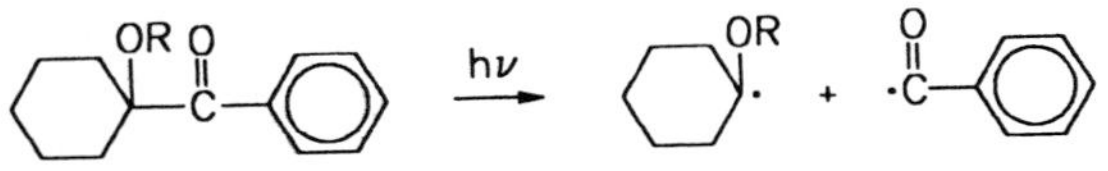

R = H, alkyl, acyl

Scheme 19

As reported in Table 16, poly(APK) exhibits [19, 20] a slightly decreased photoinitiation activity with respect to the structural model ACPK, HPK resulting one order of magnitude more active. The above findings have been interpreted assuming that the ester function in poly(APK) and ACPK has more electron-withdrawing character as compared with the hydroxy group in HPK. Accordingly, the presence in PIBOK and PABOK of an electron-donating ethereal group directly linked to the cyclohexyl phenyl ketone moiety appears to be responsible for the sharp increase of their activity (Table 16), thus reaching values very close to that found for HPK [20].

It is therefore concluded that the electronic nature of the substituent in the 1-position of the cyclohexyl ring plays an important role for the activation of this class of photoinitiators, similarly to what is found for benzoin derivatives [4, 85]. However, as the photoreactive 4-chloromethylstyrene (CMS) co-units are present in PABOK, the activity of the corresponding PIBOK/poly[(4-chloromethyl)styrene]

Table 16. UV curing in film matrix of the HDDA/BA equimolar mixture, under nitrogen, by low- and high-molecular-weight photoinitiators based on cyclohexyl phenyl ketone (CPK) moieties[a] [19, 20]

Photoinitiating system	CPK moieties[b]	CMS moieties[b,c]	Time required for 80% conversion[d] (s)
Poly(APK)	1	0	28.0
ACPK	1	0	23.0
PABOK	1	8	1.1
PIBOK	1	0	1.4
HPK	1	0	1.0
PIBOK+PCMS[c] (1:8)	1	8	1.2
PCMS[c]	0	8	2.5

[a] Polymerization conditions: see Table 1
[b] Molar % concentration in the HDDA/BA mixture
[c] CMS = 4-chloromethyl styrene units; PCMS = poly[(4-chloromethyl)styrene]
[d] Determined by IR spectroscopy

Table 17. Triplet life-time (τ) of polysiloxanes containing cyclohexyl phenyl ketone as well as alkylphenone moieties and polymerization rate (R_p) of MMA in the presence or in the absence of *N*-methyldiethanolamine [86]

Photoinitiator[a]		τ^b (ms)	10^5 R_p (mol l^{-1} s^{-1})	
Type	$\overline{M}_n$		No amine	Amine
PSCPK1	4003	4.22	6.85	4.66
PSCPK2	1967	4.13	2.87	2.94
PSAPK	3865	2.26	5.87	4.52
DEAP	–	1.71	2.15	2.35

[a] Photoinitiator concentration: $2.5 \cdot 10^{-4}$ mol l^{-1}; amine concentration: $1 \cdot 10^{-3}$ mol l^{-1} in ethyl acetate 60% MMA. UV irradiation by a 100 W high pressure Hg-W lamp, under nitrogen
[b] From phosphorescence measurements at 77 K in propan-2-ol

(PCMS) 1:8 mol/mol mixture has also been checked (Table 16). Indeed, the above mixture is found to show a slightly lower activity with respect to PABOK, thus suggesting that polymeric photoinitiators based on photosensitive moieties undergoing homolytic fragmentation may also display an activity synergism, provided that different photoreactive groups are anchored to the same macromolecular chain.

Polysiloxane photoinitiators containing terminal and side-chain alkyl phenyl ketone moieties (PSCPK1, PSCPK2 and PSAPK) have also been prepared and used for promoting the polymerization of MMA under UV irradiation [86]:

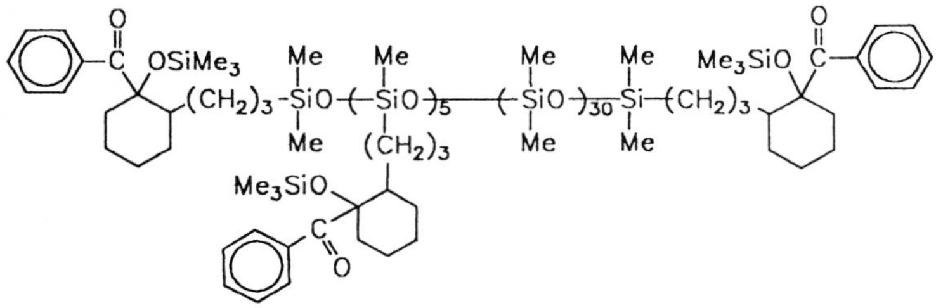

PSCPK1

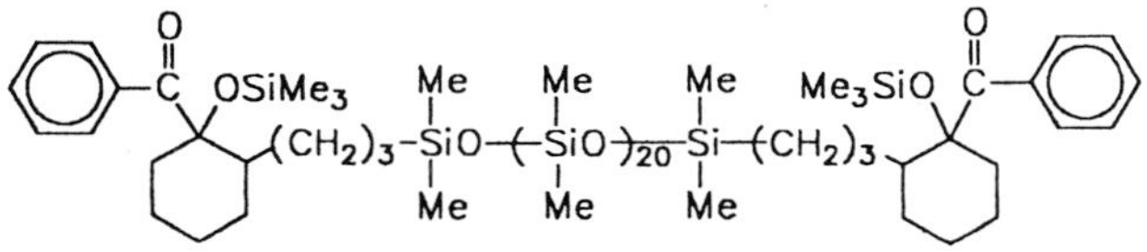

PSCPK2

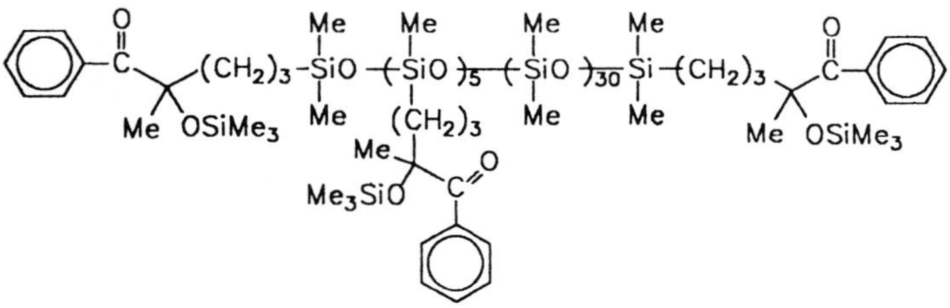

PSAPK

The photoinitiation activity of these polymeric systems has been compared with that of 2,2-diethoxyacetophenone (DEAP), although this low-molecular-weight photoinitiator cannot be strictly considered as the structural model (Table 17).

The reported data clearly indicate that the polysiloxane photoinitiators display a remarkably higher activity with respect to DEAP both in the presence and in the absence of *N*-methyldiethanolamine, which however seems to produce a variable effect depending on the type of photoinitiator. Moreover, the polymerization rate depends on both structure and molecular weight of the polysiloxane systems, an improvement of rate being observed on increasing the latter parameter. End-of-pulse transient absorption spectra of the above polymers on the microsecond time scale are assigned to the benzoyl radical produced on direct photolysis of the alkylphenone chromophores. The transient absorption spectra also correlate with the photoinduced polymerization rate of MMA, indicating that the benzoyl radical is the key initiating species. On the nanosecond time scale, laser flash photolysis of the polysiloxane initiators and the model reveals the formation of long-lived transient spectra which are also assigned to the benzoyl radical. The lack of such long-lived radicals in the case of DEAP suggests [86] that the polysiloxane units stabilize the radicals, probably through some type of intramolecular mechanism favoured by the macromolecular coiling.

An oligomeric polysiloxane photoinitiator (PSDAP) containing dialkoxy acetophenone moieties, prepared by reacting diallyloxy-acetophenone with penta-methyl-disiloxane and H-terminated dimethyl-silicone in the presence of chloro-platinic acid, has been claimed [87] to be compatible with silicone prepolymers containing methylvinyl siloxane co-units and active in their UV curing.

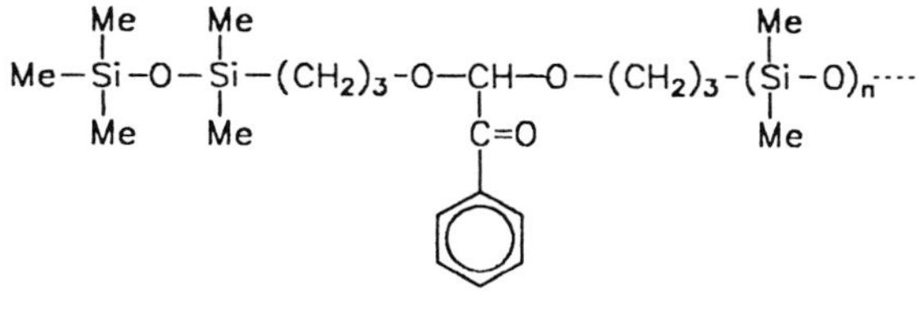

$$Me{-}Si(Me)_2{-}O{-}Si(Me)_2{-}(CH_2)_3{-}O{-}CH(C(=O)C_6H_5){-}O{-}(CH_2)_3{-}(Si(Me)_2{-}O)_n\cdots$$

PSDAP

Oligomeric benzil ketals (DTGPA), prepared by reaction of 2,2-dimethoxy-2-phenyl-acetophenone (DMPA) with triethyleneglycol, in the presence of *p*-toluen-sulphonic acid, are reported [88] to be efficient in the UV curing of styrene/unsaturated polyester formulations giving rise to coatings with very low odour.

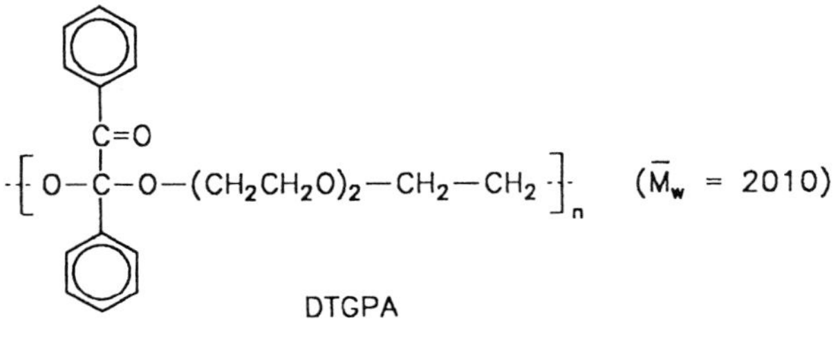

$$\left[O{-}C(C_6H_5)(C(=O)C_6H_5){-}O{-}(CH_2CH_2O)_2{-}CH_2{-}CH_2 \right]_n \quad (\bar{M}_w = 2010)$$

DTGPA

A novel class of low-odouring and non-yellowing polymeric photoinitiators, based on side-chain hydroxyalkylphenone moieties, is reported [89–91] to be very active for UV curable clear acrylic coatings. Due to low volatility and affinity for the organic phase, these photoinitiators are also claimed to be very effective in the curing of water-based acrylic emulsions in which water must be evaporated before the exposure to UV light. These commercial polymeric systems (KIP) are prepared by functionalization of α-methylstyrene prepolymers.

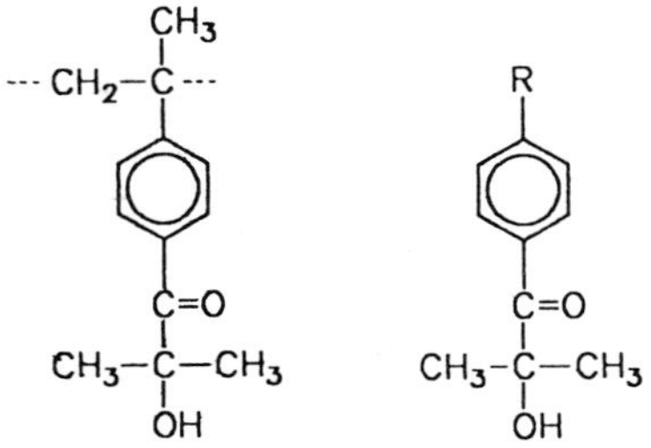

KIP

HIPK (R = H)
AIPK (R = n-$C_{12}H_{25}$)

It is well-established [74] that the low-molecular-weight structural model 2-hydroxy-2-methyl propiophenone (HIPK), upon irradiation, undergoes α-cleavage which is the main decay of the triplet state. Moreover, the absence of any aromatic substituent in the hydroxyalkyl moiety prevents the formation of yellow compounds with semiquinoid structures as those obtained [91, 92] in the case of DMPA (Scheme 20), thus allowing one to achieve uncoloured clear coatings.

However, even in the presence of vinyl monomers, an appreciable amount of the primary radical pair reacts in the cage giving rise to benzaldehyde and other carbonyl compounds [91, 93, 94] leading to coatings with acute characteristic smell (Scheme 21).

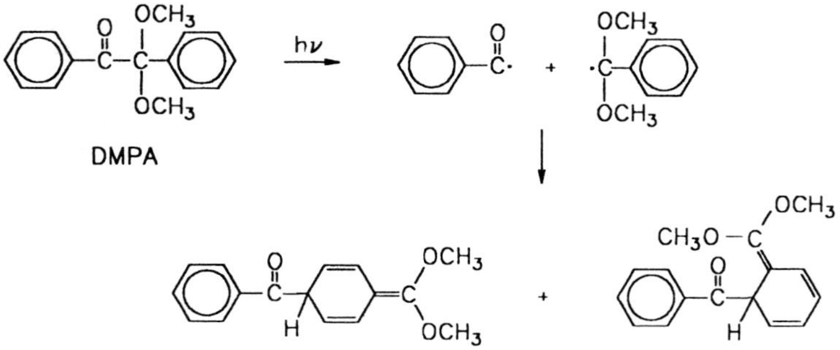

DMPA

Scheme 20

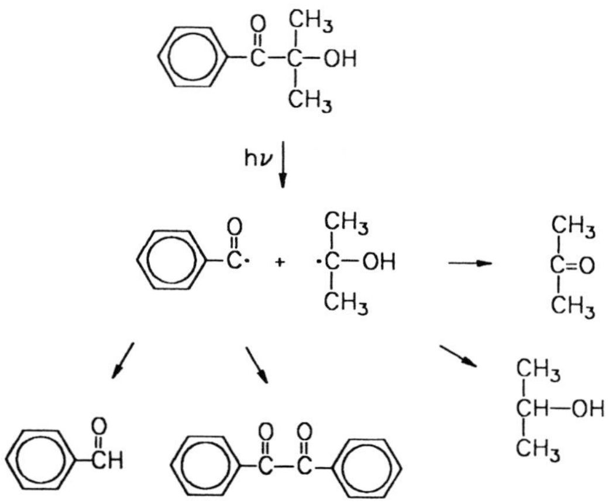

Scheme 21

KIP, by contrast, due to the formation of a polymer-anchored benzoyl radical (Scheme 22), cannot release benzaldehyde after photocleavage, thus allowing production of non-yellowing and low-odour coatings [89, 91], especially required in the field of food packaging.

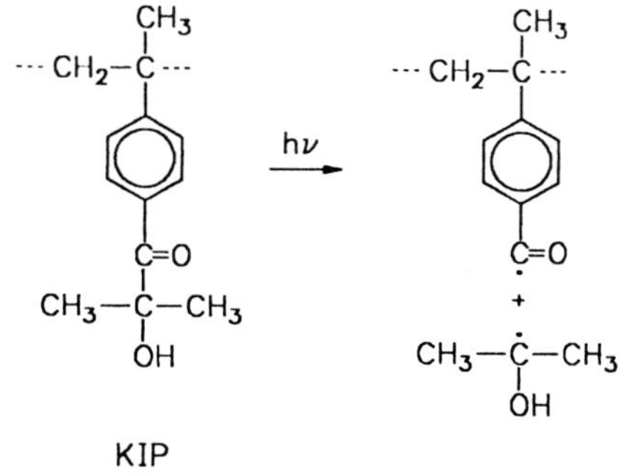

KIP

Scheme 22

Photophysical investigations by time-resolved laser spectroscopy on the above systems allowed it to be established [93] that in toluene solution and in the presence of MMA the α-cleavage quantum yield (Φ_α) in KIP is appreciably lower than in the case of HIPK, the polymeric chain reducing the efficiency of the primary radical generation (Table 18). Since it is well known [95] that the overall initiation quantum yield (Φ_i) is given by the following equation:

$$\Phi_i = \Phi_\alpha \cdot \Phi_{RM} \tag{4}$$

where Φ_{RM} represents the quantum yield of initiating monomeric radicals generation, it is concluded that Φ_{RM} is appreciably higher in the case of the polymeric system, as the MMA polymerization rate (R_p) in the presence of KIP is lower than with HIPK (Table 18).

Table 18. Polymerization rate (R_p) of MMA in toluene solution, relative quantum yields and triplet life-time ($\tau^o{}_T$) of polymeric and low-molecular-weight photoinitiators based on the hydroxyalkylphenone moiety [95]

Photoinitiating system	$\tau^o{}_T$ (ns)	R_p [a]	Φ_α [a]	Φ_i [a]	Φ_{RM} [a]
HIPK	1.4	41	0.28	1.0	3.6
AIPK	4.0	30	0.12	0.5	4.0
KIP	8.0	26	0.07	0.4	5.5

[a] Relative values with respect to DMPA, for which Φ_α, Φ_i and Φ_{RM} are arbitrarily assumed to be equal to one. $R_p \propto \Phi_i^{1/2}$

Quite similar results (Table 18) are obtained in the case of AIPK, where the long alkyl chain in the *para* position to the phenyl ring well simulates the polymeric chain in KIP. It has therefore been suggested [95] that the higher values of Φ_{RM} for KIP and AIPK against HIPK may be due either to an improved yield of radicals generated through side-reactions or to a reduction of primary radicals recombination, thus favouring the yield of initiating free radicals.

Finally, although KIP is found to be as active as HIPK in film matrix [89, 91], its better compatibility and low-odour properties make the polymeric system superior.

Acrylic and methacrylic polymeric systems, bearing side-chain hydroxyalkylphenone groups [poly(HPA) and poly(HPMA), repectively] have also been reported [96, 97] to display similar activity as the corresponding monomers in the UV curing of oligomeric epoxyacrylate/HDDA formulations. Poly(HPA), however, even if slightly more active than poly(HPMA), is found to show a lower activity as compared with HIPK, under the same experimental conditions [96].

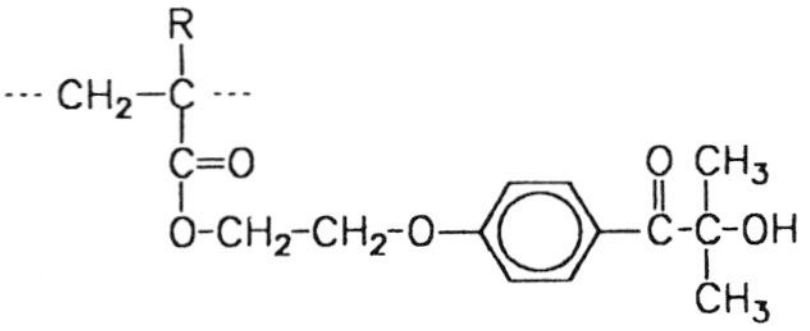

poly(HPMA) (R = CH_3)
poly(HPA) (R = H)

Copolymerizable photoinitiators based on the hydroxyalkylphenone moiety, most representatives having the general formula depicted below, have also been used [98] in acrylic, epoxyacrylate and urethaneacrylate formulations for non-yellowing and low-odour UV curable coatings for paper, metals and plastics, as well as for fast drying printing-inks. This technique is also claimed to reduce environmental pollution and save energy and materials.

$$R'-C_6H_4-C(=O)-C(CH_3)_2-OR$$

$R' = H$, $R = CH_2{=}CH{-}C(=O){-}$

$R' = CH_2{=}CH{-}COO{-}(CH_2)_2{-}O{-}$, $R = H$

$R' = CH_2{=}CH{-}COO{-}$, $R = H$

$R' = CH_2{=}CH{-}COO{-}$, $R = CH_2{=}CH{-}C(=O){-}$

More recently, new series of polymeric systems based on hydroxyalkylphenone moieties have been reported [99]. These polymers are characterized by excellent photoinitiating activity, comparable with that of the low-molecular-weight analogue HIPK, and exhibit high migration stability:

$$\cdots(CH_2-CH)_x-(CH_2-C(R')(Y))_{1-x}\cdots \quad \text{(CH bearing } C_6H_4-C(=O)-C(R)(CH_3)-OH)$$

$x = 1$ ($R = CH_3$, C_2H_5)

$0 < x < 1$ ($R = CH_3$, $R' = H$, $Y = Ph$)

$0 < x < 1$ ($R = CH_3$, $R' = CH_3$, $Y = COOCH_3$)

Moreover, their compatibility with different hydrophobic and hydrophilic UV curable formulations can easily be varied, in contrast to the corresponding low-molecular-weight photoinitiators, by choosing the appropriate comonomer in the preparation of the copolymer system.

Polymeric systems containing side-chain benzoin moieties, recently appearing [100, 102] in the patent literature, have been applied as photoinitiators to UV curing of prepolymers of a different nature. In particular, siloxane-compatible photocrosslinking initiators (PSBME) are manufactured [100] by reacting siloxanes with benzoin ether derivatives (Scheme 23) and used in UV curing of acryloxypropyl-terminated dimethylsiloxane prepolymers.

The reaction of glycidyl acrylate with α-(2-carboxyethyl)benzoin methyl ether has allowed one to obtain [101] the corresponding acrylic monomer which, upon copolymerization with different amounts of MMA, butyl methacrylate and 2-(*N*, *N*-dimethylamino)ethyl methacrylate, gives rise to polymeric photoinitiators, containing side-chain benzoin methylether moieties, for photocurable coatings:

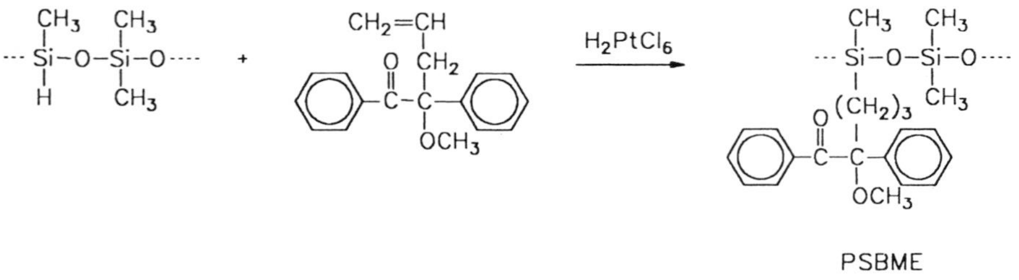

H_2PtCl_6

PSBME

Scheme 23

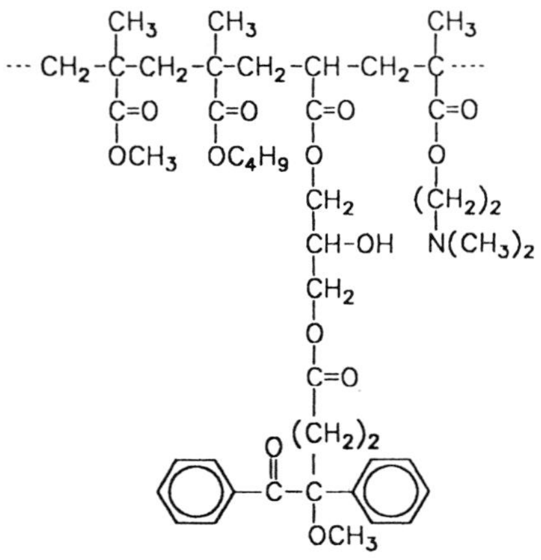

Benzoin-treated NCO-terminated polyols, derived from HO-terminated poly (ethylene adipate), toluene diisocyanate (TDI) and benzoin (Scheme 24), have been used [102] as photoinitiators of oligomeric urethaneacrylate and epoxyacrylate formulations to give crosslinked films with improved physical properties due to a reduced content of low-molecular-weight compounds.

More detailed investigations on the photoreactive behaviour of polymeric systems based on the benzoin moiety have been described [103]. Indeed, poly (benzoin acrylate) [poly(AB)] and copolymers of benzoin acrylate with styrene or MMA [poly(AB-*co*-St) and poly(AB-*co*-MMA), respectively] have been prepared and characterized. Poly(AB), when irradiated by UV light in the presence of photosensitizers such as benzophenone, *p*-benzoquinone or methyl phenyl ketone, gives a benzene-insoluble crosslinked polymer.

Poly(AB) is active [103] in the UV induced polymerization of styrene in dioxane solution. The solvent fractionation of the resulting polymeric product has been proved to be constituted by a small amount of linear and graft poly(styrene), the main product being a crosslinked polymer, according to the occurrence of termination reactions involving polymer-bound growing chain and substituted benzyl radicals.

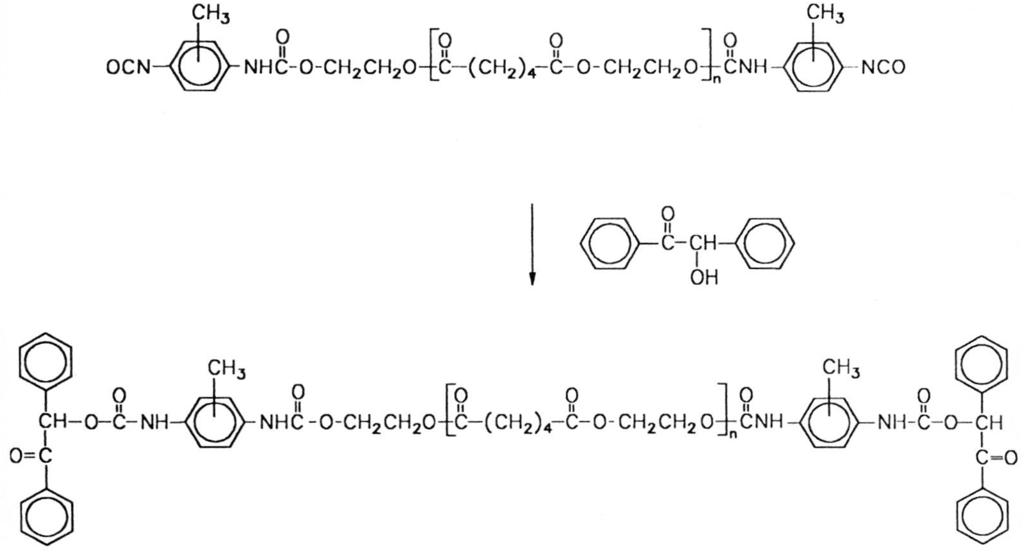

Scheme 24

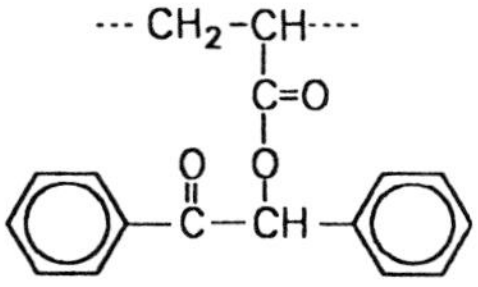

poly(AB)

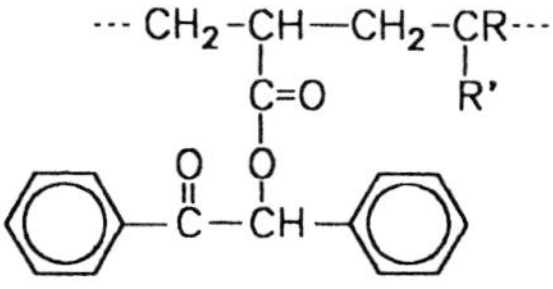

poly(AB-*co*-St) (R = H, R' = Ph)
poly(AB-*co*-MMA) (R = CH_3, R' = $COOCH_3$)

Ester derivatives of benzoin are known to display usually quite a low photoinitiation activity in the polymerization of vinyl monomers [85, 104]. By contrast, benzoin alkyl ethers are claimed to generate [17], by a photofragmentation mechanism, benzoyl and α-alkoxy benzyl radicals resulting in a much more active polymerization and crosslinking initiating species (Scheme 25). Thus, polymeric systems having the above moieties in the side chains have been prepared and their photoreactivity studied in more detail [105–107].

It has been found [105, 106] in particular that the reaction product [poly (BMEGMA-*co*-MMA)] derived from glycidyl methacrylate/methyl methacrylate copolymers [poly(GMA-*co*-MMA)] and α-(2-carboxyethyl)benzoin methyl ether (CEBME) (Scheme 26), although less efficient than CEBME alone in the UV initiated polymerization of styrene, exhibits a markedly enhanced photocrosslinking activity.

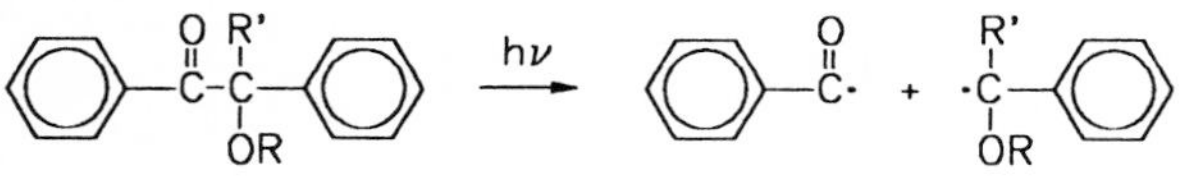

R' = H, alkyl

R = alkyl

Scheme 25

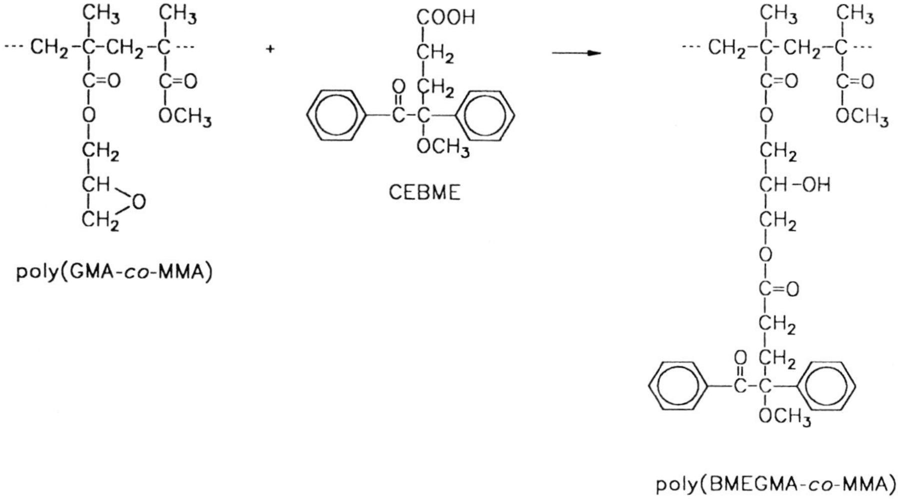

CEBME

poly(GMA-*co*-MMA)

poly(BMEGMA-*co*-MMA)

Scheme 26

Moreover, poly(α-methylolbenzoin methyl ether acrylate) [poly(MBA)] has been checked in the UV induced polymerization of styrene and compared with poly(AB) and other low-molecular-weight structural models such as benzoin, α-methylol benzoin methyl ether (MBE) and α-methylol benzoin methyl ether acetate (MBAc) [107].

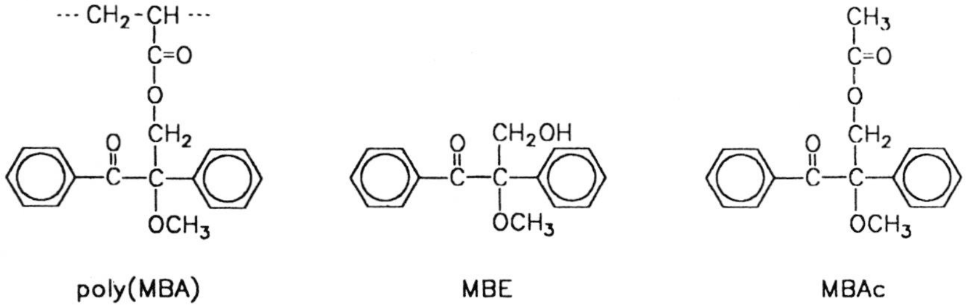

poly(MBA) MBE MBAc

The results show that poly(MBA) and poly(AB) are much more active than benzoin, MBE and MBAc, the polymerization rate of styrene being four times larger with the polymeric systems. The enhancement of photoinitiation activity by poly(MBA) is at least partially ascribed [107] to gel formation which is supposed to protect the propagating growing chain radicals from termination reactions. However, it is not possible to exclude the possibility [107] that the above behaviour could also be due to the different reactivity of benzoyl (A) and α-methoxy benzyl (B) free radicals generated by α-cleavage of the excited benzoin methyl ether moiety. In fact, the latter species remains anchored to the polymeric matrix in the case of poly(MBA), as opposed to what occurs in the corresponding low-molecular-weight analogues (Scheme 27).

Copolymers of MBA with styrene [poly(MBA-*co*-St)] containing variable amounts of MBA co-units, have also been applied [107] to the UV initiated polymerization of MMA in benzene solution.

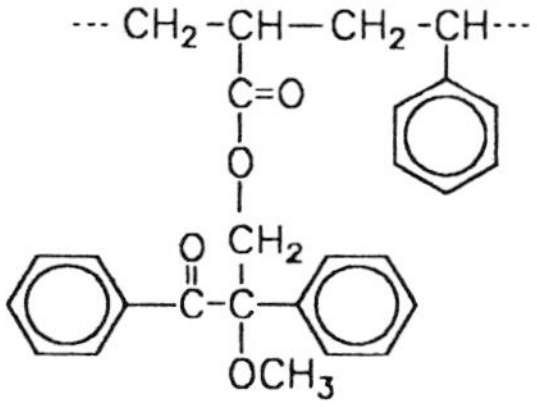

poly(MBA-*co*-St)

Large portions of the polymers obtained are insoluble in the common solvents of poly(MMA). As reported in Table 19, the conversion to poly(MMA) increases on increasing the content of MBA co-units in the copolymers, while the amount of poly(MMA) formed per mmol of MBA unit is found to decrease gradually. The authors interpret the above phenomena by assuming that the enhancement of the overall amount of MBA units causes the contemporary increase of the local concentration of primary radicals and hence the probability of termination by coupling reactions. The formation of large amounts of insoluble polymer is attributed to the crosslinking of growing graft-poly(MMA) by polymer bound α-methoxy benzyl radicals produced by α-cleavage of benzoin methyl ether moieties in MBA co-units.

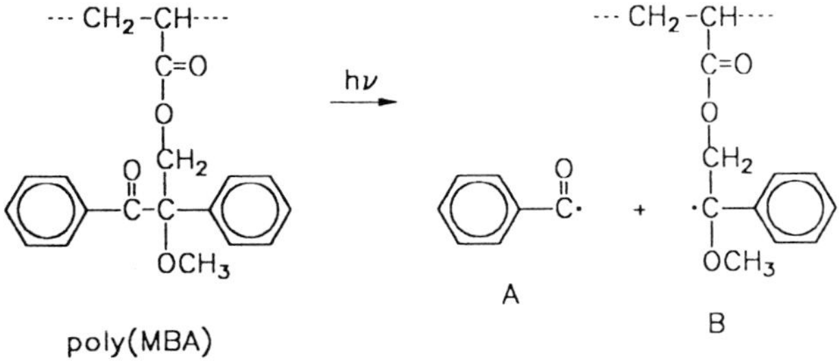

poly(MBA) $\xrightarrow{h\nu}$ A + B

Scheme 27

Table 19. Effect of the composition of poly(MBA-*co*-St)s on the photoinitiated polymerization of MMA in benzene solution [107]

Photoinitiating system[a]	MBA co-units (mol%)	Polymeric product	
		Conversion (%)	g/mmol of MBA units
Poly(MBA-*co*-St)	1.0	6	11.5
Poly(MBA-*co*-St)	11.7	26	8.4
Poly(MBA-*co*-St)	29.4	44	7.6
Poly(MBA)	100.0	52	4.9

[a] Photoinitiator concentration: 3% by weight

This occurrence indicates that photosensitive polymers can be usefully applied to photocurable polymeric systems as well as to surface protecting UV coatings.

In order to improve storage stability and compatibility of benzoin alkyl ether (BAE) photoinitiators in UV curable formulations, polymeric or oligomeric systems having BAE moieties as end-groups have been designed [108].

Three kinds of reactive BAE derivatives such as α-(2-carboxyethyl) benzoin alkyl ethers (BAE-CA), α-methylol benzoin alkyl ethers (BAE-OH) and newly synthesized α-(2-cyanato ethyl)benzoin alkyl ethers (BAE-NCO) are employed for preparing the polymeric photoinitiators.

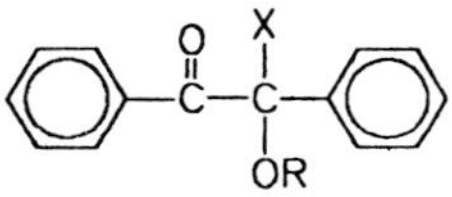

$$C_6H_5-C(=O)-C(X)(OR)-C_6H_5$$

BAE-X

BAE-CA ($X = CH_2CH_2COOH$, $R = CH_3$, i-C_4H_9)
BAE-OH ($X = CH_2OH$, $R = CH_3$, i-C_4H_9)
BAE-NCO ($X = CH_2CH_2NCO$, $R = CH_3$, i-C_4H_9)

In particular, the reaction of BAE-CA with an oligomeric epoxy resin based on bisphenol-A (BPA) gives the BAE-terminated system ER-BAE:

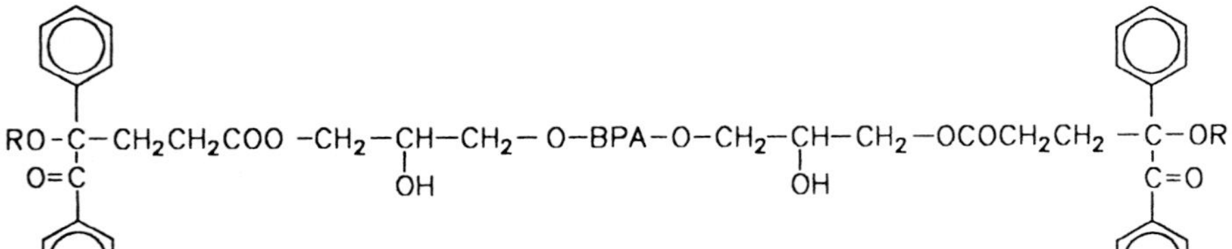

$$RO-C(C_6H_5)(C(=O)C_6H_5)-CH_2CH_2COO-CH_2-CH(OH)-CH_2-O-BPA-O-CH_2-CH(OH)-CH_2-OCOCH_2CH_2-C(C_6H_5)(C(=O)C_6H_5)-OR$$

ER-BAE

and the OH-terminated poly(tetramethylene glycol) [PTMG] is used in the reaction with BAE-NCO to give PTMG-BAE:

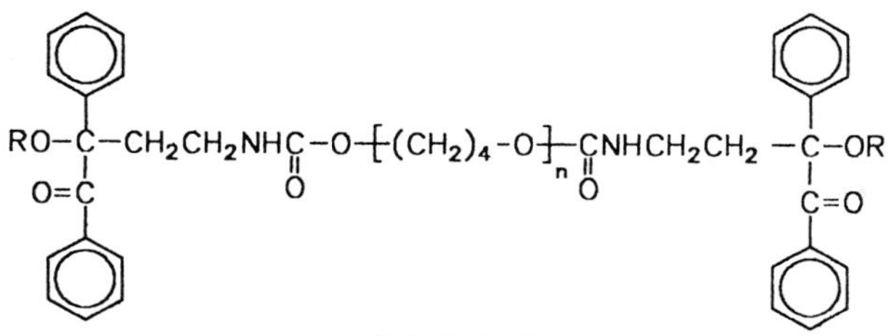

PTMG-BAE

Finally, the oligomeric PTMG-urethane modified BAE (PTMG-U-BAE) is obtained [108] through the reaction of BAE-OH with PTMG-diisocyanate, which in turn is prepared from PTMG and two equivalents of toluene diisocyanate (TDI) or hexamethylene diisocyanate (MDI):

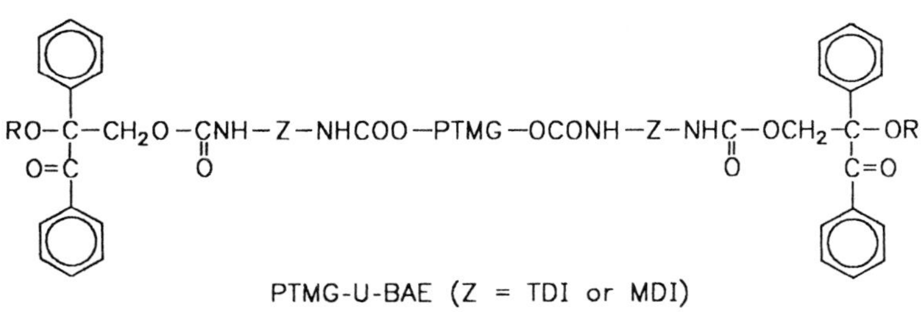

PTMG-U-BAE (Z = TDI or MDI)

The polymer-bound photoinitiators have been used to prepare photocurable formulations with epoxyacrylates, urethaneacrylates, trimethylolpropane triacrylate (TMPTA) and HDDA [108]. The above formulations have been UV irradiated on film matrix and the extent of photocuring evaluated in terms of degree of residual acrylic unsaturation at a fixed reaction time. As reported in Table 20, the polymer-bound BAE systems display an appreciably higher photoinitiation activity as compared with that of the low-molecular-weight analogue benzoin isobutyl ether (BIBE):

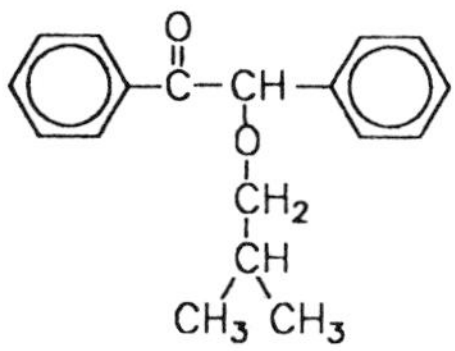

BIBE

The improved activity of the above polymeric photoinitiators has been explained [108] assuming that macroradicals generated by photofragmentation

Table 20. Photoinitiation activity, expressed as % conversion,[a] of polymer-bound BAE systems and BIBE in the UV curing of various acrylic formulations [108]

Photoinitiator[b]	Formulation		
	TMPTA	PTMG-UA[c]	PTMG-UA[c]/HDDA (3:2)
PTMG-BAE	96	70	–
BIBE	30	55	65
PTMG-U-BAE	–	–	75

[a] Determined from the amount of residual acrylic unsaturation after 120 s of irradiation time, as measured by IR spectroscopy

[b] Photoinitiator concentration: 4 equiv.% of BAE with respect to the reactive acrylate group in the formulation. Irradiation conditions: 45 mW/cm^2 by a high pressure mercury lamp

[c] PTMG urethane diacrylate obtained by reacting PTMG-diisocyanate with 2-hydroxyethyl acrylate

may be alive longer due to the retardment of recombination reactions, thus favouring the polymerization initiation.

In order to investigate in more detail the relationships between structure and photoinitiation activity in the polymeric systems based on benzoin methyl ether moieties, poly(MBA) and copolymers of MBA with menthyl acrylate [poly(MBA-*co*-MtA)s] have also been checked as photoinitiators and compared with the homopolymer of α-vinyloxymethylbenzoin methyl ether and its copolymers with menthyl vinyl ether [poly(MBVE) and poly(MBVE-*co*-MtVE)s, respectively] [61, 84].

Low-molecular-weight structural models for both the series of polymeric systems, such as MBAc and α-ethyloxymethyl benzoin methyl ether (MBEE) have also been prepared and tested in terms of photoinitiation activity [61, 84].

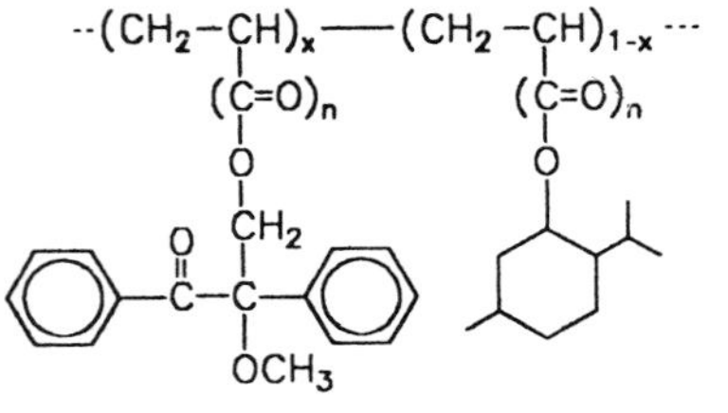

$$\cdots(CH_2-CH)_x-(CH_2-CH)_{1-x}\cdots$$

(side groups: $(C=O)_n-O-CH_2-C(OCH_3)(C_6H_5)-C(=O)-C_6H_5$ and $(C=O)_n-O$-menthyl)

poly(MBA-*co*-MtA) ($n = 1$)
poly(MBVE-*co*-MtVE) ($n = 0$)
poly(MBVE) ($n = 0$, $x = 1$)

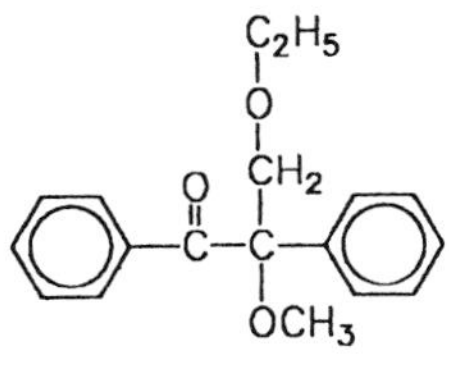

$C_6H_5-C(=O)-C(OCH_3)(C_6H_5)-CH_2-O-C_2H_5$

MBEE

The induction period values for the curing process of the HDDA/BA mixture in the presence of either polymeric or low-molecular-weight photoinitiators based on benzoin methyl ether moieties are found to be quite similar (Table 21) and one order of magnitude shorter than those reported for the systems based on *O*-acyloxime moieties (Table 15).

Table 21. UV curing in film matrix of the HDDA/BA equimolar mixture, under nitrogen, by polymeric systems bearing the benzoin methyl ether moiety (MB) at different distances from the backbone, as compared with the corresponding low-molecular-weight analogues [61, 84]

Photoinitiating system[a]		t_0[b,c] (s)	R_c[b,d] (s^{-1})
Type	MB moiety (mol%)		
Poly(MBA)	100.0	2.3	16.5
Poly(MBA-*co*-MtA)	79.0	3.6	17.6
Poly(MBA-*co*-MtA)	51.6	3.7	17.0
Poly(MBA-*co*-MtA)	36.8	3.8	17.0
Poly(MBA-*co*-MtA)	24.4	3.2	18.3
Poly(MBA-*co*-MtA)	12.2	3.4	17.6
MBAc	100.0	3.9	11.0
Poly(MBVE)	100.0	4.0	12.8
Poly(MBVE-*co*-MtVE)	81.3	4.7	15.1
Poly(MBVE-*co*-MtVE)	71.7	2.7	13.7
MBEE	100.0	4.6	10.1

[a] Photoinitiator concentration: 0.5 mol% in terms of MB moieties. Irradiation conditions: see Table 2
[b] Determined by microwave dielectrometry at 9.5 GHz in terms of ε'' (loss factor)
[c] Induction period of the curing process
[d] Polymerization rate expressed as percentage of monomer to polymer conversion over time and calculated at the half-time of the curing process

A detailed analysis of the kinetic data of Table 21 clearly shows that all the polymeric systems display higher photoinitiation activity than that of the corresponding low-molecular-weight analogues MBAc and MBEE [84]. The above results confirm that a positive "polymer effect" on activity is also present in polymeric systems working with a photofragmentation mechanism.

Considering that in the polymeric photoinitiators the local concentration of generated radicals (Scheme 27) is larger than in the corresponding low-molecular-weight structural models, a lower photoinitiation activity should be expected in the former systems, the probability of coupling reactions between free radicals being enhanced. It is therefore proposed [84] that the improvement of photoinitiation activity in the polymeric systems can be attributed to a reduced mobility of the polymer-anchored B radicals, which prevents recombination reactions, thus favouring their capability to attack acrylic monomers and then initiate the polymerization process. The higher activity of poly(MBA) and poly(MBA-*co*-MtA)s as compared with poly(MBVE) and poly(MBVE-*co*-MtVE)s, is tentatively explained [84] assuming that the latter systems are characterized by a larger conformational rigidity, as confirmed by their chiroptical properties [109]. This could favour a more pronounced tendency to give coupling reactions by the polymer-anchored B radicals. With the aim of clarifying this point, the investigation of excited-state processes and reactivity of the above systems in the polymerization of acrylic monomers in solution has been carried out [110] against the corresponding models, including α-methylolbenzoin methyl ether isobutyrate (MBI):

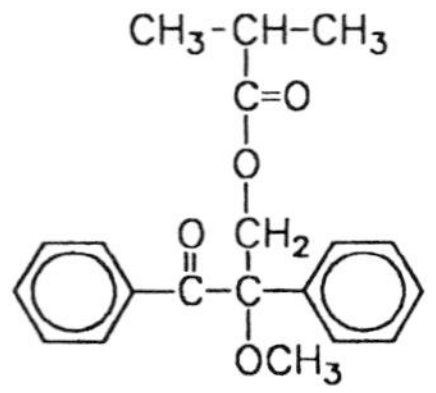

$CH_3-CH-CH_3$
$C=O$
O
$O \quad CH_2$
$C-C$
OCH_3

MBI

Indeed, photophysical studies by laser flash photolysis combined with the determination of MMA polymerization rate (R_p) in toluene solution, have allowed evaluation of the triplet state lifetime of the above systems, as well as their relative quantum yields of initiation (Φ_i) and α-cleavage (Φ_α). As reported in Table 22, Φ_α values for poly(MBA) and poly(MBVE) are appreciably lower than those for MBI and MBEE, respectively. On this basis, the polymeric photoinitiators would be expected to display lower activity than the models in the polymerization of acrylic monomers. On the contrary, poly(MBA) and poly(MBVE), together with the related copolymers, show higher values of R_p and hence Φ_i, in the UV initiated polymerization of MMA in toluene solution. These findings, therefore, confirm the previously obtained results in film matrix, where a HDDA/BA equimolar mixture was used as curing formulation (Table 21).

As with the previous reported conclusion concerning the polymeric photoinitiators based on hydroxyalkyl acetophenone moiety (KIP), in this case Φ_{RM} (Eq. 4) also is much higher in the polymeric systems than in the models (Table 22), indicating that the B radicals anchored to the polymer backbone are less prone to give radical-radical combination, thus favouring their reaction with the acrylic monomers.

Further investigations on polymeric systems containing side-chain benzoin methyl ether and benzoin moieties confirm the enhanced reactivity of the polymeric photoinitiators [111, 112]. Indeed, copolymers of MBA with methyl acrylate and methyl methacrylate [poly(MBA-*co*-MA) and poly(MBA-*co*-MMA), respectively] and the corresponding copolymers based on α-methylolbenzoin acry-

Table 22. Rate of polymerization and relative quantum yields in the polymerization of MMA in toluene solution[a] [110]

Photoinitiator	R_p	Φ_i	Φ_α	Φ_{RM}
MBEE	0.80	0.64	0.50	1.28
MBI	0.75	0.56	0.45	1.24
Poly(MBVE)	1.20	1.44	0.35	4.11
Poly(MBA)	1.80	3.24	0.25	12.96
Poly(MBA-*co*-MtA)	1.40	1.96	–	–
Poly(MBVE-*co*-MtVE)	1.60	2.56	–	–

[a] Values relative to 2,2-dimethoxy-2-phenyl acetophenone (DMPA), set as equal to one

late and methacrylate [poly(MBHA-*co*-MA) and poly(MBHMA-*co*-MMA)] have been prepared and their efficiency checked both in the photoinitiated polymerization of MMA in solution and in the cure of acrylic formulations in film matrix, against the corresponding low-molecular-weight structural models MBI, MBHI and MBHP.

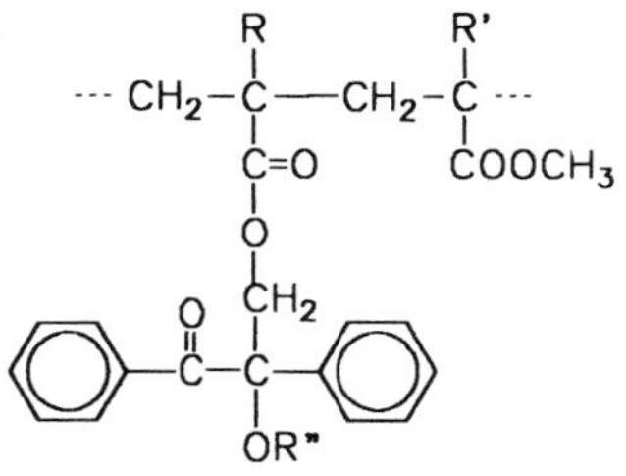

poly(MBA-*co*-MA) (R = R' = H, R" = CH_3)
poly(MBA-*co*-MMA) (R = H, R' = R" = CH_3)
poly(MBHA-*co*-MA) (R = R' = R" = H)
poly(MBHMA-*co*-MMA) (R = R' = CH_3, R" = H)
poly(MBHMA-*co*-MA) (R = CH_3, R' = R" = H)

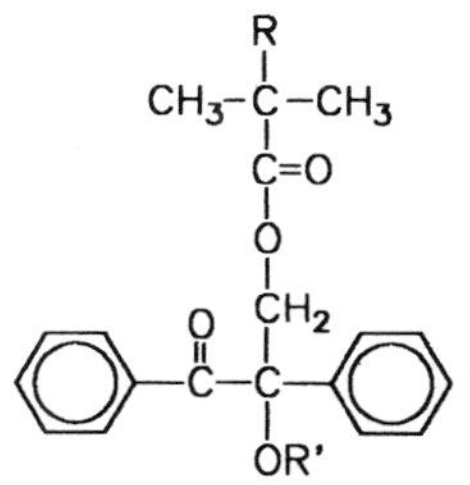

MBI (R = H, R' = CH_3)
MBHI (R = R' = H)
MBHP (R = CH_3, R' = H)

As reported in Tables 23 and 24, the polymeric systems display higher efficiency than the corresponding models in both types of UV initiated processes.

Moreover, the polymers based on benzoin moieties are less active than the corresponding benzoin methyl ether derivatives. The greater reactivity of the polymeric initiators has been attributed by the authors to their role as chain terminators. In fact, the polymer chain may well shield the polymer-bound substituted benzyl radicals, thereby diminishing their termination efficiency

Table 23. Photoinitiated polymerization of methyl methacrylate (MMA) in benzene solution, using polymeric photoinitiators based on benzoin methyl ether and benzoin moieties (BE) as well as their low-molecular-weight analogues[a] [111]

Photoinitiator		Polymer yield (g/mmol of BE)
Type	BE[b] (mol%)	
Poly(MBHA-*co*-MA)	48	9.6
MBHI	100	8.0
Poly(MBHMA-*co*-MMA)	49	9.0
Poly(MBHMA-*co*-MA)	64	7.5
MBHP	100	6.1
Poly(MBA-*co*-MA)	68	17.2
Poly(MBA-*co*-MMA)	37	15.0
MBI	100	6.1

[a] 5:1 (v:v) benzene–MMA solution. Photoinitiator concentration: 10^{-2} mol l^{-1} in terms of photoreactive groups
[b] Determined by ^1H NMR analysis

Table 24. UV curing in film matrix of UVE74/TEGDA mixture using polymeric photoinitiators based on benzoin methyl ether and benzoin moieties (BE) and their low-molecular-weight analogues[a] [111]

Photoinitiator		Curing rate[b]
Type	BE[c] (mol%)	
Poly(MBHA-*co*-MA)	48	8
MBHI	100	19
Poly(MBHMA-*co*-MMA)	49	11
MBHP	100	21
Poly(MBA-*co*-MA)	68	4
MBI	100	11

[a] Comonomers mixture: 73.4% of triethylene glycol (TEGDA) and 26.6% of UVE74 (Setacure AP570). Belt speed: 8 m/min. Photoinitiator concentration: 0.1 mol l^{-1} in terms of photoreactive groups. Thickness of liquid film: 50μm
[b] Number of passes under UV lamps required to cure the acrylic formulation
[c] Determined by ^{1}H NMR analysis

with respect to the low-molecular-weight analogues. The lower photoinitiation activity of poly(MBHA-*co*-MA) and poly(MBHMA-*co*-MMA) with respect to poly(MBA-*co*-MA) and poly(MBA-*co*-MMA) has been explained on the basis of the different capacity of the polymer-bound substituted benzyl radicals to act as chain terminators and scavengers of benzoyl primary radicals in the two systems. In fact, in the polymers based on benzoin moieties, the benzyl-type radicals may react with propagating and benzoyl primary radicals by two different mechanisms involving hydrogen atom transfer and radical-radical combination, the latter route being possible only in the polymeric systems bearing side-chain benzoin methyl ether moieties (Scheme 28).

It is well established [113, 114] that both aromatic and aliphatic tertiary amines can be applied in the photoinitiated polymerization of multifunctional acrylates to accelerate the curing process, even in the presence of benzoin derivatives. However, since amines, in addition to oxygen scavengers, behave as chain transfer agents, this last process predominates, with a strong reduction in the average molecular weight of the polymer chains, particularly when the amine concentration is rather high.

One way of minimizing the detrimental effect caused by chain transfer on decreasing molecular weight and crosslink density of UV curable coatings has been successfully applied [115] by using copolymerizable methacrylic monomers containing the tertiary amine moiety as components of reactive acrylic formulations in the presence of benzoin ether derivatives.

In this context, polymeric photoinitiators bearing side-chain benzoin methyl ether and tertiary amine moieties in the same macromolecule, have been prepared and the effect of amine on the photoinitiation activity investigated [116–118].

In particular, copolymers of MBA with 2-(*N*,*N*-dimethylamino)ethyl acrylate, 2-(*N*,*N*-diethylamino)ethyl acrylate, 3-(*N*,*N*-dimethylamino)propyl acrylate and 3-(*N*,*N*-diethylamino)propyl acrylate [poly(MBA-*co*-DMEA)s, poly(MBA-*co*-DEEA)s, poly(MBA-*co*-DAPA)s and poly(MBA-*co*-DEPA)s, respectively] at

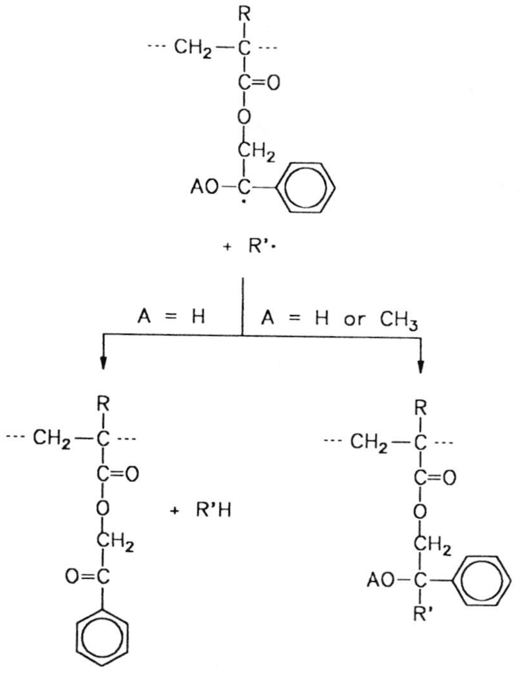

R' = growing chain or benzoyl primary radicals **Scheme 28**

different compositions have been checked and compared with the corresponding mixtures of poly(MBA) with poly(DMEA), poly(DEEA), poly(DAPA) and poly(DEPA). The mixtures of low-molecular-weight model compounds MBI, DMEI, DEEI, DAPI and DEPI have been similarly investigated (Table 25).

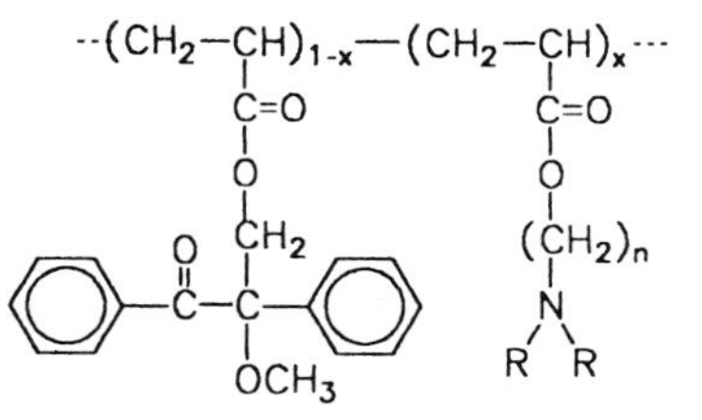

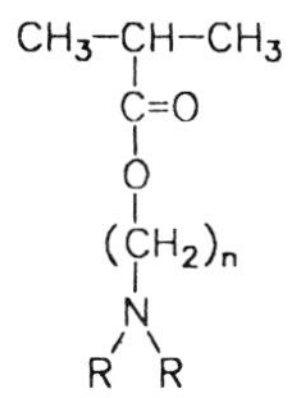

poly(MBA-*co*-DMEA) ($n = 2$, $R = CH_3$)
poly(MBA-*co*-DEEA) ($n = 2$, $R = C_2H_5$)
poly(MBA-*co*-DAPA) ($n = 3$, $R = CH_3$)
poly(MBA-*co*-DEPA) ($n = 3$, $R = C_2H_5$)
poly(DMEA) ($n = 2$, $R = CH_3$, $x = 1$)
poly(DEEA) ($n = 2$, $R = C_2H_5$, $x = 1$)
poly(DAPA) ($n = 3$, $R = CH_3$, $x = 1$)
poly(DEPA) ($n = 3$, $R = C_2H_5$, $x = 1$)

DMEI ($n = 2$, $R = CH_3$)
DEEI ($n = 2$, $R = C_2H_5$)
DAPI ($n = 3$, $R = CH_3$)
DEPI ($n = 3$, $R = C_2H_5$)

Table 25. UV curing in film matrix of the HDDA/BA equimolar mixture, under nitrogen, by polymeric and low-molecular-weight photoinitiators based on benzoin methyl ether (MB) and tertiary amine (AM) moieties [116–118]

Photoinitiating system[a]			t_0 [b,c] (s)	$t_{1/2}$ [b,d] (s)	Rc_{max} [b,e] (s^{-1})
Type	MB (mol%)	AM (mol%)			
Poly(MBA-*co*-DMEA)	44	56	0.5	3.9	17
Poly(MBA)/poly(DMEA)	44	56	0.9	4.6	16
Poly(MBA)/DMEI	44	56	1.1	4.8	17
MBI/DMEI	44	56	1.2	6.7	11
Poly(MBA-*co*-DEEA)	52	48	1.5	4.9	17
Poly(MBA)/poly(DEEA)	50	50	1.5	4.6	16
Poly(MBA)/DEEI	51	49	2.4	4.9	14
MBI/DEEI	46	54	2.3	7.3	11
Poly(MBA-*co*-DAPA)	52	48	0.9	4.3	16
Poly(MBA)/poly(DAPA)	51	49	1.9	5.3	15
Poly(MBA)/DAPI	50	50	1.3	4.7	15
MBI/DAPI	50	50	2.6	7.3	12
Poly(MBA-*co*-DEPA)	55	45	0.4	3.9	17
Poly(MBA)/poly(DEPA)	55	45	0.6	4.4	16
Poly(MBA)/DEPI	55	45	1.1	5.4	14
MBI/DEPI	55	45	0.6	6.4	10

[a] Photoinitiator concentration 0.5 mol% in terms of MB moieties. Irradiation conditions: see Table 2
[b] Determined by microwave dielectrometry at 9.5 GHz in terms of ε'' (loss factor)
[c] Induction period of the curing process
[d] Half-time of the process
[e] Maximum polymerization rate expressed as percentage of monometer to polymer conversion over time

All the above copolymers and mixtures of poly(MBA) with low- and high-molecular-weight tertiary amine compounds display an appreciably higher activity, as measured by the maximum of polymerization rate (Rc_{max}), with respect to the corresponding mixtures involving the low-molecular-weight photoinitiator MBI. These results confirm previous observations [84] concerning the homopolymers of MBA and MBVE as well as their copolymers with MtA and MtVE as compared with MBAc and MBEE (Table 21).

All the MBA/*N,N*-dialkylamino acrylate copolymers behave similarly to poly(MBA-*co*-MtA)s (Tables 21 and 25), thus suggesting that the replacement of MtA by *N,N*-dialkylamino acrylate co-units does not markedly affect the photoinitiation activity of the system. Accordingly, BMI/*N,N*-dialkylamino isobutyrates mixtures exhibit substantially the same activity as MBI alone [118]. Similar results have previously been obtained for 2,2-dimethoxy-2-phenyl acetophenone (DMPA), when additioned with diethylmethylamine, in the UV initiated polymerization of *n*-butyl methacrylate [113]. However, a remarkable shortening of the induction period (t_o) of UV curing is observed for all the polymeric photoinitiators in the presence of tertiary amines as compared with the low-molecular-weight MBI/*N,N*-dialkylamino isobutyrates systems, the maximum effect resulting in the case of MBA/*N,N*-dialkylamino acrylate copolymers (Table 25).

The above data have been interpreted assuming that the side-chain tertiary amino groups in the copolymers behave as hydrogen donors towards the adjacent polymer-anchored benzylic radicals generated by the photofragmentation of benzoin methyl ether moieties of MBA co-units. The resulting alkylamino radicals may then react with the oxygen present in traces in the acrylic formulation, thus giving rise to peroxy radicals which completely consume the oxygen through a chain mechanism involving adjacent amino groups (Scheme 29):

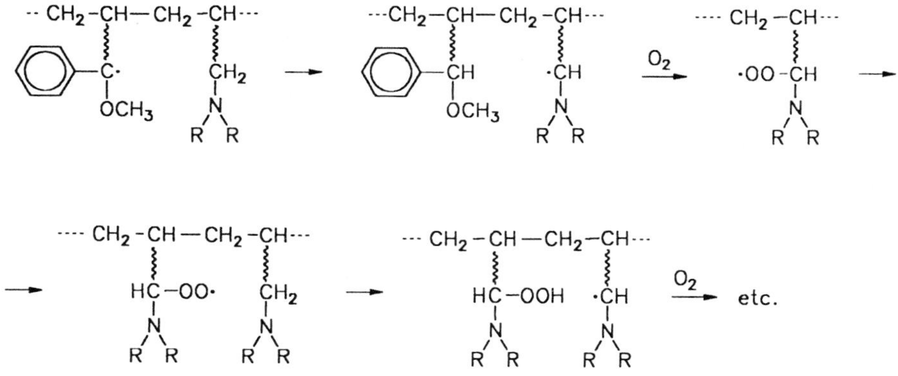

Scheme 29

This picture is also consistent with the fact that, in general, t_o becomes shorter on increasing the content of tertiary amine co-units in the copolymer [118]. The observation that Rc_{max} decreases in poly(MBA-*co*-DAPA) and poly(MBA-*co*-DEPA) on increasing the content of tertiary amine co-units and that the minimum activity is observed in poly(MBA-*co*-DMEA) and poly(MBA-*co*-DEEA) in correspondence with the highest content of tertiary amine co-units [118] can easily be explained. In fact, due to the mechanism proposed in Scheme 29, as soon as the traces of oxygen are consumed, the residual amine co-units, in excess with respect to MBA units, continue the conversion of the substituted benzyl-type polymeric radicals into the alkylamino radicals, which are known to display lower reinitiation constants for acrylic monomers.

In conclusion, the polymeric photoinitiators bearing side-chain benzoin methyl ether and tertiary amine moieties appear very promising for fast UV curable coatings because they combine the high activity of the photoreactive polymers with the oxygen scavenging effect typical of amine functional groups. Moreover, the anchorage of amine moieties to the photoreactive polymer matrix should also give some other advantages in terms of ultimate properties of the resulting coatings such as no reduction of crosslink density as well as no appreciable emission of bad odours, due to the lack of volatility of the polymeric photoinitiator.

All the low- and high-molecular weight photoinitiators, working with a photofragmentation mechanism, examined up to now usually fail in the UV curing of white pigmented coatings, as they absorb in the same spectral region as

TiO_2, the most used component for this type of formulation. Quite recently, low-molecular-weight photoinitiators, based on the acylphosphinoxide moiety, have been introduced [119]. They are claimed to work with a Norrish I type fragmentation mechanism [120, 121] and exhibit an absorption band centered at about 380 nm with a tail over 400 nm [76, 119]. Such absorption characteristics make this class of photoinitiators particularly suitable for the UV curing of TiO_2-pigmented coatings [122] as well as of thick-walled glass fibre-reinforced polyesters [123]. In this context, polymeric photoinitiators bearing side-chain acyl diphenylphosphinoxide moieties have also been prepared and checked in the UV curing of both HDDA/BA (1:1) and TiO_2-pigmented epoxyacrylate formulations [124].

In particular, the homopolymer of methacryloyl chloride [poly(MAC)] and its copolymers with MMA [poly(MAC-*co*-MMA)] have been functionalized to give poly(MAPO) and poly(MAPO-*co*-MMA)s (Scheme 30):

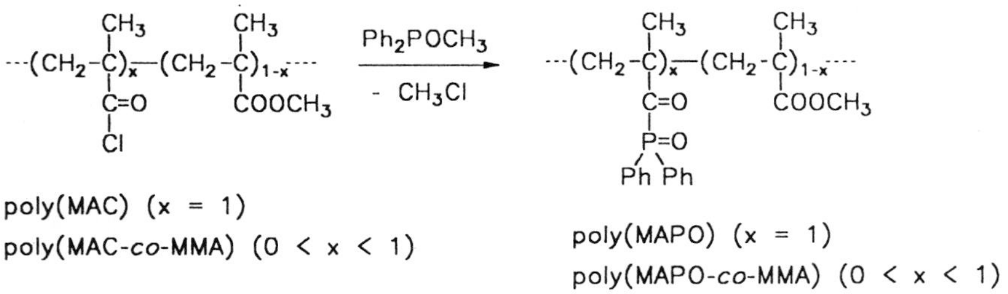

$$\cdots(CH_2\text{-}C(CH_3)(C(=O)Cl))_x\text{-}(CH_2\text{-}C(CH_3)(COOCH_3))_{1-x}\cdots \xrightarrow[-\ CH_3Cl]{Ph_2POCH_3} \cdots(CH_2\text{-}C(CH_3)(C(=O)P(=O)Ph_2))_x\text{-}(CH_2\text{-}C(CH_3)(COOCH_3))_{1-x}\cdots$$

poly(MAC) (x = 1)
poly(MAC-*co*-MMA) (0 < x < 1)

poly(MAPO) (x = 1)
poly(MAPO-*co*-MMA) (0 < x < 1)

Scheme 30

The low-molecular-weight analogue pivaloyl diphenylphosphinoxide (PIVPO) has also been prepared in order to compare its photoinitiation activity with that of the polymeric systems:

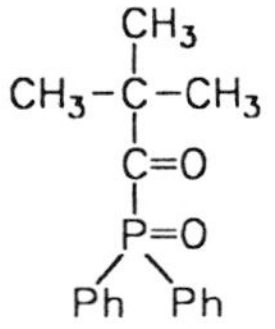

$$CH_3\text{-}C(CH_3)_2\text{-}C(=O)\text{-}P(=O)Ph_2$$

PIVPO

As reported in Table 26, poly(MAPO) and poly(MAPO-*co*-MMA) systems are found [124] to display a higher photoinitiation activity in clear coating formulations, expressed in terms of half-time of the curing process, essentially due to their shorter induction period, as compared with the structural model PIVPO. The polymeric systems, therefore, appear to be preferable for their higher productivity in clear coatings formulations.

Table 26. UV curing in film matrix of HDDA/BA equimolar mixture, under nitrogen, by polymeric and low-molecular-weight photoinitiators based on the acyldiphenylphosphinoxide (APO) moiety [124]

Photoinitiating system[a]		t_0 [b,c] (s)	$t_{1/2}$ [b,d] (s)	Rc_{max} [b,e] (s^{-1})
Type	APO moiety[f] (mmol/g)			
Poly(MAPO)	1.78	12.6	28.2	3.7
Poly(MAPO-*co*-MMA)	1.97	15.0	24.6	6.2
Poly(MAPO-*co*-MMA)	1.53	14.4	27.6	4.2
Poly(MAPO-*co*-MMA)	1.01	7.8	18.0	6.1
Poly(MAPO-*co*-MMA)	0.20	12.6	32.4	3.0
PIVPO	3.49	23.4	35.4	5.1

[a] Photoinitiator concentration: 0.1 mol% of APO moieties. Irradiation conditions: see Table 2
[b] Determined by microwave dielectrometry at 9.5 GHz in terms of ε'' (loss factor)
[c] Induction period of the curing process
[d] Time required to obtain 50% conversion of the HDDA/BA mixture
[e] Maximum polymerization rate expressed as percentage of monomer to polymer conversion over time
[f] As determined by UV spectroscopy

Table 27. UV curing of a TiO_2-pigmented epoxyacrylate formulation in the air by polymeric and low-molecular-weight photoinitiators based on the acyldiphenylphosphinoxide (APO) moiety [124]

Photoinitiating system[a]		Curing rate[b]	Pendulum hardness[c] (s)	W.I.[d]
Type	APO moiety (mmol/g)			
Poly(MAPO)	1.78	4	135	76.8
Poly(MAPO-*co*-MMA)	1.97	3	140	73.3
PIVPO	3.49	3	145	76.2

[a] Photoinitiator concentration: 2 wt% in terms of APO moieties
[b] Expressed as number of passes required at a conveyor belt speed of 8 m/min and under constant irradiation conditions, in order to obtain a non-tacky surface coating
[c] Expressed as oscillation time of the Koenig pendulum after three passes under constant irradiation conditions
[d] Whiteness index, by colorimetry, of the final coating

The above polymeric and low-molecular-weight photoinitiators, when employed in the UV curing of a TiO_2-pigmented epoxyacrylate formulation, exhibit substantially the same activity within experimental error (Table 27). In this case no substantial gain of productivity is obtained with the polymeric photoinitiators, the ultimate properties of the resulting coatings being similar or slightly poorer compared with those obtained with PIVPO.

However, it is worth noting that these polymeric systems display [106] a significantly higher stability than PIVPO, both under daylight exposure and hydrolytic conditions, thus allowing storage problems to be overcome when applied to UV curable coatings.

More recently, taking into account that in the case of low-molecular-weight acyldiphenylphosphinoxides the best performances, in terms of stability and photoinitiation activity, are observed when the acyl group is linked to a phenyl ring [125, 126], polymeric photoinitiators bearing side-chain benzoyldiphenylphosphinoxide moieties have been prepared [127]. In particular, the homopolymer of 4-vinylbenzoic acid and its copolymers with MMA at different compositions [poly(VBA) and poly(VBA-*co*-MMA)s, respectively] have been transformed in poly(VBPO) and poly(VBPO-*co*-MMA)s by treatment with thionyl chloride followed by functionalization with methoxydiphenylphosphine, according to Scheme 31:

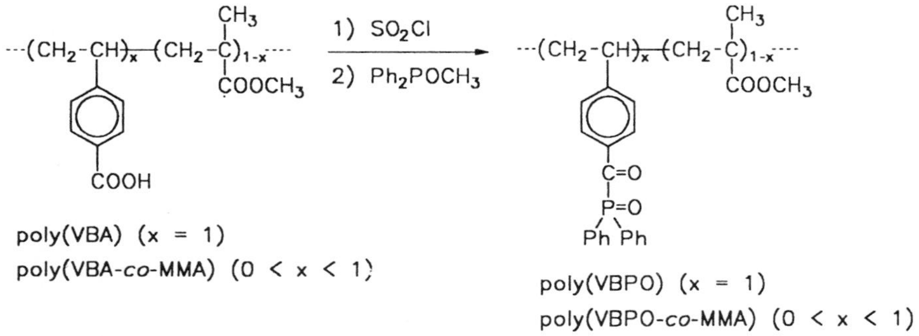

$$\cdots(CH_2-CH)_x-(CH_2-C(CH_3)(COOCH_3))_{1-x}\cdots \text{ [side chain: } C_6H_4-COOH] \xrightarrow[2)\ Ph_2POCH_3]{1)\ SO_2Cl} \cdots(CH_2-CH)_x-(CH_2-C(CH_3)(COOCH_3))_{1-x}\cdots \text{ [side chain: } C_6H_4-C(=O)-P(=O)Ph_2]$$

poly(VBA) (x = 1)
poly(VBA-*co*-MMA) (0 < x < 1)

poly(VBPO) (x = 1)
poly(VBPO-*co*-MMA) (0 < x < 1)

Scheme 31

The comparison of the kinetic data reported in Tables 26 and 28 clearly shows that poly(VBPO-*co*-MMA)s are more active in the photoinitiated polymerization of the HDDA/BA equimolar mixture than poly(MAPO-*co*-MMA)s, notwithstanding their uncomplete solubility in the acrylic formulation. In particular, the remarkable shortening of the induction period (t_0) causes a sharp increase of the overall polymerization efficiency in poly(VBPO-*co*-MMA)s, as indicated by $t_{1/2}$ values which are about one third of those for poly(MAPO-*co*-MMA)s.

In order to obtain complete solubility of poly(VBPO) and poly(VBPO-*co*-MMA)s in the HDDA/BA equimolar mixture, UV curing experiments have also been repeated on the same acrylic formulation diluted with chloroform and the results compared with those found for the corresponding low-molecular-weight structural model compound 4-isopropyl-benzoyldiphenylphosphinoxide (IBPO).

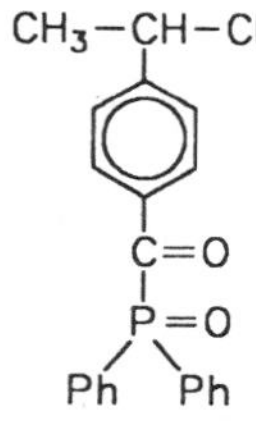

$$CH_3-CH(CH_3)-C_6H_4-C(=O)-P(=O)Ph_2$$

IBPO

Table 28. UV curing in film matrix of HDDA/BA equimolar mixture, under nitrogen, by poly (VBPO-*co*-MMA)s [127]

Photoinitiator[a]		t_0 [b,c] (s)	$t_{1/2}$ [b,d] (s)	Rc_{max} [b,e] (s^{-1})
Type	VBPO co-units[f] (mmol/g)			
Poly(VBPO-*co*-MMA)	0.70	0.6	9.1	9.4
Poly(VBPO-*co*-MMA)	1.17	2.6	9.5	9.2
Poly(VBPO-*co*-MMA)	1.83	2.9	8.4	11.0
Poly(VBPO-*co*-MMA)	2.07	1.4	8.9	9.4

[a] Photoinitiator concentration: 0.1 mol% of VBPO moieties (uncomplete solubility). Irradiation conditions: see Table 2
[b] Determined by microwave dielectrometry at 9.5 GHz, in terms of ε'' (loss factor)
[c] Induction period of the curing process
[d] Time required for obtaining 50% conversion of the HDDA/BA mixture
[e] Maximum polymerization rate, expressed as percentage of monomer to polymer conversion over time
[f] Determined by UV spectroscopy

Indeed, the kinetic parameters (Table 29) are substantially the same, thus indicating a practically equal productivity of poly(VBPO) and poly(VBPO-*co*-MMA)s, as compared with IBPO. Moreover, the photoinitiation activity of the polymers appears independent of the content of VBPO co-units, and suggests that the individual photoreactive co-units behave as isolated moieties.

Table 29. UV curing, under nitrogen, of HDDA/BA (1:1)/$CHCl_3$ mixture by polymeric and low-molecular-weight photoinitiators based on the benzoyldiphenylphosphinoxide (BPO) moiety[a] [127]

Photoinitiator[b]		t_0 [c,d] (s)	$t_{1/2}$ [c,e] (s)	Rc_{max} [c,f] (s^{-1})
Type	BPO moiety[g] (mmol/g)			
IBPO	2.87	0.5	7.3	7.9
Poly(VBPO)	2.28	0.5	8.2	7.3
Poly(VBPO-*co*-MMA)	2.07	1.3	9.7	6.2
Poly(VBPO-*co*-MMA)	1.93	1.0	8.0	7.7
Poly(VBPO-*co*-MMA)	1.83	1.0	6.0	7.9
Poly(VBPO-*co*-MMA)	1.17	1.7	9.0	7.2

[a] Mixture consisting of 1 part of HDDA/BA (1:1) and 2 parts of $CHCl_3$ (wt/wt)
[b] Photoinitiator concentration: 0.1 mol% of BPO moieties. Irradiation conditions: see Table 2
[c] Determined by microwave dielectrometry at 9.5 GHz, in terms of ε'' (loss factor)
[d] Induction period of the curing process
[e] Time required for obtaining 50% conversion of the HDDA/BA mixture
[f] Maximum polymerization rate, expressed as percentage of monomer to polymer conversion over time
[g] Determined by UV spectroscopy

Polysilanes having the general formula $(-SiR_1R_2-)_n$ strongly absorb in the UV region between 295 and 400 nm, depending on the nature of the substituents R_1 and R_2 [128]. High- and low-molecular-weight silyl radicals can be formed according to Scheme 32:

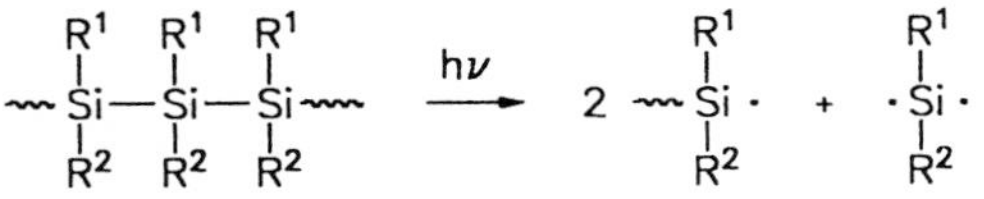

$$\sim Si(R^1)(R^2)-Si(R^1)(R^2)-Si(R^1)(R^2)\sim \xrightarrow{h\nu} 2 \sim Si(R^1)(R^2)\cdot + \cdot Si(R^1)(R^2)\cdot$$

Scheme 32

These radicals are able to initiate the polymerization of acrylates, methacrylates and styrene [129]. In particular, poly(methyl phenyl silylene) [poly(MPSi)] is found to be the most effective, in accordance with a very high quantum yield of silyl radical formation.

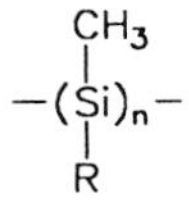

$$-(Si(CH_3)(R))_n-$$

poly(MPSi) (R = Ph)
poly(MASi) (R = $OCOCH_3$)

Poly(methyl acetoxy silylene) [poly(MASi)] is also claimed [130] to initiate the polymerization of cyclohexyl methacrylate upon UV irradiation and to crosslink the resulting polymer by air exposure. Finally, the pyridinium salt of partially chloromethylated poly(methyl phenyl silylene) (Q-PMPSi) is reported [131] to initiate the polymerization, in aqueous solution, of hydrophilic vinyl monomers such as methacrylic acid, acrylamide, 2-hydroxyethyl methacrylate and 1-vinyl-2-pyrrolidone.

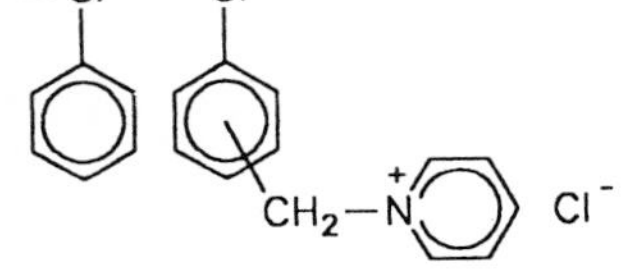

$$\sim Si(CH_3)(C_6H_5)-Si(CH_3)(C_6H_4CH_2-N^+C_5H_5)\sim \quad Cl^-$$

Q-PMPSi

Recently, telechelic photoinitiators with alkoxythiocarbonyl sulfide groups have been reported [132] to be active in the photoinitiated polymerization of MMA. In particular, the photochemical efficiencies of BPX, BMX and poly(MMA) terminated by one or two isopropoxythiocarbonyl sulfide groups (PMMA-X) have been compared in the polymerization of MMA. The above systems, upon irradiation, give photodissociation reactions as depicted in Scheme 33:

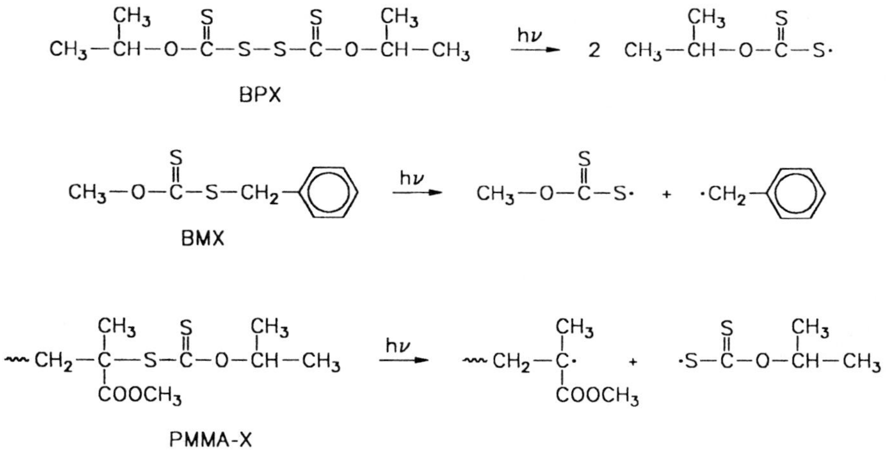

$$CH_3-CH(CH_3)-O-C(=S)-S-S-C(=S)-O-CH(CH_3)-CH_3 \xrightarrow{h\nu} 2\ CH_3-CH(CH_3)-O-C(=S)-S\cdot$$

BPX

$$CH_3-O-C(=S)-S-CH_2-C_6H_5 \xrightarrow{h\nu} CH_3-O-C(=S)-S\cdot + \cdot CH_2-C_6H_5$$

BMX

$$\sim CH_2-C(CH_3)(COOCH_3)-S-C(=S)-O-CH(CH_3)-CH_3 \xrightarrow{h\nu} \sim CH_2-C\cdot(CH_3)(COOCH_3) + \cdot S-C(=S)-O-CH(CH_3)-CH_3$$

PMMA-X

Scheme 33

Quantum yields are found to follow the order: BPX > BMX > difunctional poly(MMA) > monofunctional poly(MMA). Moreover, the photoinitiation efficiency increases with increasing the molecular weight of the functional poly(MMA).

The use of a low-molecular-weight sulphur-containing photoinitiator, such as *S*-benzoyl *O*-ethyl xanthate (BEX) is described for the polymerization of MMA [133]. Laser flash photolysis of BEX reveals [134] that the primary photoprocess involves a C(=O)–S cleavage to generate both the benzoyl and the ethoxythiocarbonylthiyl radicals which initiate the polymerization (Scheme 34).

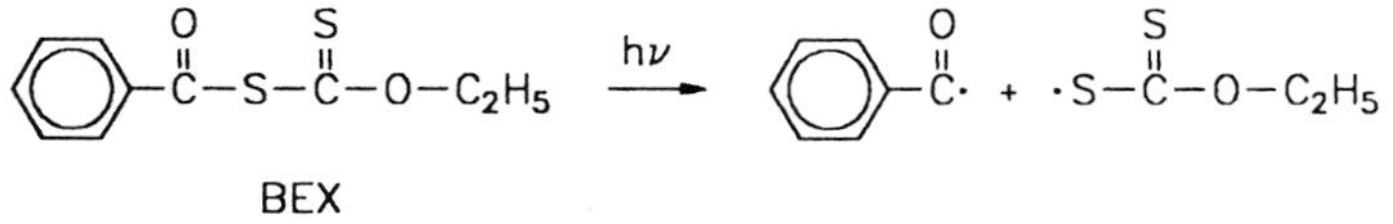

$$C_6H_5-C(=O)-S-C(=S)-O-C_2H_5 \xrightarrow{h\nu} C_6H_5-C\cdot(=O) + \cdot S-C(=S)-O-C_2H_5$$

BEX

Scheme 34

Poly(*S*-benzoyl *O*-ethylmethacrylate xanthate) resin [poly(BEMX)], obtained by functionalization of crosslinked 2-hydroxyethyl methacrylate/ethyleneglycol dimethacrylate beads, results [135], upon irradiation at 400 nm, in a heterogeneous photoinitiator for the polymerization of MMA and styrene more effective than the low-molecular-weight analogue BEX.

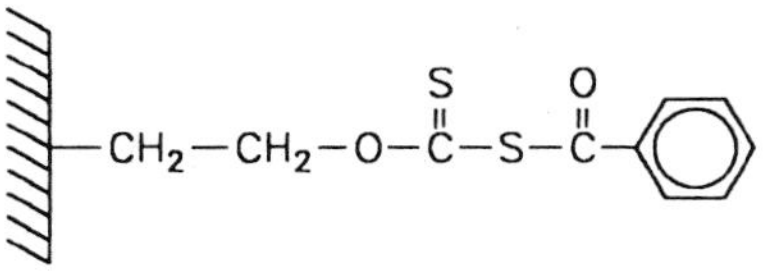

$$\text{(resin)}-CH_2-CH_2-O-C(=S)-S-C(=O)-C_6H_5$$

poly(BEMX)

This behaviour is explained assuming that, when the thiocarbonylthiyl radicals are immobilized on the resin, their termination efficiency is lower, thus favouring the propagation reaction. However, when high monomer conversions are reached, termination of the homopolymer chains by the polymer-bound thiocarbonylthiyl radicals becomes predominant and a high graft yield is obtained.

5 Special Applications of Polymeric Photoinitiators

This section deals with the use of photoreactive polymeric systems as promoters of UV induced polymerization and crosslinking of monomers and oligomers employed in peculiar applications.

Photoinitiators containing in the same macromolecule photosensitive groups operating with different radical generation mechanisms are also considered, particularly when these photoreactive moieties are responsible for synergistic effects.

As far as special applications are concerned, one example is given by acrylic and methacrylic polymers bearing side-chain hydroxyalkylphenone moieties, such as poly(HPA) and poly(HPMA), previously described in Sect. 4. They are claimed [96, 97] to produce a two-layer system consisting of a base layer of the polymeric photoinitiator, obtained by spin coating of its solution over the substrate surface, and a top layer of a UV-curable coating resin. This procedure is emphasized as a new method for surface chemical modification and for obtaining very thin photostructurable layers. In fact, by using the polymeric photoinitiator, after exposure to UV light, an insoluble crosslinked film displaying a strong adhesion to the substrate surface is obtained, whereas in the presence of a low-molecular-weight analogue such as HIPK, a polymeric film without any characteristic structure is given [96, 97].

It is noteworthy that the final thickness of the coating is not much affected by the depth of the coating resin layer, but depends essentially on the thickness of the polymeric photoinitiator layer, which in turn is affected by the concentration of the solution of the photoreactive polymer used for the spin coating process (Table 30). Interdiffusion between the solid photoinitiator layer and the liquid resin on the top may play a crucial role besides the photochemical and radical processes. A further advantage of the two-layer system process is the need for a much smaller amount of the polymeric photoinitiator with respect to the conventional UV curing, where a low-molecular-weight system in employed [96].

Another interesting procedure for obtaining a two-layer system involves [97] the formation of a base layer of the polymeric photoinitiator, through a spin coated film onto different substrates, followed by their immersion, under UV irradiation, into multifunctional acrylic formulations dissolved in solvents unable to remove the base layer from the substrate. By this method, a top layer of crosslinked acrylic resin is obtained, having strong adhesion to the substrate. The replacement of the polymeric photoinitiator by a low-molecular-weight analogue

Table 30. Thickness of photoinitiator layer and coating as a function of poly(HPA) concentration [96]

Poly(HPA) concentration[a] (%)	Thickness (μm)	
	Photoinitiator layer	Coating[b]
0.5	0.035	12.7
1.0	0.075	13.8
2.0	0.130	14.8
5.0	0.250	16.6
10.0	0.670	17.9
15.0	1.00	18.5

[a] In acetone
[b] Epoxyacrylate/HDDA 4:1 (w/w)

does not produce the formation of a two-layer system but the gelation of the whole acrylic formulation, as the photoinitiator is dissolved and hence diffuses into the reaction medium.

A special application of polymeric photoinitiators based on side-chain polyhaloacyl aromatic groups is claimed [136] for obtaining lamination adhesives between different polymeric films. Indeed, poly(α-methylstyrene) has been transformed in the corresponding trichloroacetyl derivative [poly(MSCA)] by a Friedel-Crafts reaction with acetyl chloride and successive chlorination (Scheme 35). Trimethylolethane dimethacrylate (TMEDMA) has then been interposed between poly(MSCA)-coated cellophane and oriented poly(propylene) films to give, upon UV irradiation, polymeric laminates (Scheme 35).

Polymer coatings on cellulose or silica have also been produced [137] by immobilizing over their surface reactive photoinitiators based on hydroxyalkylphenone moieties [cell-GEHMP and SiO_2-$Si(EO)_nHMP$, respectively] (Scheme 36) able to initiate, upon irradiation, the polymerization of acrylate monomers.

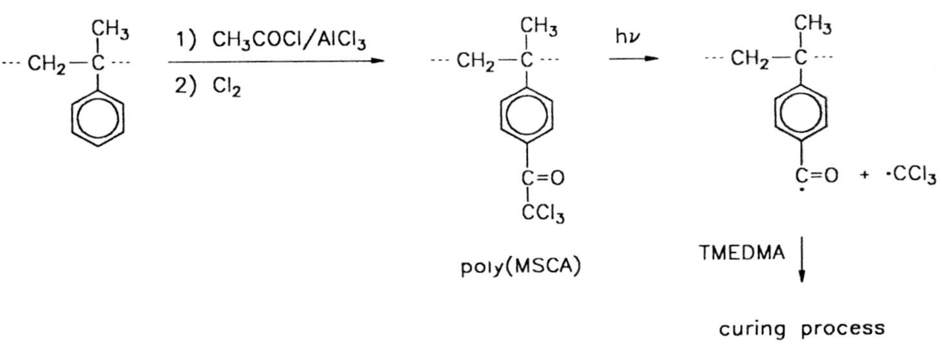

1) $CH_3COCl/AlCl_3$
2) Cl_2
poly(MSCA)
hν
$\cdot CCl_3$
TMEDMA
curing process

Scheme 35

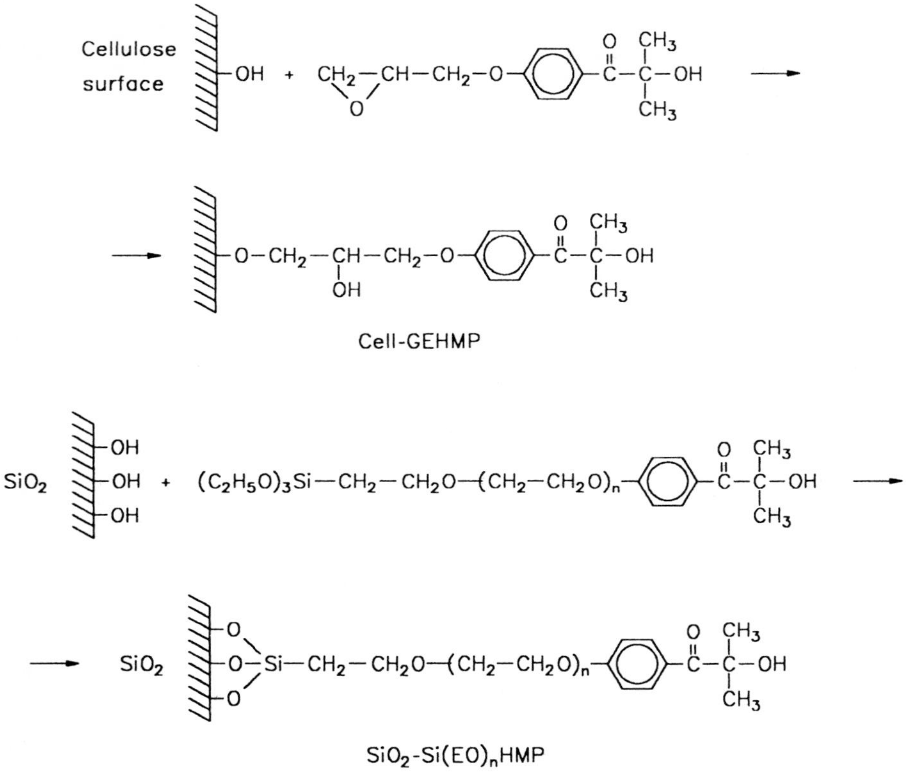

Cellulose surface —OH + CH_2—CH—CH_2—O—C_6H_4—C(=O)—C(CH_3)$_2$—OH (epoxide) →

→ Cellulose —O—CH_2—CH(OH)—CH_2—O—C_6H_4—C(=O)—C(CH_3)$_2$—OH

Cell-GEHMP

SiO_2 (—OH)$_3$ + $(C_2H_5O)_3Si$—CH_2—CH_2O—$(CH_2$—$CH_2O)_n$—C_6H_4—C(=O)—C(CH_3)$_2$—OH →

→ SiO_2 (—O)$_3$Si—CH_2—CH_2O—$(CH_2$—$CH_2O)_n$—C_6H_4—C(=O)—C(CH_3)$_2$—OH

SiO_2-Si(EO)$_n$HMP

Scheme 36

Indeed, photopolymerization of acrylamide in aqueous solution on cell-GEHMP gives cellulose with the surface covered by a poly(acrylamide) gel. Similarly, UV initiated polymerization of 2-hydroxyethyl acrylate, 2-(dimethylamino) ethyl methacrylate and *N*-vinylpyrrolidone carried out with the above functionalized SiO_2 results in the formation of a surface polymer coating.

Another special application of polymeric photoinitiators is represented by the synthesis of block copolymers which can be realized by several routes with different efficiency.

The photoreduction of aromatic ketones by polymeric systems having tertiary amine end groups provides an elegant way for the preparation of block copolymers with high efficiency [138]. The method consists of the synthesis of the bifunctional azo-derivative 4,4′-azobis (*N*,*N*-dimethylaminoethyl-4-cyano pentanoate) (ADCP), successively used as free radical thermal initiator for the preparation of tertiary amine-terminated poly(styrene).

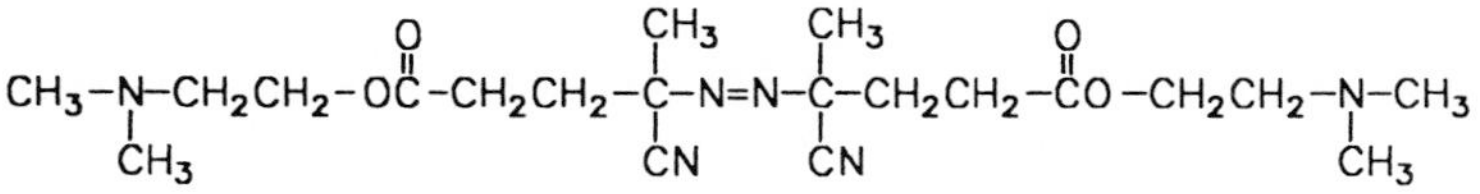

$CH_3-N(CH_3)-CH_2CH_2-OC(=O)-CH_2CH_2-C(CH_3)(CN)-N{=}N-C(CH_3)(CN)-CH_2CH_2-C(=O)O-CH_2CH_2-N(CH_3)-CH_3$

ADCP

The resulting poly(styrene), which is expected to have tertiary amine groups attached at each end of the polymer chain, due to the well established termination mechanism by radical-radical combination, are used, under UV irradiation, in conjunction with 9-fluorenone as a photo-redox system for the free radical polymerization of MMA to yield MMA/St/MMA block copolymers (Scheme 37):

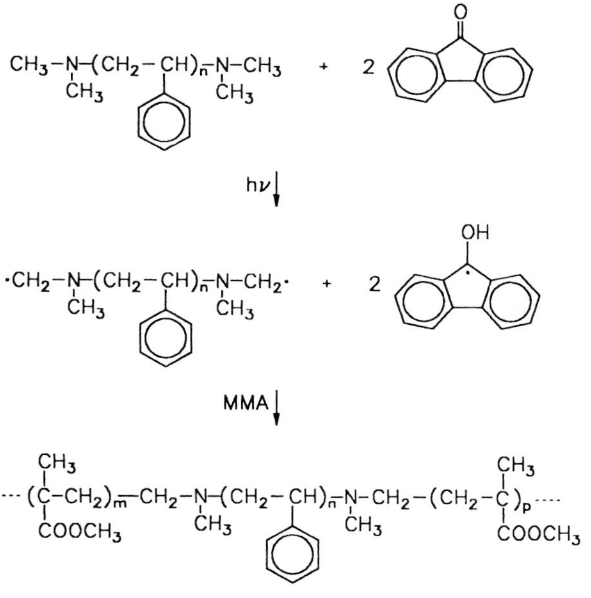

$CH_3-N(CH_3)-(CH_2-CH(C_6H_5))_n-N(CH_3)-CH_3 \; + \; 2$ 9-fluorenone

$h\nu \downarrow$

$\cdot CH_2-N(CH_3)-(CH_2-CH(C_6H_5))_n-N(CH_3)-CH_2\cdot \; + \; 2$ (9-hydroxyfluorenyl radical, OH)

$MMA \downarrow$

$\cdots(C(CH_3)(COOCH_3)-CH_2)_m-CH_2-N(CH_3)-(CH_2-CH(C_6H_5))_n-N(CH_3)-CH_2-(CH_2-C(CH_3)(COOCH_3))_p\cdots$

MMA/St/MMA block copolymers

Scheme 37

Solubility characteristics of the polymeric product allow us to exclude the presence of homopoly(styrene). Moreover, taking into account that semipinacol-type radicals, as those produced by photoreduction of fluorenone, are not significantly active for initiating MMA polymerization [8], the presence of homopoly(MMA) can also be ruled out.

A different approach to obtain styrene/MMA block copolymers involves the use of bifunctional initiators containing thermal-labile azo and photo-labile benzoin moieties [139]. As a first step, prepolymers are prepared by thermal decomposition of the azo moiety in the presence of one monomer. These photoactive prepolymers, containing benzoin end groups, are successively employed for the block copolymerization under UV irradiation of a different monomer.

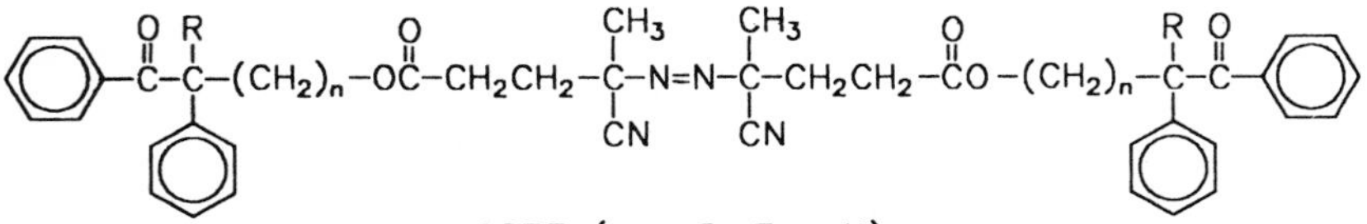

ACPB (n = 0, R = H)
ABME (n = 1, R = OCH_3)

Indeed, 4,4′-azobis-(4-cyanopentanoyl)-bis-benzoin (ACPB) and 4,4′-azobis-(4-cyanopentanoyl)-bis-(α-methylolbenzoin methyl ether) (ABME) have been used in the synthesis of poly(styrene)s possessing benzoin or benzoin methyl ether end groups (Scheme 38):

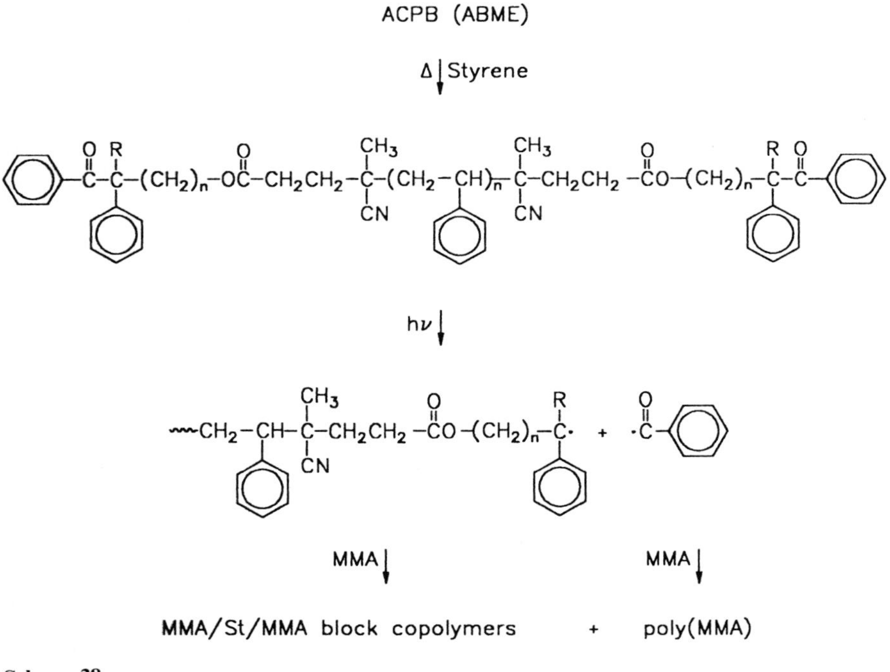

ACPB (ABME)

Δ ↓ Styrene

hν ↓

MMA ↓ MMA ↓

MMA/St/MMA block copolymers + poly(MMA)

Scheme 38

However, as seen in Scheme 38, block copolymers are accompanied by a relevant amount (30–87%) of homopoly(MMA), due to the contemporary formation of benzoyl radicals by photofragmentation. Minor amounts (1–20%) of homopoly(styrene) are also found, due to the recombination of polymeric benzyl-type primary radicals.

An inverted sequence of the same procedure has also been used [139] to prepare the same three-block copolymers. Indeed, thermal polymerization of MMA by ABME gives rise to a polymer mainly containing only one benzoin methyl ether moiety per macromolecule, since growing MMA radicals terminate mostly by disproportionation. Thus, terminally photoactive poly(MMA) is used to obtain the photoinitiated block copolymerization of styrene. In this case, a 90% yield of block copolymers is obtained, appreciably higher than in the preceding method, fully consistent with the usual assumption that the termination in styrene polymers occurs by combination. In fact, coupling of the growing styryl radicals with the less reactive poly(MMA)-bound methoxy benzyl radicals also contributes to the formation of block copolymers.

Different vinyl monomers have also been reported to give block copolymers, in combination with other bifunctional initiators, provided that the two photolabile groups differ sufficiently in the selective absorption of the light to permit a two-step synthesis. Indeed, an azo-oligoperoxyester has been prepared [55] by reaction of hydrogen peroxide with 4,4′-azobis-(4-cyanopentanoic acid chloride):

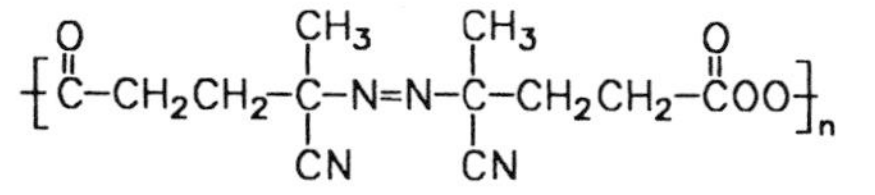

$$\left[-\overset{O}{\overset{\|}{C}}-CH_2CH_2-\underset{CN}{\overset{CH_3}{C}}-N=N-\underset{CN}{\overset{CH_3}{C}}-CH_2CH_2-\overset{O}{\overset{\|}{C}}OO- \right]_n$$

$n = 8\text{-}10$

Poly(styrene)s containing acylperoxide groups are thus obtained by selective photolysis of the azo moieties at 350 or 371 nm. These prepolymers are successively used as macronitiators for the free radical polymerization of vinyl chloride at 70 °C. Styrene/vinyl chloride block copolymers are thus produced [55] by the above two-step route, although relevant amounts (50–60%) of poly(styrene) and poly(vinyl chloride), due to both low peroxide content (~ 0.6 groups per macromolecule of polystyrene) and chain transfer with solvent and monomer, are also present.

These findings clearly indicate that monomers and solvents having a low capability for giving chain transfer have to be used in order to minimize the contemporary formation of homopolymers.

Similar photochemical essays are performed with di-*tert*-butyl 4,4′-azo-bis-(4-cyanoperoxy pentanoate) (BACPP) [55], obtained with a high degree of purity.

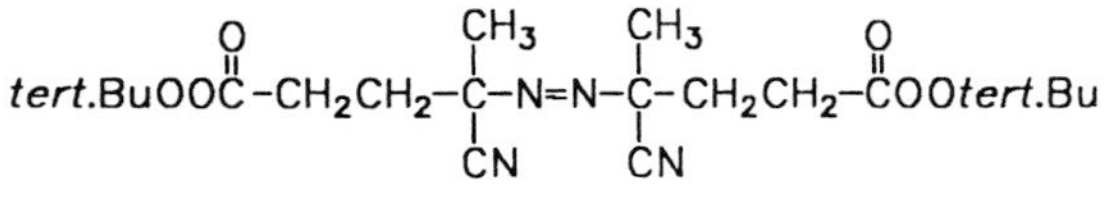

$$\textit{tert}.BuOO\overset{O}{\overset{\|}{C}}-CH_2CH_2-\underset{CN}{\overset{CH_3}{C}}-N=N-\underset{CN}{\overset{CH_3}{C}}-CH_2CH_2-\overset{O}{\overset{\|}{C}}OO\textit{tert}.Bu$$

BACPP

The good solubility of BACPP in styrene, as compared with the previous azo-oligoperoxyester, gives rise, by photolysis at 367 nm, to a well defined polymeric initiator consisting of a telechelic poly(styrene) with two perester end groups. On heating at 95 °C in the presence of MMA, styrene/MMA block copolymers are obtained in high yield without any residual homopoly(styrene) [55]. This photochemical pathway therefore appears preferable to the thermal and redox methods previously reported for the synthesis of ω,ω'-bis-peroxy-poly(styrene)s [140].

Another synthetic route, capable of giving block copolymers, involves the use of a trichloroacetyl-azo initiator. Indeed, terminal carbon-centred macroradicals are formed upon irradiation of polymers having $-CX_3$, $-CH_2X_2$, and $-CH_3X$ end groups (X = halogen atom) in the presence of carbonyl compounds of transition metals, preferably the manganese and rhenium carbonyls, $Mn_2(CO)_{10}$ and $Re_2(CO)_{10}$ [141] (Scheme 39):

$$Mn_2(CO)_{10} + 2X_3C{-}R \xrightarrow{h\nu} 2Mn(CO)_5X + 2X_2\dot{C}{-}R$$

Scheme 39

A polymeric initiator with terminal halogen containing groups can be prepared by using a bifunctional initiator 4,4′-azobis(4-cyanopentanoic acid-trichloroacetyl-amide) (ACPT) [142]. When this polymer is irradiated at 436 nm in the presence of $Mn_2(CO)_{10}$ and the appropriate monomer, block copolymers are obtained [143], according to Scheme 40:

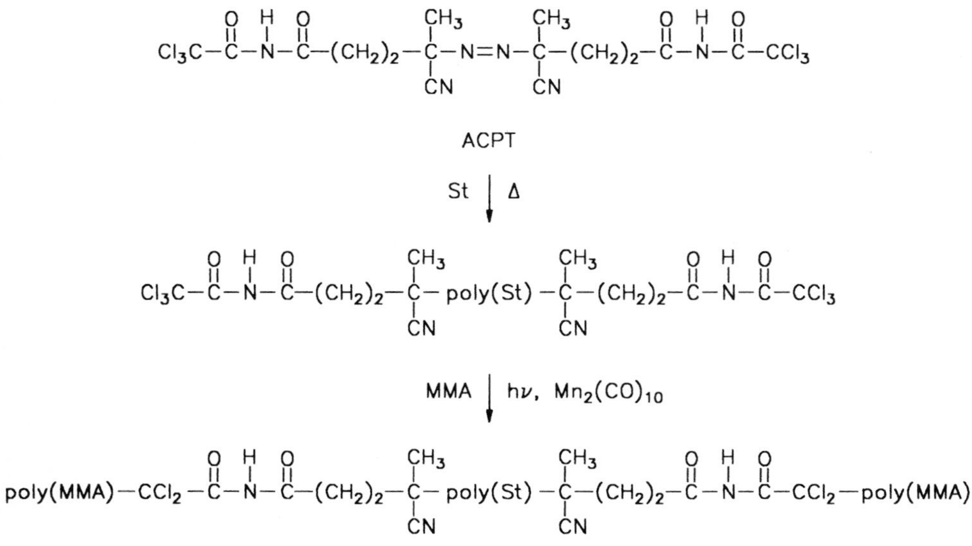

$$Cl_3C{-}C(=O){-}NH{-}C(=O){-}(CH_2)_2{-}C(CH_3)(CN){-}N{=}N{-}C(CH_3)(CN){-}(CH_2)_2{-}C(=O){-}NH{-}C(=O){-}CCl_3$$

ACPT

$$\downarrow \text{St}, \Delta$$

$$Cl_3C{-}C(=O){-}NH{-}C(=O){-}(CH_2)_2{-}C(CH_3)(CN){-}\text{poly(St)}{-}C(CH_3)(CN){-}(CH_2)_2{-}C(=O){-}NH{-}C(=O){-}CCl_3$$

$$\downarrow \text{MMA}, h\nu, Mn_2(CO)_{10}$$

$$\text{poly(MMA)}{-}CCl_2{-}C(=O){-}NH{-}C(=O){-}(CH_2)_2{-}C(CH_3)(CN){-}\text{poly(St)}{-}C(CH_3)(CN){-}(CH_2)_2{-}C(=O){-}NH{-}C(=O){-}CCl_2{-}\text{poly(MMA)}$$

Scheme 40

The above procedure allows one to obtain MMA-St-MMA three-block copolymers with efficiency in the 40–70% range.

Block copolymers can also be prepared [144] by using benzoin-modified oligourethanes as photoinitiators in the MMA polymerization, according to Scheme 41:

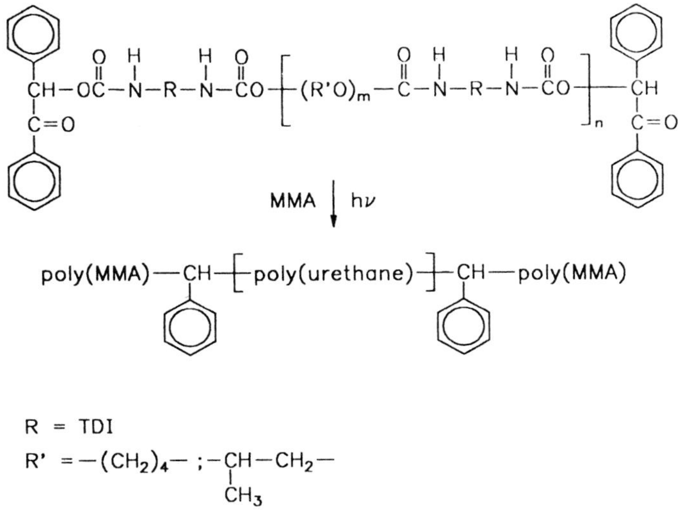

$$\mathrm{Ph{-}CH(C(=O)Ph){-}O\overset{O}{\overset{\|}{C}}{-}\overset{H}{N}{-}R{-}\overset{H}{N}{-}\overset{O}{\overset{\|}{C}}O{-}\left[(R'O)_m{-}\overset{O}{\overset{\|}{C}}{-}\overset{H}{N}{-}R{-}\overset{H}{N}{-}\overset{O}{\overset{\|}{C}}O\right]_n{-}CH(Ph){-}C(=O)Ph}$$

$$\xrightarrow[h\nu]{MMA}$$

$$\mathrm{poly(MMA){-}CH(Ph){-}\left[poly(urethane)\right]{-}CH(Ph){-}poly(MMA)}$$

R = TDI

R' = $-(CH_2)_4-$; $-CH(CH_3)-CH_2-$

Scheme 41

Following the above procedure, thermoplastic block copolymers are obtained, their properties depending on the nature of the oligoether used in the preparation of the oligourethane photoinitiator. Indeed, the packing density, tensile strength, and relative elongation of the block copolymers increase on increasing the molecular weight of the oligoether and on replacing oligopropylene glycol with oligotetramethylene glycol. However, considering that low molecular weight benzoyl primary radicals are also obtained upon irradiation, the block copolymers should be accompanied by a significant amount of homopoly(MMA).

Di-block copolymers may also be formed by using dithiocarbamate free radicals. Indeed, copolymers containing poly(styrene) and poly(hydroxyethyl methacrylate) blocks have been obtained by a two-step procedure [145]. Firstly, styrene is photopolymerized in the presence of benzyl *N,N*-diethyldithiocarbamate (BDC) by a living radical mechanism [146]. In fact, as the benzyl and thiyl radicals, formed by the photofragmentation of BDC, participate mainly in the initiation and termination reactions respectively, polystyrene with a dithiocarbamate end group is thus obtained. The successive UV irradiation of this polymer, in the presence of hydroxyethyl methacrylate (HEMA), gives rise to the di-block copolymer, according to Scheme 42.

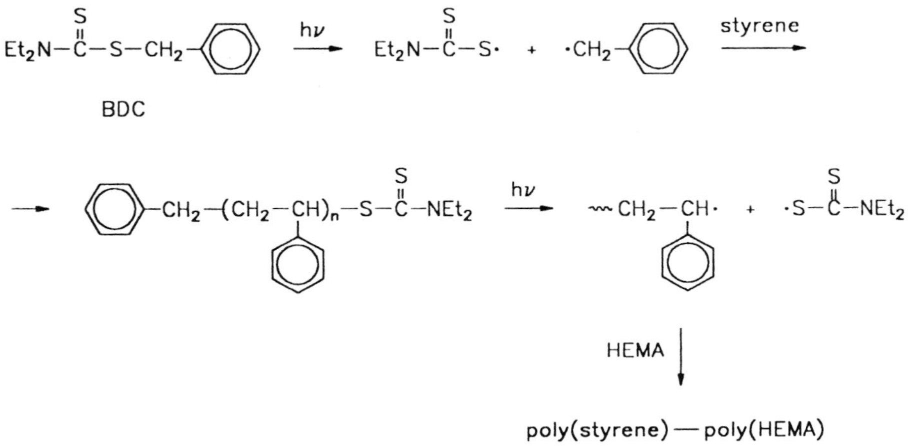

$$Et_2N-\overset{S}{\overset{\|}{C}}-S-CH_2-C_6H_5 \xrightarrow{h\nu} Et_2N-\overset{S}{\overset{\|}{C}}-S\cdot + \cdot CH_2-C_6H_5 \xrightarrow{\text{styrene}}$$

BDC

$$\longrightarrow C_6H_5-CH_2-(CH_2-\underset{C_6H_5}{CH})_n-S-\overset{S}{\overset{\|}{C}}-NEt_2 \xrightarrow{h\nu} \sim CH_2-\underset{C_6H_5}{CH}\cdot + \cdot S-\overset{S}{\overset{\|}{C}}-NEt_2$$

$$\downarrow \text{HEMA}$$

poly(styrene) — poly(HEMA)

Scheme 42

Analogously, di-block copolymers consisting of one block derived from a single monomer and another block, containing randomly distributed repeating units afforded by two different comonomers, are reported [147]. In particular, poly(St-*co*-MMA)-*b*-poly(VAc) and poly(St-*co*-MMA)-*b*-poly[acrylic acid (AA)] have been prepared with 55–92% blocking selectivity. Moreover di-block copolymers having alternating structures from different comonomers are also obtained using the same photochemical procedure [147]. Indeed, poly[St-*alt*-diethyl fumarate (DEF)]-*b*-poly[maleic anhydride (MAn)-*alt*-isobutyl vinyl ether (IBVE)] and poly[St-*alt*-diisopropyl fumarate (DiPF)]-*b*-poly(MAn-*alt*-IBVE) have been synthesized with 63–80% blocking selectivity.

Dithiocarbamate-functionalized polymers of styrene and MMA at both ends are prepared [148] by thermal free radical initiation with tetraethylthiuram disulfide which is known [146] to behave as initiator, chain transfer agent and terminator (iniferter). Successive photolysis of the terminal dithiocarbamate end groups, in the presence of another vinyl monomer, allows one to obtain three-block copolymers (Scheme 43).

Poly(MMA)-poly(St)-poly(MMA) and poly(BA)-poly(St)-poly(BA) have thus been obtained with good blocking selectivity. This is particularly high in the latter case as the dithiocarbamate radical is substantially unable to initiate the polymerization of acrylates, BA reacting only with the polystyryl radical.

According to a similar procedure, block copolymers are claimed to be prepared by photodecomposition of bis-dithiocarbamates in the presence of an acrylic ester to give a polymeric intermediate which is successively photodecomposed in the presence of another reactive monomer [149].

The difunctional photoiniferter such as *p*-xylylene bis(*N*,*N*-diethyldithiocarbamate) (XDC) has led to several ABA-type three-block copolymers by successive photoinitiated polymerizations [147].

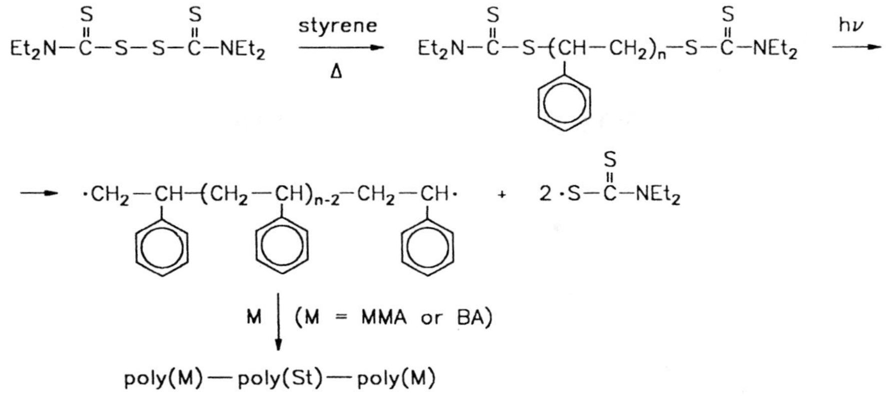

$$Et_2N-\overset{S}{\overset{\|}{C}}-S-S-\overset{S}{\overset{\|}{C}}-NEt_2 \xrightarrow[\Delta]{styrene} Et_2N-\overset{S}{\overset{\|}{C}}-S\!-\!(\underset{C_6H_5}{CH}-CH_2)_n\!-\!S-\overset{S}{\overset{\|}{C}}-NEt_2 \xrightarrow{h\nu}$$

$$\longrightarrow \cdot CH_2-\underset{C_6H_5}{CH}-(CH_2-\underset{C_6H_5}{CH})_{n-2}-CH_2-\underset{C_6H_5}{CH}\cdot \;+\; 2\cdot S-\overset{S}{\overset{\|}{C}}-NEt_2$$

$$\Big\downarrow M \quad (M = MMA \text{ or } BA)$$

$$\text{poly(M)}-\text{poly(St)}-\text{poly(M)}$$

Scheme 43

$$\begin{matrix}Et\\Et\end{matrix}\!>N-\overset{S}{\overset{\|}{C}}-S-CH_2-C_6H_4-CH_2-S-\overset{S}{\overset{\|}{C}}-N\!<\!\begin{matrix}Et\\Et\end{matrix}$$

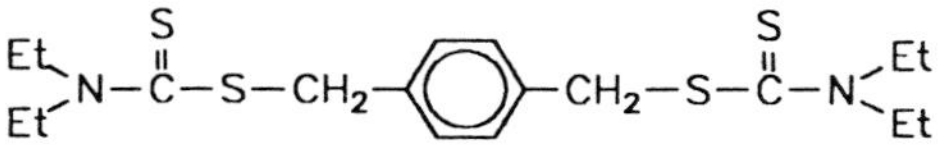

XDC

Indeed, poly(VAc)-poly(St-*co*-MMA)-poly(VAc), and poly(M_3-*alt*-M_4)-poly (M_1)-poly(M_3-*alt*-M_4) and poly(M_3-*alt*-M_4)-poly(M_1-*alt*-M_2)-poly(M_3-*alt*-M_4), where M_1, M_2, M_3 and M_4 are St, DEF or DiPF, IBVE, and MAn, respectively, were obtained.

Copolymers consisting of polysiloxane and poly(MMA) blocks may also be prepared starting from a dithiocarbamate-containing polysiloxane as photoinitiator in the presence of MMA, according to Scheme 44:

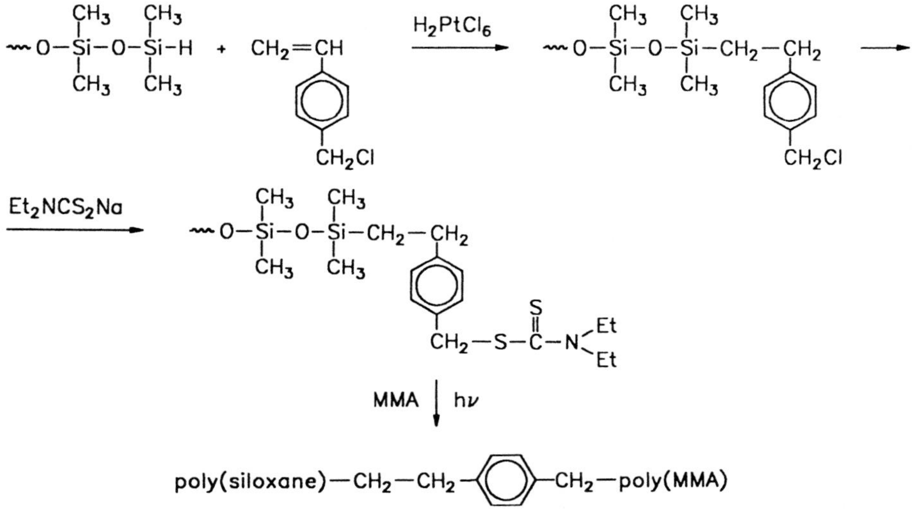

$$\sim O-\underset{CH_3}{\overset{CH_3}{Si}}-O-\underset{CH_3}{\overset{CH_3}{Si}}-H \;+\; CH_2=\underset{C_6H_4CH_2Cl}{CH} \xrightarrow{H_2PtCl_6} \sim O-\underset{CH_3}{\overset{CH_3}{Si}}-O-\underset{CH_3}{\overset{CH_3}{Si}}-CH_2-\underset{C_6H_4CH_2Cl}{CH_2} \longrightarrow$$

$$\xrightarrow{Et_2NCS_2Na} \sim O-\underset{CH_3}{\overset{CH_3}{Si}}-O-\underset{CH_3}{\overset{CH_3}{Si}}-CH_2-\underset{C_6H_4-CH_2-S-\overset{S}{\overset{\|}{C}}-N(Et)_2}{CH_2}$$

$$\Big\downarrow MMA \quad h\nu$$

$$\text{poly(siloxane)}-CH_2-CH_2-C_6H_4-CH_2-\text{poly(MMA)}$$

Scheme 44

By this procedure it is possible to synthesize [150] block copolymers, having thermoplastic elastomeric properties, with a micro-domain morphology and glass transition temperatures of -120 and $105\,^{\circ}\mathrm{C}$, corresponding to polysiloxane and poly(MMA) blocks, respectively.

The preparation of block copolymers by combination of thermally radical and photoinduced cationic polymerization processes has also been reported [151]. Indeed, styrene/cyclohexene oxide (CHO) copolymers have been synthesized by using a bifunctional azobenzoin initiator such as ABME, previously described, through a two-step procedure. In the first step, thermal free radical polymerization of styrene in the presence of the above azobenzoin initiator gives poly(styrene) prepolymers with benzoin photoactive end groups, as reported in Scheme 38. These prepolymers, upon photolysis and subsequent oxidation to the corresponding carbocations in the presence of 1-ethoxy-2-methylpyridinium hexafluoro phosphate ($EMP^{+}PF_6^{-}$), finally give block copolymers by cationic polymerization of cyclohexene oxide (Scheme 45).

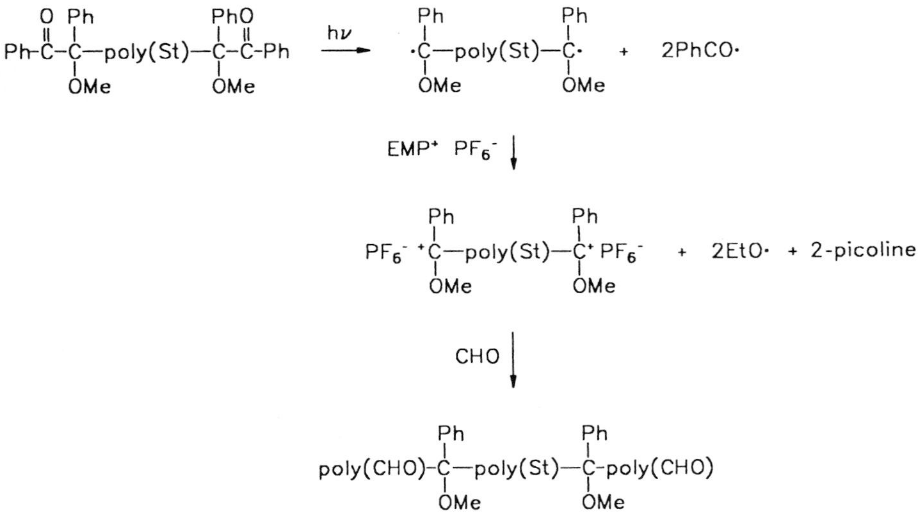

Ph–C(=O)–C(Ph)(OMe)–poly(St)–C(Ph)(OMe)–C(=O)–Ph —hν→ ·C(Ph)(OMe)–poly(St)–C(Ph)(OMe)· + 2PhCO·

↓ $EMP^{+}\ PF_6^{-}$

PF_6^{-} ⁺C(Ph)(OMe)–poly(St)–C⁺(Ph)(OMe) PF_6^{-} + 2EtO· + 2-picoline

↓ CHO

poly(CHO)–C(Ph)(OMe)–poly(St)–C(Ph)(OMe)–poly(CHO)

Scheme 45

A reversed sequence of the same procedure has also been followed [151] by using ABME in conjunction with $EMP^{+}PF_6^{-}$ and CHO in order to obtain, under UV irradiation, poly(CHO) containing an azo linkage in the main chain. As the photolysis of the azo group at 350 nm irradiation is not achievable due to the higher absorption coefficient of the residual benzoin methyl ether end groups, poly(CHO) is thermally decomposed in the presence of styrene, to give mainly three-block copolymers (Scheme 46).

The formation of ABA-type block copolymers is expected, due to the well known termination mechanism of growing poly(styrene) by radical-radical combination. This procedure, therefore, can be used to prepare AB- or ABA-type

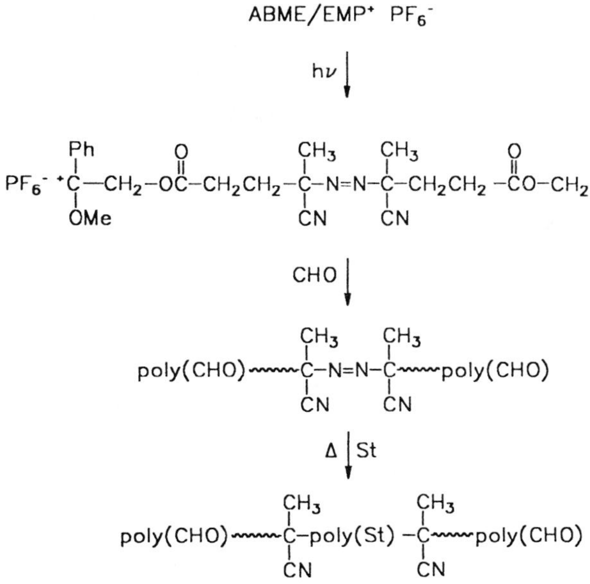

ABME/EMP^+ PF_6^-

$\downarrow h\nu$

$PF_6^-\ {}^+C(Ph)(OMe)-CH_2-OC(=O)-CH_2CH_2-C(CH_3)(CN)-N{=}N-C(CH_3)(CN)-CH_2CH_2-C(=O)O-CH_2-C^+(Ph)(OMe)\ PF_6^-$

$\downarrow$ CHO

poly(CHO)~~~$C(CH_3)(CN)-N{=}N-C(CH_3)(CN)$~~~poly(CHO)

$\downarrow$ Δ St

poly(CHO)~~~$C(CH_3)(CN)$–poly(St)–$C(CH_3)(CN)$~~~poly(CHO)

Scheme 46

block copolymers depending on the kinetic behaviour of the radically polymerizable monomer involved in the second step.

Bifunctional initiators such as bis-(2-glycylaminophenyl) disulfide (GAPDS) and 2,2′-azo-2,2′-dimethyldipropiono hydrazide (AMPH) have been used [152] to obtain hydrocarbon chain polymeric initiators with amino end groups able to promote the polymerization of aminoacid *N*-carboxy anhydrides (NCA) (Scheme 47).

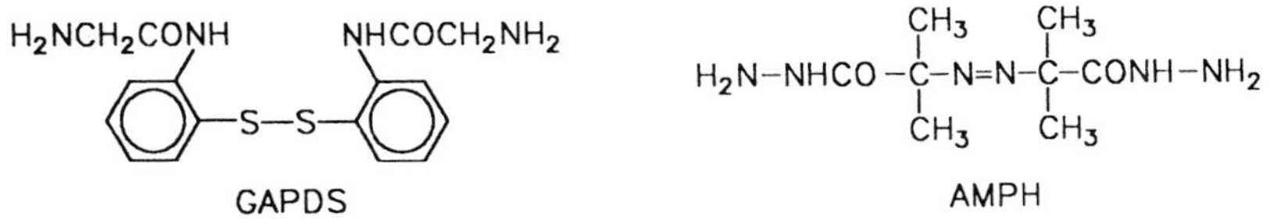

$H_2NCH_2CONH-C_6H_4-S-S-C_6H_4-NHCOCH_2NH_2$

GAPDS

$H_2N-NHCO-C(CH_3)_2-N{=}N-C(CH_3)_2-CONH-NH_2$

AMPH

GAPDS and AMPH can also be employed, first in the synthesis of the corresponding polypeptide initiators containing –S–S– or N=N groups and then submitted to photolysis, thus giving the free radical polymerization of vinyl monomers (Scheme 48).

In both cases, copolymers containing hydrocarbon chain and polypeptide blocks of ABA and AB structures are obtained.

Similarly, the incorporation of diaryldisulfide moieties on a cellulosic prepolymer and its irradiation in the presence of styrene or chloroprene is reported [153] to give multiphase three-block copolymers (Scheme 49).

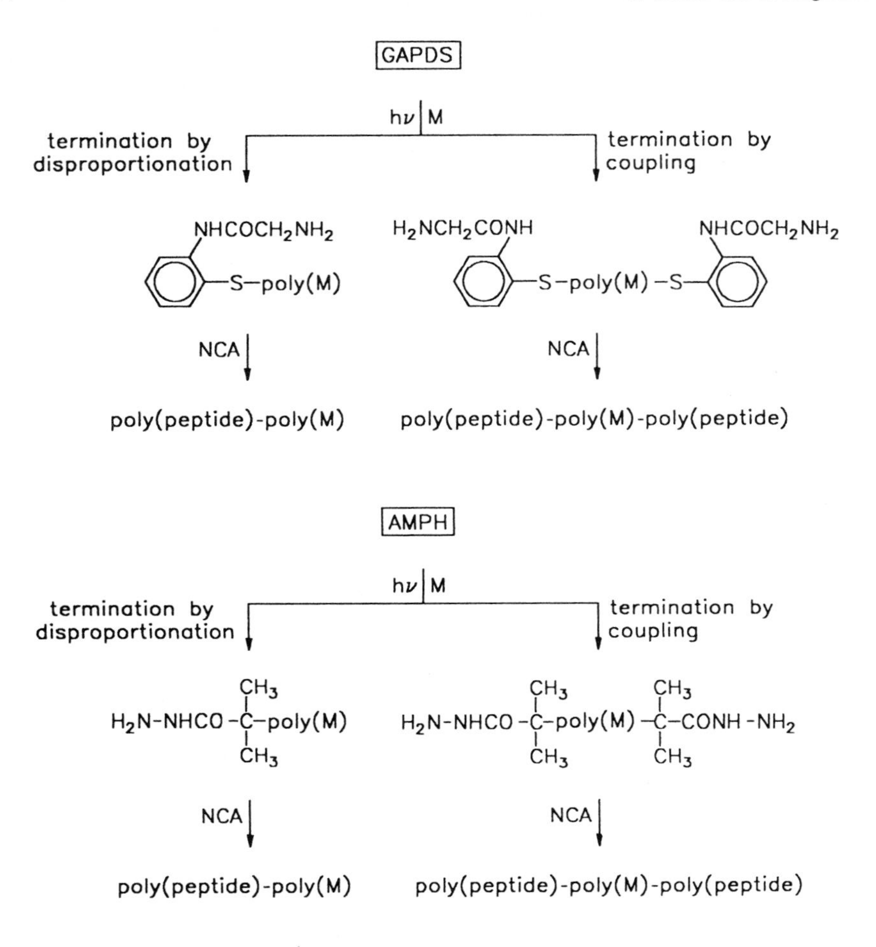

Scheme 47

The cellulose triester macroinitiator is prepared by a three-step procedure involving first the cleavage by HBr of the cellulose triester, then the transformation by hydrolysis of the resulting Br-terminated derivative into the corresponding l-hydroxy-cellulose triester and finally its coupling with a suitable disulfide, such as bis-(4-isocyanatophenyl) disulfide, to give the final initiator. This method is claimed [153] to give a thermoplastic elastomer having a polychloroprene central block with only minor contamination by homopolymers.

Block copolymers are also reported [55, 75, 154, 155] to be prepared by photolysis, in the presence of a vinyl monomer, of light sensitive groups incorporated

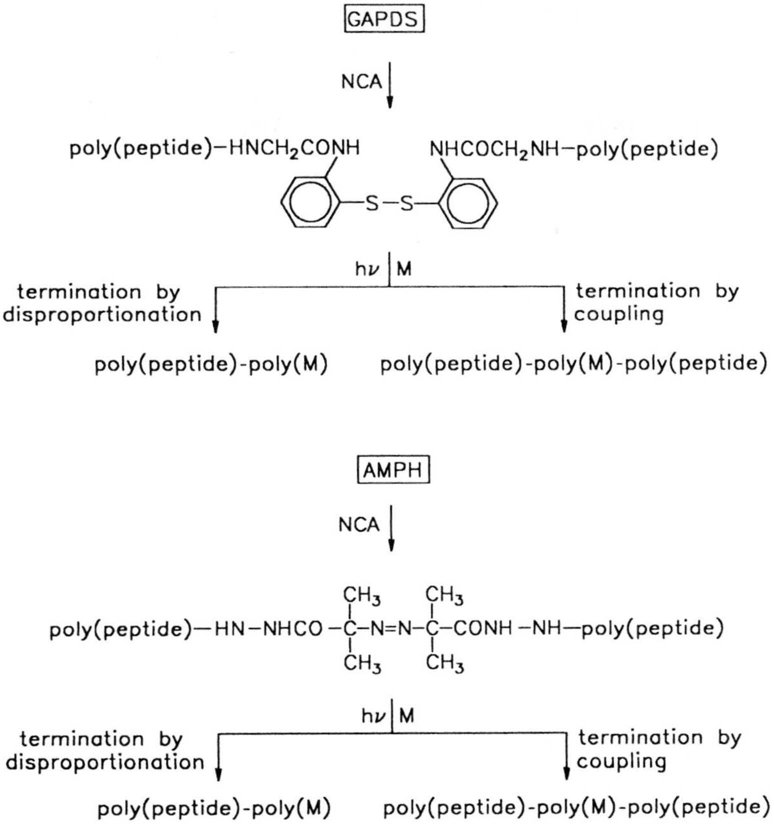

GAPDS

NCA ↓

poly(peptide)–$HNCH_2CONH$–C_6H_4–S–S–C_6H_4–$NHCOCH_2NH$–poly(peptide)

hν | M

termination by disproportionation ↓ poly(peptide)-poly(M)

termination by coupling ↓ poly(peptide)-poly(M)-poly(peptide)

AMPH

NCA ↓

poly(peptide)–HN–NHCO–$C(CH_3)_2$–N=N–$C(CH_3)_2$–CONH–NH–poly(peptide)

hν | M

termination by disproportionation ↓ poly(peptide)-poly(M)

termination by coupling ↓ poly(peptide)-poly(M)-poly(peptide)

M and NCA as in Scheme 47

Scheme 48

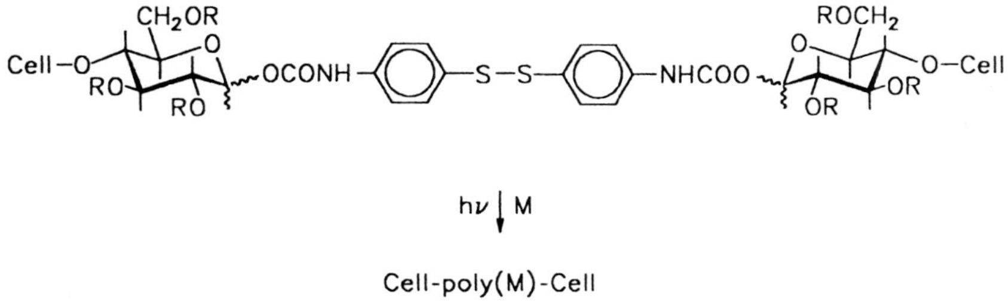

Cell–O–(glucopyranose: CH_2OR, RO, RO)~OCONH–C_6H_4–S–S–C_6H_4–NHCOO~(glucopyranose: $ROCH_2$, OR, OR)–O–Cell

hν ↓ M

Cell-poly(M)-Cell

R = CH_3CO-, C_2H_5CO-/CH_3CO-, C_3H_7CO-

Cell = cellulose triester, M = styrene, chloroprene

Scheme 49

into a condensation polymer, thus producing by free radical polymerization the polyvinyl block. Depending on the termination mode, that is by disproportionation and/or by coupling, the structure of the resulting block copolymers can be CB-VB and/or CB-VB-CB (where CB and VB indicate polycondensate and polyvinyl blocks, respectively) in the case where the polymeric precursor contains only one photolabile moiety (X) per macromolecule. When the number of photodissociable groups is more than one, the block copolymers are also constituted by VB-CB-VB three-blocks and $(VB\text{-}CB)_n$ multi-blocks (Scheme 50).

Indeed, copolyesters having incorporated *O*-acylated oximino ketone groups are obtained [75] by interfacial condensation of different acid dichlorides, such as terephthaloyl-, isophthaloyl- and sebacoyl chlorides, with small amounts of 1-(4′-hydroxyphenyl)-1,2-propanedione-2-oxime (HPPO) or *p*-hydroxyphenyl glyoxal aldoxime (HPGA) in the presence of a large excess of bisphenol-A:

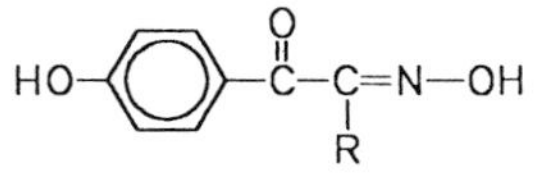

HPPO (R = CH_3)
HPGA (R = H)

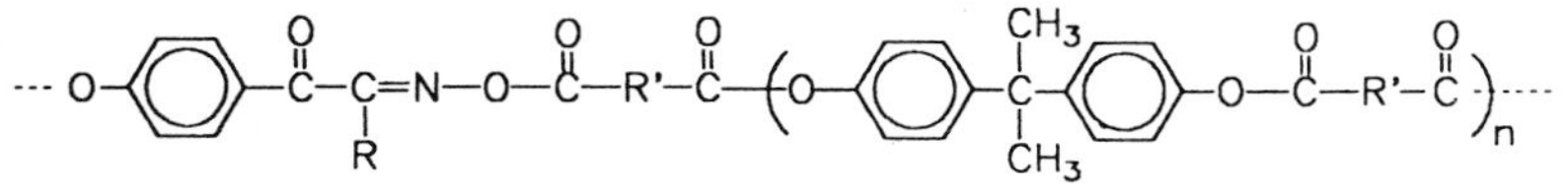

R = H, CH_3; R' = $(CH_2)_8$, *m*-phenylene, *o*-phenylene

The above copolyesters, when irradiated at 365 nm in the presence of vinyl or acrylic monomers, give rise to block copolymers (Scheme 51).

The polyadipates of tetrachloro bisphenol-A depicted below, containing small amounts of HPPO, are similarly used [154] as prepolymers to give, upon UV irradiation, macroradicals able to initiate the polymerization of styrene and produce block copolymers.

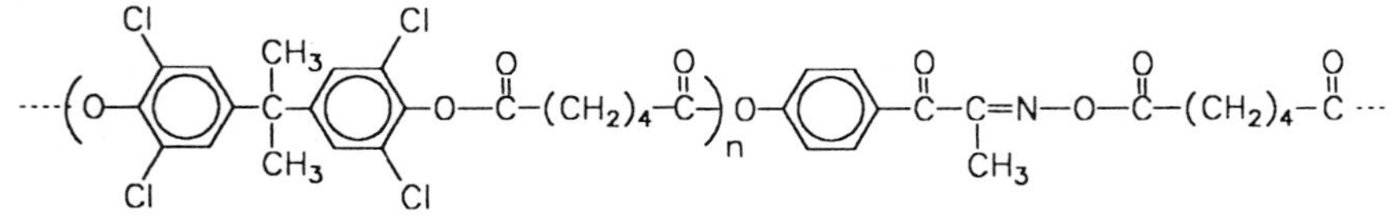

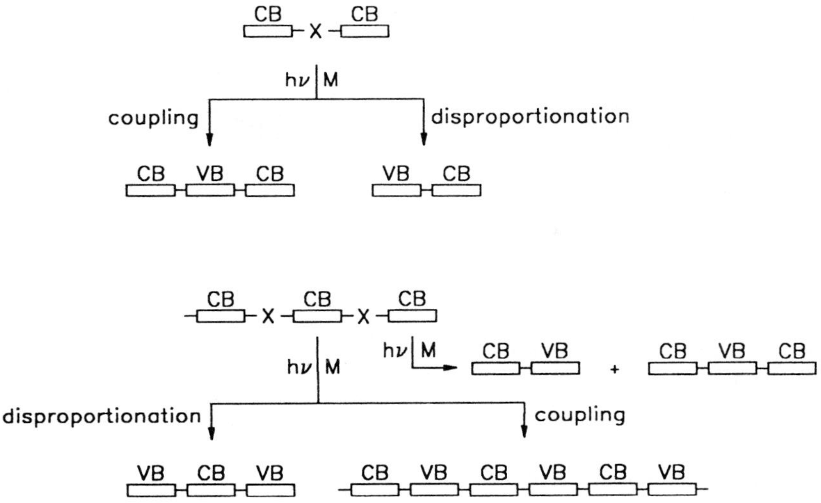

Scheme 50

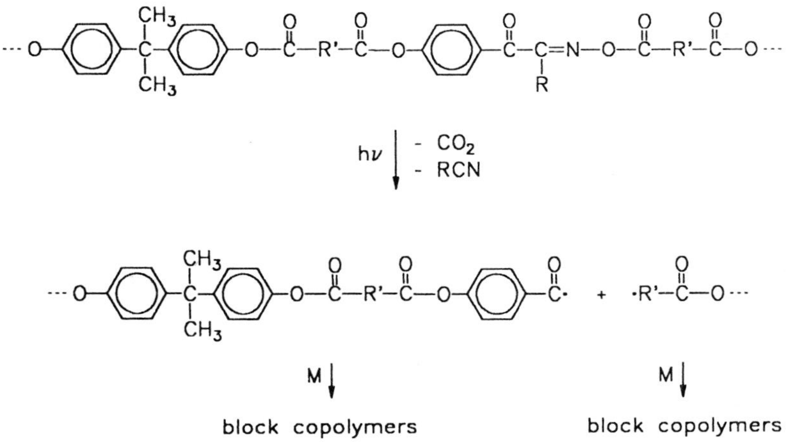

Scheme 51

An analogous polycondensate photoinitiator is obtained [55, 155] by incorporation of 4,4′-dihydroxybenzoin methyl ether into poly(bisphenol-A carbonate):

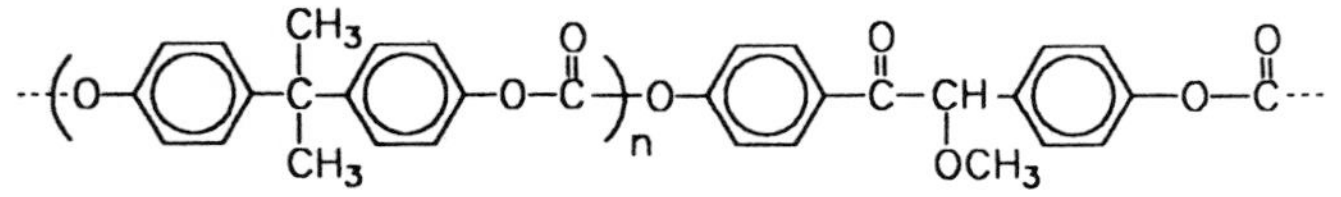

The same procedure is then applied to the above polycarbonate in order to obtain block copolymers through photolysis of the main chain benzoin methyl ether moieties in the presence of a vinyl monomer such as MMA, ethyl acrylate or acrylonitrile.

In conclusion, these photochemical methods allow one to obtain many types of block copolymer in which polycondensate- and polyvinyl-block lengths can be modulated, respectively, by the content of photolabile moiety in the prepolymers and by the polymerization conditions.

A further interesting feature of the polymeric photoinitiators is based on the possibility of anchoring different photosensitive moieties to the same macromolecule in order to provide a synergistic effect on activity due to their interaction along the polymer chain. In this context, the synthesis of several copolymeric photoinitiators bearing side-chain thioxanthone and hydroxyalkylphenone or morpholino ketone moieties has been reported recently [156]. In particular, the free radical copolymerization of 2-[2-(acryloyloxy)ethylthio] thioxanthone (AETX) with either [4-(2-acryloyloxyethoxy)phenyl]-2-hydroxy-2-propyl ketone (HPA) or [4-(2-acryloyloxyethylthio)phenyl]-2-(*N*-morpholino)-2-propyl ketone (APMK) [poly(AETX-*co*-HPA) and poly(AETX-*co*-APMK), respectively] and of 4-[2-(methacryloyloxy)ethoxycarbonyl] thioxanthone (METX) with APMK [poly(METX-*co*-APMK)] has been performed.

The above systems are used as initiators for the UV curing of different coating formulations based on mono- and multifunctional acrylates.

Table 31. UV curing activity, as determined by Koenig pendulum hardness, of a TiO_2-pigmented polyester acrylate formulation in the presence of polymeric and low-molecular-weight initiators based on thioxanthone and hydroxypropiophenone moieties, as a function of irradiation conditions[a]

Photoinitiating system[b]		Conveyor belt speed (m/min)			
Type	Hydroxypropiophenone moiety (mol%)	5	10	15	20
		Koenig pendulum hardness (s)			
Poly(AETX-*co*-HPA)	91	124	83	71	58
Poly(AETX-*co*-HPA)	50	133	95	81	62
HIPK/ITX[c]	91	92	66	53	41

[a] Formulation composition: 75 wt. parts of polyester acrylate (Laromer EA81), 25 wt. parts of HDDA and 30 wt. parts of TiO_2. UV source: two middle pressure mercury lamps (80 W/cm) at 10 cm distance from the substrate. Formulation thickness on a glass plate: 24 μm

[b] Photoinitiator concentration: 3 wt% with respect to the formulation, in combination with 2 wt% of *N*-methyldiethanolamine

[c] HIPK: 2-hydroxy-2-methyl propiophenone. ITX: 2-isopropyl thioxanthone

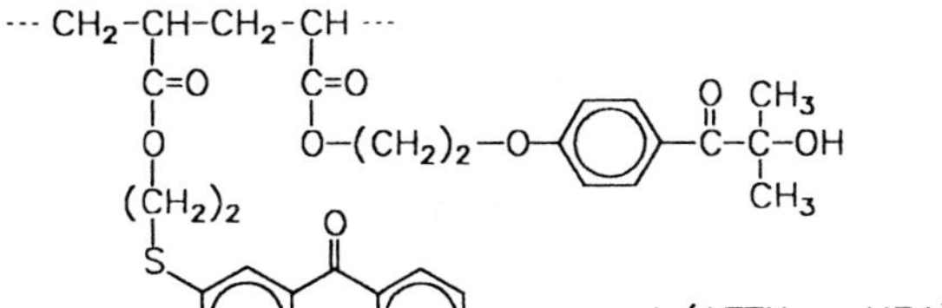

poly(AETX-*co*-HPA)

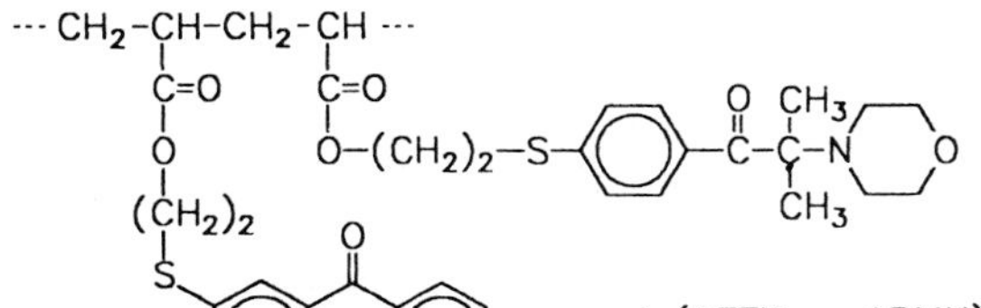

poly(AETX-*co*-APMK)

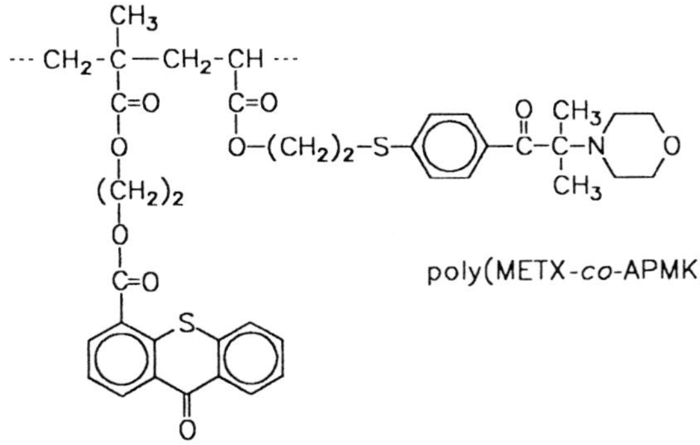

poly(METX-*co*-APMK)

A special emphasis is given [156] to poly(AETX-*co*-HPA) which, combined with *N*-methyldiethanolamine, displays a higher photoinitiation activity, as compared with the corresponding mixture of the low-molecular-weight analogues HIPK and ITX, in the UV curing of a TiO_2-pigmented coating formulation, based on a polyester acrylate (Table 31).

6 Final Remarks

The results reported in the present overview dealing with the recent developments of fundamental and applied research as well as the potential applications allow one to predict that polymeric photoinitiators will, in the near future, become the object of increasing interest, particularly in the area of UV curable surface coatings. In fact, coating technology has become a mature branch of science and cannot expand to meet the needs of the many required specific applications unless new materials are developed having tailor-made and uniform molecular structure. In this context, polymeric systems, because of their high synthetic flexibility and peculiar structural properties, are the best candidates in the field of photoinitiators to reach this important target. An obvious development of these systems will be represented by the extension of the anchorage to macromolecular chains of a great variety of photoreactive moieties, already tested as low-molecular-weight photoinitiators.

In our opinion, however, the most exciting results should be obtained by new polymeric systems containing multifunctional photosensitive moieties. This class of photoinitiator, some preliminary examples of which are reported in the preceding section, should be able, in principle, to cover a wide range of technological applications, depending on the intrinsic photophysical and photochemical properties of each single moiety and its capability of mutual interaction. The prevalence of this last effect could in fact largely modify the overall photophysical behaviour of the polymeric photoinitiators with a potential synergism in terms of curing rate in industrial applications.

This approach, for instance, could be usefully applied in the area of pigmented coatings. In fact, for this kind of applications, the range of UV absorption of the photoinitiator has to be located in the spectral window offered by the pigment. As a consequence, many conventional photoinitiators fail to cure pigmented formulations. In this frame, polymeric systems properly designed for containing specific moieties able to act as photosensitizing and photoinitiating species, respectively, could overcome the crucial problem related to this important industrial process. The suitable photosensitizer has to absorb the light at longer wavelengths than the pigment and efficiently transfer its excitation energy to the photoinitiating moiety in order to generate the primary radicals able to promote the curing process. For doing this, it is necessary that the triplet state energy level of the photoinitiating moiety is lower than that of the sensitizer. This has recently been achieved [157, 158] through the combination of low-molecular-weight systems consisting of substituted thioxanthones, which display an intense absorption band with a maximum around 385 nm and triplet state energies in the 61–63 Kcal/mole range, and photoinitiators based on α-morpholinoacetophenones having substantially the same level of triplet state energy or slightly lower.

The synergistic behaviour of the above systems has been explained in terms of both triplet-triplet energy transfer and photoreduction of the thioxanthone derivative by the amino group of the morpholino moiety [157–161].

The modulation of composition and distribution of the two photoreactive moieties along a polymer chain, readily obtainable by copolymerization, could therefore allow us to reach the most favourable situation in terms of interacting distance, thus giving rise to polymeric systems which could be successfully applied in the area of pigmented coatings such as white laquers, silk screen inks and printing inks.

7 References

1. Hutchinson J, Ledwith A (1974) Adv Polym Sci 14: 49
2. Ledwith A (1976) J Oil Col Chem Assoc 59: 157
3. Davis MJ, Doherty J, Godfrey AA, Green PN, Young JRA, Parrish MA (1978) J Oil Col Chem Assoc 61: 256
4. McGinnis VD (1979) Photogr Sci Eng 23: 124
5. Hammond GS, Moore WM (1959) J Am Chem Soc 81: 6334
6. Block H, Ledwith A, Taylor AR (1971) Polymer 12: 271
7. Braun D, Becker KH (1971) Makromol Chem 147: 91
8. Hutchinson J, Lambert MC, Ledwith A (1973) Polymer 14: 250
9. Breslow R (1972) Chem Soc Rev 1: 553
10. Breslow R, Baldwin S, Flechtner T, Kalicky P, Liu S, Washburn W (1973) J Am Chem Soc 95: 3251
11. Sumitomo H, Nobutoki K, Susaki K (1971) J Polym Sci A-1 9: 809
12. David C, Demarteau W, Geuskens G (1969) Polymer 10: 21
13. Sànchez G, Weill G, Knoesel R (1978) Makromol Chem 179: 131
14. Gleria M, Minto F, Flamigni L, Bortolus P (1987) Macromolecules 20: 1766
15. Gibson HW, Bailey FC, Chu JYC (1979) J Polym Sci Polym Chem Ed 17: 777
16. Barton J, Capek I, Šušoliak O, Juranicovà V (1978) Makromol Chem 179: 2997
17. Heine H-G, Rosenkranz H-J, Rudolph H (1972) Angew Chem Int Ed 11: 974
18. Carlini C, Ciardelli F, Donati D, Gurzoni F (1983) Polymer 24: 599
19. Carlini C, Dal Canto S, Donati D, Moroni A (1983) Pitture e Vernici 59: 73
20. Carlini C (1986) British Polym J 18: 236
21. Carlini C, Ciardelli F, Rolla PA in: Patsis AV (ed) Proceedings of XIIIth Internat Conference in Organic Coatings Science and Technology, 7–11 July 1987, Athens; Proc p 79
22. Carlini C, Toniolo L, Rolla PA, Barigelletti F, Bortolus P, Flamigni L (1987) New Polym Mat 1: 63
23. Kamachi M, Kikuta Y, Nozakura S (1979) Polym J 11: 273
24. David C, Naegelen V, Piret W, Geuskens G (1975) Eur Polym J 11: 569
25. Inokuti M, Hirayama F (1965) J Chem Phys 43: 1978
26. David C, Baeyens-Volant D, Geuskens G (1976) Eur Polym J 12: 71
27. Flamigni L, Barigelletti F, Bortolus P, Carlini C (1984) Eur Polym J 20: 171
28. Flamigni L, Bortolus P, Barigelletti F, Carlini C in: Marchetta C (ed) Proceedings of Internat Workshop on "Future Trends in Polymer Science and Technology. Polymers: Commodities or Specialities?", 8–13 Oct 1984, Capri (Italy); Available from CNR, Rome, Vol 2, p 164
29. De Schryver FC, Boens N (1976) J Oil Col Chem Assoc 59: 171
30. Ledwith A, Bosley JA, Purbrick MD (1978) J Oil Col Chem Assoc 61: 95
31. Fouassier JP (1983) J Chim Phys 80: 339
32. Shaefer CG, Peters KS (1980) J Am Chem Soc 102: 7566
33. Cohen SG, Parola A, Parsons GH Jr (1973) Chem Rev 73: 141
34. Berner G, Kirchmayr R, Rist G (1978) J Oil Col Chem Assoc 61: 105
35. Ghosh P, Ghosh R (1981) Eur Polym J 17: 817
36. Miyasaka H, Morita K, Kamada K, Nagata T, Kiri M, Mataga N (1991) Bull Chem Soc Jpn 64: 3229
37. Clarke SR, Shanks RA (1980) J Macromol Sci Chem A14: 69
38. Ledwith A (1977) Pure & Appl Chem 49: 431
39. Davis GA, Carapellucci PA, Szoc K, Gresser JD (1969) J Am Chem Soc 91: 2264
40. Cohen SG, Parsons G (1970) J Am Chem Soc 92: 7603
41. Davidson RS, Lambeth PF (1967) Chem Commun 1265
42. Bachmann WE (1933) J Am Chem Soc 55: 391
43. Bartholomew RF, Davidson RS, Howell MJ (1971) J Chem Soc C 2804
44. Ledwith A, Purbrick MD (1973) Polymer 14: 521
45. Sandner MR, Osborn CL, Trecker DJ (1972) J Polym Sci A-1 10: 3173
46. Catalina F, Peinado C, Sastre R, Mateo JL, Allen NS (1989) J Photochem Photobiol A: Chem 47: 365
47. Allen NS, Catalina F, Peinado C, Sastre R, Mateo JL, Green PN (1987) Eur Polym J 23: 985
48. Catalina F, Peinado C, Madruga EL, Sastre R, Mateo JL (1990) J Polym Sci Part A: Polym Chem 28: 967

49. Allen NS, Peinado C, Lam E, Kotecha JL, Catalina F, Navaratnam S, Parsons BJ (1990) Eur Polym J 26: 1237
50. Catalina F, Peinado C, Mateo JL, Bosch P, Allen NS (1992) Eur Polym J 28: 1533
51. Allen NS, Hurley JP, Rahman A, Follows GW, Waddel I (1993) Eur Polym J 29: 1155
52. Kinstle JF, Watson SL Jr (1975) J Radiat Curing 2(2): 7
53. Ciardelli F, Ruggeri G, Aglietto M, Angiolini D, Carlini C, Bianchini G, Siccardi G, Bigogno G, Cioni L (1989) J Coatings Technol 61: 77
54. Smets G, Hamouly SNE, Oh TJ (1984) Pure & Appl Chem 56: 439
55. Smets G (1985) Polym J 17: 153
56. Smets G, Oh TJ (1986) ACS Polym Prepr 27: 80
57. Tsubakiyama K, Kuzuba M, Yoshimura K, Yamamoto M, Nishijima Y (1991) Polym J 23: 781
58. Smets G, Oh TJ (1986) J Polym Sci Part C: Polym Lett 24: 229
59. Mateo JL, Manzarbeitia JA, Sastre R (1987) J Photochem Photobiol A: Chem 40: 169
60. Mateo JL, Bosch P, Catalina F, Sastre R (1993) J Polym Sci Part A: Polym Chem 31: 153
61. Carlini C, Ciardelli F, Rolla PA, Tombari E, Li Bassi G, Nicora C in: Radtech Europe '89 on Radiation Technologies, 9–11 Oct 1989. Florence. Available from Radtech Europe, Fribourg; Conference Papers p 369
62. Sengupta PK, Modak SK (1985) Makromol Chem 186: 1593
63. Hageman HJ (1989) In: Allen NS (ed) Photopolymerisation and photoimaging science and technology. Elsevier Science Publishers, p 38
64. Davidson RS, Hageman H, Lewis S (1993) Radcure coatings and inks. Aspects of photoinitiation, 19–20 October 1993, Egham. Available from Paint Research Association, Teddington, Middlesex (UK); Conference Papers p 137
65. Ghosh P, Mitra PS (1975) J Polym Sci Polym Chem Ed 13: 921
66. Ghosh P, Mitra PS (1976) J Polym Sci Polym Chem Ed 14: 981
67. Ghosh P, Banerjee AN (1977) J Polym Sci Polym Chem Ed 15: 203
68. Ghosh P, Mitra PS (1977) J Polym Sci Polym Chem Ed 15: 1743
69. Ghosh P, Ghosh TK (1984) J Polym Sci Polym Chem Ed 22: 2295
70. Fouassier JP (1990) Progr Org Coatings 18: 229
71. Chinmayanandam RB, Melville HW (1954) Trans Faraday Soc 50: 73
72. Heine H-G (1972) Tetrahedron Lett 4755
73. Sandner MR, Osborn CL (1974) Tetrahedron Lett 415
74. Eichler J, Herz CP, Naito I, Schnabel W (1980) J Photochem 12: 225
75. Delzenne GA, Laridon U, Peeters H (1970) Eur Polym J 6: 933
76. Baxter JE, Davidson RS, Hageman HJ, McLauchlan KA, Stevens DG (1987) Chem Commun 73
77. Gaur HA, Groenenboom CJ, Hageman HJ, Hakvoort GTM, Oosterhoff P, Overeem T, Polman RJ, Van der Werf S (1984) Makromol Chem 185: 1795
78. Fouassier JP, Lougnot DJ, Scaiano JC (1989) Chem Phys Lett 160: 335
79. Gupta SN, Thijs L, Neckers DC (1981) J Polym Sci Polym Chem Ed 19: 855
80. Gupta I, Gupta SN, Neckers DC (1982) J Polym Sci Polym Chem Ed 20: 147
81. Schultz AR (1960) J Polymer Sci 47: 267
82. Naito I, Schnabel W (1984) Polymer J 16: 81
83. Naito I, Ueki T, Tabara S, Tomiki M, Kinoshita A (1986) J Polym Sci Part A: Polym Chem 24: 875
84. Angiolini L, Carlini C (1990) Chimica Industria 72: 124
85. Lewis FD, Hoyle CH, Magyar JG (1975) J Org Chem 40: 488
86. Allen NS, Hardy SJ, Jacobine AF, Glaser DM, Catalina F, Navaratnam S, Parsons BJ (1991) J Photochem Photobiol A: Chem 62: 125
87. Fabrizio LF, Lin SOS, Jacobine AF (Loctite Corp) (1985) US 4,536,265; Chem Abs (1986) 104: 6677p
88. Huesler R, Kirchmayr R, Rutsch W, Rembold M (Ciba-Geigy AG) (1989) Eur Pat Appl EP 304,886; Chem Abs (1990) 112: 119595u
89. Li Bassi G, Nicora C, Cadonà L, Carlini C (Fratelli Lamberti S.p.A. and Consiglio Nazionale delle Ricerche) (1985) Eur Pat Appl EP 161,463; Chem Abs (1986) 104: 169042s
90. Li Bassi G (1986) Pitture e Vernici 62: 30
91. Li Bassi G, Cadonà L, Broggi F in: Radcure Europe '87, 4–7 May 1987. Munich. Available from Assoc. Finish. Processes SME, Deaborn, MI; Conference Proc p 3–15
92. Kirchmayr R, Berner G, Huesler R, Rist G (1982) Farbe Lack 88: 910
93. Phan XT (1986) J Radiat Curing 13: 11

94. Phan XT (1986) J Radiat Curing 13: 18
95. Fouassier JP, Lougnot DJ, Li Bassi G, Nicora C (1989) Polymer Commun 30: 245
96. Koehler M, Ohngemach J in: Radcure Europe '87, 4–7 May 1987. Munich. Available from Assoc. Finish. Processes SME, Deaborn, MI; Conference Proc p 3–1
97. Köhler M, Poetsch E, Ohngemach J, Dorsch D, Eidenschink R, Greber G (Merck GmbH) (1988) Eur Pat Appl EP 258,719; Chem Abs (1988) 109: 139179c
98. Köhler M, Ohngemach J, Poetsch E (Merck GmbH) (1987) Ger Pat 3,534,645; Chem Abs (1987) 107: 134811b
99. Klos R, Gruber H, Greber G (1991) J Macromol Sci Chem A28: 925
100. Lin SOS, Jacobine AF (Loctite Corp) (1985) US 4,534,838; Chem Abs (1986) 104: 51572m
101. Hatton KB, Irving E, Walsche JMA, Mallaband A (Ciba-Geigy AG) (1989) Eur Pat Appl EP 302,831; Chem Abs (1989) 111: 8004k
102. Schmidle CJ (Thiokol Corp) (1978) Belg Pat 860,306; Chem Abs (1978) 89: 25324j
103. Kurusu Y, Nishiyama H, Okaware M (1967) Kogyo Kagaku Zasshi 70: 593; Chem Abs (1968) 68. 4015k
104. Rudolph H, Rosenkranz HJ, Heine H-G (1975) Appl Polym Symp 26: 157
105. Shim JS, Park NG, Kim UY, Ahn K-D (1984) Polymer (Korea) 8: 34; Chem Abs (1984) 100: 210503a
106. Hong SI (1990) Makromol Chem Macromol Symp 33: 213
107. Ahn K-D, Ihn KI, Kwon I-C (1986) J Macromol Sci Chem A23: 355
108. Ahn K-D, Kwon I-C, Choi H-S (1990) J Photopolym Sci Technol 3: 137
109. Angiolini L, Carlini C, Tramontini M, Altomare A (1990) Polymer 31: 212
110. Fouassier JP, Ruhlmann D, Zahouily K, Angiolini L, Carlini C, Lelli N (1992) Polymer 33: 3569
111. Hageman H, Davidson RS, Lewis S in: Radtech Europe '91, on Creating Tomorrow's Technology, 29 Sept–2 Oct 1991. Edinburgh (UK); Available from Radtech, Fribourg, Conference Proc p 691
112. Davidson RS (1993) J Photochem Photobiol A: Chem 69: 263
113. Allen NS, Catalina F, Green PN, Green WA (1986) Eur Polym J 22: 49
114. Hoyle CE, Kim K-J (1987) J Appl Polym Sci 33: 2985
115. Hoyle CE, Keel M, Kim K-J (1988) Polymer 29: 18
116. Carlini C, Angiolini L, Lelli N, Ciardelli F, Rolla PA in: XXth FATIPEC Congress, 17–21 Sept 1990. Nice. Available from EREC, Puteaux (France); Proc p 413
117. Di Battista P, Giaroni P, Li Bassi G, Angiolini L, Carlini C, Lelli N in: Radtech Europe '91 on Creating Tomorrow's Technology, 29 Sept–2 Oct 1991. Edinburgh (UK); Available from Radtech, Fribourg, Conference Proc p 655
118. Angiolini L, Caretti D, Carlini C, Lelli N (1993) Polym Adv Technol 4: 375
119. Lechtken P, Buethe I, Hesse A (Basf AG) (1980) Ger Pat DE 2,830,927; Chem Abs (1980) 93: 46823u
120. Sumiyoshi T, Schnabel W, Henne A, Lechtken P (1985) Polymer 26: 141
121. Baxter JE, Davidson RS, Hageman HJ, Overeen T (1987) Makromol Chem Rapid Commun 8: 311
122. Jacobi M, Henne A (1985) Polym Paint Colour J 175: 636
123. Nickolaus W, Hesse A, Scholz D (1980) Plastverarbeiter 31: 723
124. Angiolini L, Caretti D, Carlini C, Lelli N, Rolla PA (1993) J Appl Polym Sci 48: 1163
125. Lechtken P, Buethe I, Jacobi M, Trimborn W (Basf AG) (1980) Ger Pat DE 2,909,994; Chem Abs (1981) 94: 103560c
126. Jacobi M, Henne A in: Radcure '83, May 9–11, 1983. Lausanne (Switzerland), Available from SME, Deaborn, MI, Technical Papers FC83-256
127. Angiolini L, Caretti D, Carlini C (1994) J Appl Polym Sci 51: 133
128. Miller R, Michl J (1989) Chem Rev 89: 1359
129. West R, Wolf AR, Peterson DJ (1986) J Rad Curing 35
130. Eck H, Rengstl A, Jira R, Fleischmann G (Wacker-Chemie GmbH) (1989) Ger Pat DE 3,814,429; Chem Abs (1990) 112: 159751y
131. Kminek I, Yagci Y, Schnabel W (1992) Polym Bull 29: 277
132. Timpe HJ, Ali S (1991) Acta Polym 42: 243
133. Ajayaghosh A, Das S, George MV (1993) J Polym Sci A: Polym Chem 31: 653
134. Weir D, Ajayaghosh A, Muneer M, George MV (1990) J Photochem Photobiol A: Chem 52: 425

135. Ajayaghosh A, Francis R, Das S (1993) Eur Polym J 29: 63
136. Reiter RH, Rosen G (Sun Chemical Corp) (1976) US Pat 3,992,363; Chem Abs (1977) 86: 55143e
137. Koelher M, Ohngemach J (1989) Polym Mat Sci Eng 60: 1
138. Yagci Y, Hizal G, Tunca Ü (1990) Polym Commun 31: 7
139. Yagci Y, Önen A (1991) J Macromol Sci Chem A28: 129
140. Piirma I, Chou LP (1979) J Appl Polym Sci 24: 2051
141. Banford CH (1974) in: Jenkins AD, Ledwith A (ed) Reactivity, mechanism and structure in polymer chemistry, John Wiley, New York, p 52
142. Yagci Y, Muller M, Schnabel W (1991) Macromol Reports A28 (Suppl 1): 37
143. Yagci Y (1993) Radcure coatings and inks. Aspects of photoinitiation: Conference papers, 19–20 October 1993. Egham. Available from Paint Research Association, Teddington, Middlesex (UK), p 151
144. Maslyuk AF, Kercha SF, Sopina IM (1989) Ukr Khim Zh 55: 758; Chem Abs (1990) 112: 139872k
145. Himori S (Mitsubishi Petrochemical Co Ltd) (1988) Eur Pat Appl EP 296,850; Chem Abs (1989) 110: 232282m
146. Otsu T, Yoshida M (1982) Makromol Chem Rapid Commun 3: 127
147. Otsu T, Kuriyama A (1985) Polym J 17: 97
148. Turner SR, Blevins RW (1990) Macromolecules 23: 1856
149. Otsu T, Himori S, Kiriyama T (Mitsubishi Petrochemical Co Ltd) (1988) Eur Pat Appl EP 286,376; Chem Abs (1989) 110: 136755x
150. Himori S (Mitsubishi Petrochemical Co Ltd) (1990) Eur Pat Appl EP 386,615; Chem Abs (1991) 114: 7427a
151. Yagci Y, Önen A, Schnabel W (1991) Macromolecules 24: 4620
152. Vlasov GP, Rudkovskaya GD, Ovsyannikova LA (1982) Makromol Chem 183: 2635
153. Mezger T, Cantow H-J (1983) Makromol Chem Rapid Commun 4: 313
154. Lanza E, Berghmans H, Smets G (1973) J Polym Sci Polym Phys Ed 11: 95
155. Doi T, Smets G (1989) Macromolecules 22: 25
156. Köhler M, Dorsch D, Ohngemach J, Greber G (Merck GmbH) (1989) Eur Pat Appl EP 341,560; Chem Abs (1990) 112: 159156h
157. Dietliker K, Rembold MW, Rist G, Rutsch W, Sitek F in: Radcure Europe '87, 4–7 May 1987. Munich. Conference Proc p 3–37
158. Dietliker K (1993) Radiation curing in polymer science and technology, Vol. II Photoinitiating systems, Fouassier JP, Rabek JF (eds). Elsevier Science Publ, London and New York, Chapt 3, p. 155
159. Meier K, Zweifel H (1986) J Photochem 35: 353
160. Fouassier JP, Lougnot DJ (1990) in: ACS Symposium Series 147, Hoyle CE, Kinstle JF (Eds) Am Chem Soc. Washington, p 59
161. Fouassier JP, Burr D (1991) Eur Polym J 27: 657

Editor: Prof. G. Wegner
Received 30 May 1994

Metal Complex in Polymer Membrane as a Model for Photosynthetic Oxygen Evolving Center

R. Ramaraj
School of Chemistry, Madurai Kamaraj University, Madurai - 625 021, Tamil Nadu, India

M. Kaneko
Department of Chemistry, Ibaraki University, Mito, Ibaraki 310, Japan

Photosynthesis is attracting attention as an important model of artificial photochemical conversion system in relevant to solar energy conversion for new energy resources. In the photosynthesis, dioxygen evolution is the most important process which provides electrons to the whole photochemical system. Several proposals have been put forward to elucidate the mechanism in the dioxygen formation from two water molecules and four molecules of one-electron oxidation catalyst. The protein part of the oxygen evolving center plays an important role for the catalysis. However, these mechanisms remain the most obscure part of plant photosynthesis. In order to construct artificial photosynthesis for the future energy source, it is important to utilize heterogeneous system such as polymer aggregates. The present authors have established new and active water oxidation catalysts as models for the photosynthesis especially by using heterogeneous polymer systems.

This review article mainly summarizes the work done on artificial water oxidation processes using polymer membranes, and the mechanism of the dioxygen evolution will be discussed. In the model water oxidation systems studied, the multielectron transfer catalytic metal complexes such as Mn and Ru are oxidized by chemical, electrochemical and photochemical methods to produce reactive higher oxidation states which oxidize two water molecules to liberate dioxygen both in homogeneous and heterogeneous polymer membrane systems. Structural reorganization of the catalytic molecules in the polymer membrane during dioxygen evolution is also described. Visible light splitting of water has been achieved by a system composed of a semiconductor photoanode modified with a polymer membrane incorporating water oxidation catalysts.

Advances in Polymer Science Vol. 123

1 Introduction

Solar energy conversion by artificial photosynthesis is becoming a more and more important and urgent research subject as an alternative new energy resource since the global environmental problems such as CO_2 increase came to be serious. Photosynthesis is an excellent model for achieving artificial solar energy conversion, for which water oxidation is the most important primary step[1-3] to provide electrons to the whole system to produce high energy compounds. In photosynthesis, the four-electron process of water oxidation to liberate dioxygen catalyzed at Mn-protein complex is coupled with an one-electron process of photoexcitation occurring at chlorophyll. The difficulty in creating non-biological model systems for photoinduced water oxidation lies in the requirement to couple the four-electron oxidation of two molecules of water (Eq. 1) with the one-electron chemistry of the photoredox process[4].

$$2\,H_2O \longrightarrow O_2 + 4\,H^+ + 4\,e^{-1} \tag{1}$$

In artificial photosynthesis, attempts are currently being made to molecular units that mimic the function of the oxygen evolving center (OEC), a key constituent of the plant photosystem II. It is generally agreed that the OEC is constituted by at least four manganese ions acting as a storage center for the four redox equivalents required to bring about the release of dioxygen from two water molecules. The OEC has been shown to operate by stepwise accumulation of positive charges and release of protons. Structural features of the OEC complex are now under intensive investigation[1-3]. To realize the four-electron water oxidation process and also to understand the mechanism involved in this process, one approach is to design a catalyst system such as a polymer membrane[1-5]. The major problem in this area of research is to establish an efficient and stable oxygen evolving catalyst. Currently, efforts are being made to design a model where the involvement of more than two metal centers leads to the formation of an active multielectron transfer site in a microheterogeneous phase rather than in a homogeneous solution.

In photosynthesis, a series of redox reactions takes place by which an electron is transported from one terminal to another terminal through electron transfer between neighboring redox centers. The reactions in photosynthesis are carried out in an immobilized state[1-3]. When electrons are extracted from water in photosynthesis dioxygen is liberated as a byproduct. Attempts to construct a model system have been made in terms of the use of microheterogeneous reaction environments such as micelle, bilayer, etc.[4]. However, the utilization of a polymer membrane system has not been explained extensively except for photogalvanic cells[5,6]. The functions of polymeric materials in artificial oxygen evolving center are summarized as follows: (1) to provide catalytic center by holding metal ions in an active structure, (2) to provide heterogeneous matrix necessary for achieving unidirectional charge flow in photoconversion systems and (3) to stabilize the active catalyst structure. This third point is especially important since high oxidation state complex species capable of oxidizing water are strong oxidants which can oxidize the ligand of the other complex molecules leading to so-called self-decomposition of the complex.

One of the outstanding properties of polymer membranes is their multiphase structure with a microheterogeneous environment[5,6]. Electrochemical properties of the incorporated molecules could be influenced by such a microenvironment, but no direct study has been reported concerning water oxidation processes. In water oxidation, a higher oxidation state metal complex is an active species, but it is not yet clear if this high oxidation state complex can be stably present in the polymer membrane to achieve water oxidation. It is expected that such active complex species can be present in the membrane if no electron-donating group exists in the microenvironment and if the high oxidation state complex can be stabilized by an anionic environment.

The polymer membrane coated electrode is one of the most promising approaches since these polymer membrane coatings provide a catalytic center with designed properties and lead to direct processes with electrons. Among the electron processes, the involvement and transformation of the redox centers in the charge transport and the utilization of electrons at the membrane are the most fundamental ones and they are explained in this article.

2 Photosynthetic Oxygen-Evolving Center Composed of a Mn-Protein Complex

In green plants there are two different photosystems (PS I and PS II) (Fig. 1). Photosystem I (PS I) produces a strong reductant and a weak oxidant, whereas photosystem II (PS II) generates a strong oxidant required for the oxidation of water to dioxygen. The PS II reaction is initiated by light absorption in the pigment complexes of PS II, followed by excitation energy transfer to the reaction center where the primary photochemical reaction takes place. During the last decade the general organization scheme of the reaction center complex has been established; but the ability of green plants to produce molecular oxygen from water using sun light energy is still a great mystery which remains to be explained at the molecular level. The basic principles of the functional organization scheme of photosynthetic water oxidation became apparent after the reports of Joliot and Kok[7]. In the photosynthetic oxygen evolving center, the oxidation state is represented by the so-called S-state model (Fig. 2) (S = the redox state of the oxygen evolving system and also to express the charge accumulation process). In this scheme P680 is chlorophyll *a* molecule; Z is electron donor and Q is electron acceptor in PS II. It is envisaged that holes are accumulated as S_4 state after successive four-electron oxidation of S_0 by photoinduced charge separation occurring successively at the photoreaction center. The important problem of water oxidation via four-step one-electron redox sequence, within

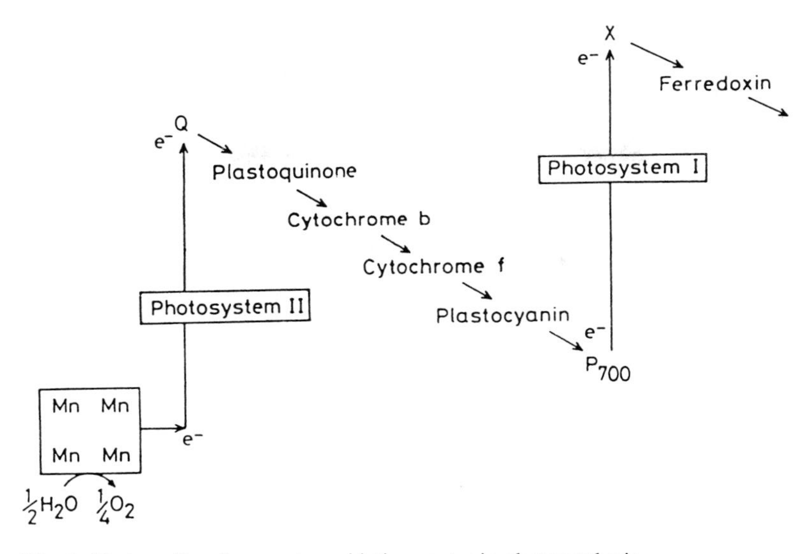

Fig. 1. Electron flow from water oxidation center in photosynthesis

$$4\,Z \quad P680 \quad Q \xrightarrow{4h\nu} 4\,Z \quad P680^{+}\ Q^{-} \longrightarrow$$
$$4\,Z^{+} \quad P680 \quad Q^{-} \longrightarrow$$

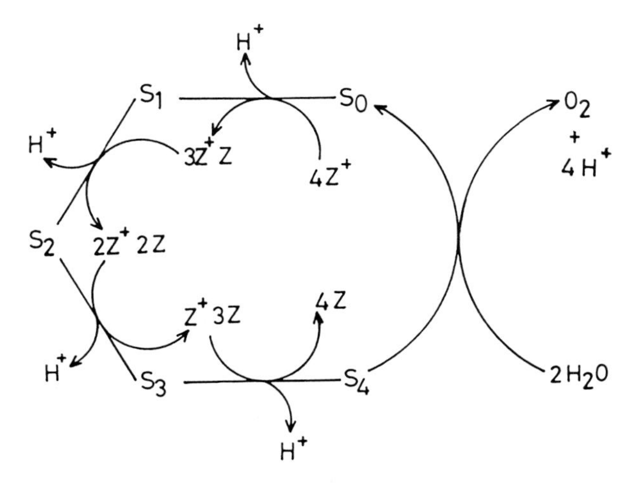

Fig. 2. S-state model showing the redox state of the oxygen evolving system

biological organisms, is how to stabilize the reactive intermediates in order to prevent rapid oxidative destruction. Their goal can be achieved by using appropriate redox groups which are intercalated into a specific protein matrix acting as enzyme. There are at least three different pools of bound manganese but only one of them was inferred to be directly involved in water oxidation[8,9]. Despite extensive studies, the exact number of functional manganese ions has not yet been determined unambiguously. Despite considerable variability in oxygen evolution rate there remains good agreement that four Mn ions are present in PS II membrane having optimal activity. The requirement of four Mn/Ps II for optimal activity would not necessarily imply that all four Mn are involved in water oxidation.

Photosynthetic water oxidation has been proposed based on the early experimental data. Three different models are considered: (i) involvement of manganese-water system in the redox reactions, (ii) involvement of another ligand or component other than manganese-water system and (iii) the redox reactions via non-water ligand system. The water oxidizing enzyme complex is located within a membrane called thylakoids and exposed to its inner side. Significant progress was achieved after the isolation of thylakoid membrane and of PS II particles which remained highly active in oxygen evolution. This study has been highlighted by major advances in our understanding of the molecular components essential for the photosynthetic process[1-3, 7-9]. These advances lead to a more rational design and synthesis of a variety of complex molecular systems aimed at increasing the efficiency of charge transfer and charge separation in the molecular level[10, 11]. The other major development culminated in the recent crystallization and direct structure determination of the *in vivo* bacterial photosynthetic centers although it does not contain OEC[12]. The photosynthetic water oxidation occurs at an enzyme that contains a special chlorophyll *a* for the generation of sufficiently oxidizing redox equivalents and a manganese center that catalyses the redox sequence leading from water to dioxygen. There are considerable supports for the binding of at least four Mn ions to the oxygen evolution center with two Mn ions involved in the catalytic site and the other two Mn ions in the oxidation of this site. The special chlorophyll *a* and the catalytic manganese center are incorporated into protein matrices and are coupled with other redox centers. Although the general function and organization of the oxygen evolving center is now available, yet the many details of the molecular mechanism and its control are still missing.

3 Models for Photosynthetic Oxygen Evolving Center

In the photosynthesis, a four-electron process of water oxidation to give dioxygen is coupled with an one-electron process of photoexcitation occurring at chlorophyll with the mediation of Mn complex[1-3]. The water oxidation reaction utilizing two water molecules is driven by four holes generated sequentially by charge separation following light absorption at a special chlorophyll *a* in photosystem II. The holes and water molecules are located in a protein and the water oxidation reaction can take place if the electron transfer from water to the oxidation center is suitably mediated by catalyst which interacts with two water molecules. To realize the four-electron water oxidation and also to understand the mechanism involved in this process, one approach is to design a catalyst system in a condensed system such as polymer membrane. The major problem in this area of research is to establish an efficient and stable oxygen evolving catalyst.

3.1 Metal Oxide Catalysts

The early system developed for water oxidation based on an electron acceptor (A), water oxidation catalyst (Cat.)[4,13–19] and the one-electron oxidant (S^+) generated by visible light irradiation oxidizes water with dioxygen formation according to Fig. 3. PtO_2 and RuO_2 can be used as catalysts for water oxidation reaction as shown in Fig. 3. The oxidation of water to dioxygen requires the cooperation of four oxidizing equivalents of sufficient redox power ($E_0 = 0.83$ V vs. SCE at pH = 7) and this has been difficult to achieve in the laboratory although green-plant photosynthesis solved the problems long ago. However, the recent works[13–19] have established that metal oxides such as PtO_2, IrO_2 and RuO_2 in powdered or colloidal form are capable of mimicking the functions of the manganese catalyst in photosynthesis. The redox catalyst facilitates coupling of the one-electron reduction of a suitable electron acceptor (S^+) to water oxidation process according to Fig. 3. In order for the production of dioxygen to be cyclic with respect to S^+ it is necessary to bring about the one-electron oxidation of S; if this can be achieved by a photochemical process then the overall photooxidation of water to oxygen can be realized. This has been

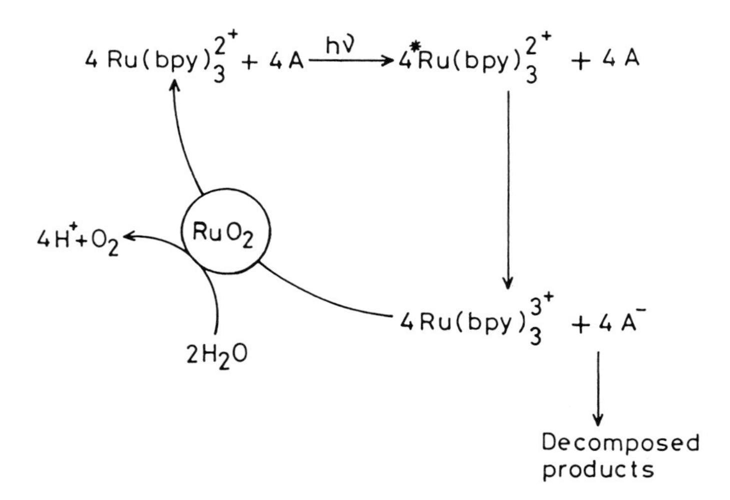

Fig. 3. Photolytic water oxidation using a heterogeneous catalyst

achieved only with the $[Ru(bpy)_3]^{2+/3+}$ couple since the photophysical and photochemical properties of the other oxidants preclude their use as photosensitizers[18, 20, 21]. It was reported that oxygen was evolved with a strong one-electron oxidant $[Ru(bpy)_3]^{3+}$ at pH 8 without any catalyst[22]; but it was found that oxygen evolution did not occur with $[Ru(bpy)_3]^{3+}$ in the absence of a catalyst[17, 19, 22–24]. The irradiation of sensitizer $[Ru(bpy)_3]^{2+}$ in the presence of a sacrificial electron acceptor (such as $S_2O_8^{2-}$, $[Co(NH_3)_5Cl]^{2+}$, Mn^{IV} pyrophosphate) and water oxidation catalyst RuO_2/MnO_2 resulted in continuous oxygen formation (Eqs. 2–4).

$$[Ru(bpy)_3]^{2+} \xrightarrow{h\nu} {}^*[Ru(bpy)_3]^{2+} \quad (2)$$

$$^*[Ru(bpy)_3]^{2+} + A \longrightarrow [Ru(bpy)_3]^{3+} + A^- \quad (3)$$

$$4\,[Ru(bpy)_3]^{3+} + 2\,H_2O \longrightarrow 4\,[Ru(bpy)_3]^{2+} + 4\,H^+ + O_2 \quad (4)$$

where $A = S_2O_8^{2-}$, $[Co(NH_3)_5Cl]^{2+}$.

A different reaction scheme was proposed[25] using Co(III) and Fe(III) hydroxides in combination with $[Ru(bpy)_3]^{3+}$ considering the adsorption of the oxidant $[Ru(bpy)_3]^{3+}$ onto the surface OH groups of the catalyst, which requires definite acid-base properties to be able to participate in the cation exchange in the specific pH region (Eqs. 5–7):

$$M^{III} - OH \rightleftharpoons M^{III} - O^- + H^+ \xrightarrow{[Ru(bpy)_3]^{3+}} M^{III} - O^- \ldots Ru^{3+}_{ads} \quad (5)$$

$$M^{III} - O^- \ldots Ru^{3+}_{ads} \longrightarrow M^{II} \ldots O^*_{ads} \ldots Ru^{2+}_{ads} \quad (6)$$

$$2\; M^{II} \ldots O^*_{ads} \ldots Ru^{2+}_{ads} \xrightarrow{2\,H_2O} 2\,M^{II}(H_2O) + O_2 + 2\,Ru^{2+} \quad (7)$$

where M^{III} and $M^{II} \ldots O^*_{ads} \ldots Ru^{2+}_{ads}$ represent catalytically active cation of the Co^{III} or Fe^{III} hydroxides and active form of oxygen together with the reduced forms of the oxidant and the catalyst respectively, and Ru^{2+} is $[Ru(bpy)_3]^{2+}$.

Eventhough oxygen evolution is realized using model systems of heterogeneous oxide catalysts, the mechanisms of all the reactions mentioned are quite complex, and the literatures contains an abundance of different and frequently conflicting opinions. These solid catalysts are heterogeneous in nature and their redox chemistries are difficult to study. They are non-specific and active for hydrogen evolution as well. It is necessary to design and study a chemical model of the photosynthetic oxygen evolution in homogeneous solution, through which it is possible to understand the mechanism of dioxygen formation from two water molecules.

3.2 *Model Mn Complexes*

Manganese is an essential element forming the active sites of a number of metalloproteins[1-3]. Chemists have been investigating the coordination chemistry of bi- and polynuclear manganese complexes which contains O and N donor atoms and a variety of bridging O/N ligands. A large number of such complexes have been synthesized and characterized[10]. The electronic and spectroscopic properties of these model compounds are very similar to those of the biomolecules. This has lead to a deeper understanding of the structure and the function of the metalloproteins. The metalloproteins with two or more manganese atoms per subunit are interesting inorganics since the manganese atoms are relatively close to one another.

It is reasonable to choose manganese compounds as a model for water oxidation to mimic photosynthetic oxygen evolving center. The previous attempts to realize such water oxidation included oxo-bridged dinuclear manganese complexes and manganese porphyrins[26, 27]. Calvin[26] reported that the di-μ-oxo-bridged binuclear manganese complex, $[(bpy)_2Mn^{III}(\mu - O)_2Mn^{IV}(bpy)_2]^{3+}$, was able to decompose water photolytically, but the initial claim was later withdrawn[28]. The present authors[29,30] found that the dinuclear manganese complex oxidizes water with the formation of visible oxygen bubbles when it was suspended in water as a heterogeneous catalysts in the presence of an oxidant such as Ce(IV) ion, although the same complex can not oxidize water when used in a homogeneous aqueous solution. It is obvious that two or four molecules of the dinuclear manganese complex may be oxidized by four Ce(IV) ions to produce four-electron depletion center on the solid particle (Fig. 4). This oxidizing center then leads to the four-electron oxidation of two molecules of weakly bound water with subsequent evolution of dioxygen. It has already been suggested[25, 31-33] that the microheterogeneous environment around the manganese centers and the close proximity of the manganese atoms to one another are responsible for water-splitting in photosystem II. This is supported by this finding[29, 30] that the dinuclear manganese complex is catalytically active only in the heterogeneous form.

McAuliffe et al.[34-36] have found that a large number of manganese complexes of Schiff base ligands evolve oxygen upon irradiation in an aqueous solution in the presence of p-benzoquinone. Sawyer et al.[37] have described a series of dinuclear manganese complexes of Schiff base ligands which in highly oxidized form oxidize water to oxygen. Oxygen electrode was mainly used for the detection of dissolved oxygen in solution analogous to the method employed earlier[26, 27]. However, oxygen electrode is sensitive to temperature changes caused by photolysis. The present authors[38] have found that the Schiff base complex of manganese becomes insoluble upon addition of Ce(IV) ions, and the insoluble complex oxidizes water by the Ce(IV) oxidant with the formation of visible oxygen gas bubbles. The role of heterogeneity of the present catalyst system would be to bring the oxygen atoms of two water molecules into close proximity to form an O–O bond which leads to the generation of oxygen.

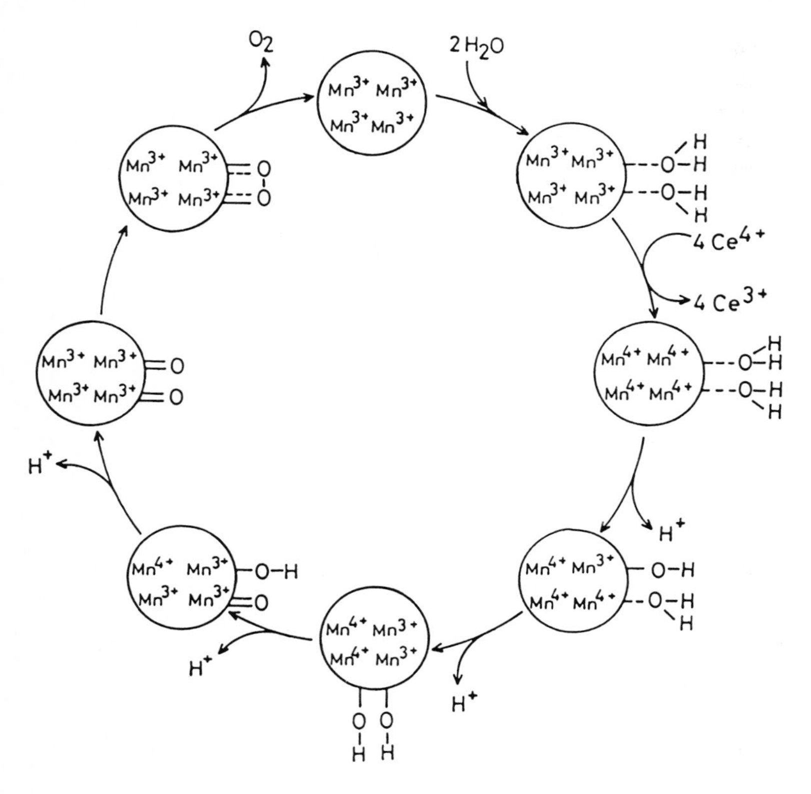

Fig. 4. A schematic depiction of heterogeneous water oxidation mechanism

3.3 Model Ru Complexes

The major problem encountered in earlier works is the lack of an efficient, stable, and fast-acting catalyst capable of mediating the oxidation of water to dioxygen by a strong oxidizing agent (Eq. 8),

$$4\,\mathrm{Ox} + 2\,\mathrm{H_2O} \longrightarrow 4\,\mathrm{Red} + 4\,\mathrm{H^+} + \mathrm{O_2} \tag{8}$$

Recently many groups have become interested in developing a stable and efficient homogeneous oxygen evolving catalyst using multielectron transfer metal complexes. Meyer et al.[39] first developed homogeneous oxygen evolving catalyst dinuclear ruthenium complex $[(bpy)_2(H_2O)RuORu(H_2O)(bpy)_2]^{4+}$ (pby = 2,2′-bipyridine). This work was rapidly followed by other groups[29, 30, 38–56] and a number of examples of homogeneous water oxidation catalysts have been reported.

The dinuclear ruthenium complexes are stable in the Ru^{III}–Ru^{III} oxidation state and easily form Ru^{V}–Ru^{V} upon electrochemical or chemical oxidation[39, 30, 46]. Recently, the existence of higher oxidation state Ru^{V}–Ru^{V}

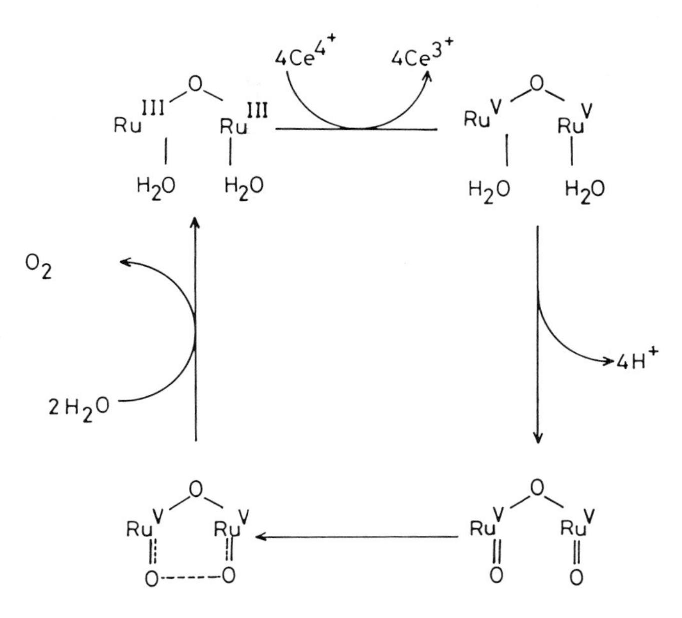

Fig. 5. Water oxidation redox cycle using the dimer ruthenium complex

complex has been characterized by Resonance Raman spectroscopy[56]. The higher oxidation state of the ruthenium complex (Ru^{V}–Ru^{V}) realizes four-electron water oxidation to evolve oxygen in a homogeneous solution with the formation of Ru^{III}–Ru^{III} as shown in Fig. 5. This reaction scheme (Fig. 5) supports the fact that the loosely bound water molecules in the protein bound manganese would lead to the water oxidation. It was also reported[16, 48, 50] that the substituted 2,2′-bipyridine complexes acts as water oxidation catalysts in combination with $[Ru(bpy)_3]^{2+}$ complex photochemically. It was reported that during photolysis $[Ru(bpy)_3]^{2+}$ underwent decomposition to produce a dimer ruthenium complex which catalysed water oxidation in the subsequent reaction (Eqs. 9 and 10).

$$4\,[Ru(bpy)_3]^{2+} + A \xrightarrow{h\nu} 4\,[Ru(bpy)_3]^{3+} + A^- \quad (9)$$

$$4\,[Ru(bpy)_3]^{3+} + 2\,H_2O \xrightarrow{Ru^{III}-Ru^{III}} O_2 + 4\,H^+ + 4\,[Ru(bpy)_3]^{2+} \quad (10)$$

where A = electron acceptor.

It is noted that the linearly oxo-bridged trinuclear ruthenium complex $[NH_3)_5Ru-O-Ru(NH_3)_4-O-Ru(NH_3)_5]^{6+}$ (Ru^{III}–Ru^{IV}–Ru^{III}, called Ru-red) is a better catalyst for water oxidation than the dimer ruthenium complexes in homogeneous solution[46,47]. The Ru^{III}–Ru^{IV}–Ru^{III} is oxidized electrochemically

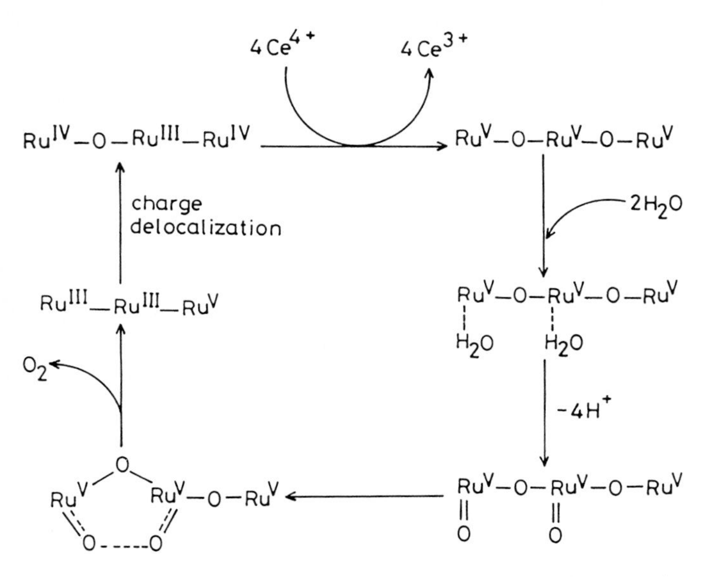

Fig. 6. Water oxidation redox cycle using the trimer ruthenium complex

or chemically to produce Ru^V–Ru^V–Ru^V and this higher oxidation state trimer catalysed water oxidation efficiently to evolve oxygen as shown in Fig. 6. In a polynuclear metal complex, the charge delocalization and stabilization as well as close proximity of the metal centers explain the mechanisms of oxygen formation through O–O bond formation from loosely bound two water molecules.

The particular concern was that most of these oxygen evolving homogeneous catalysts decomposed irreversibly with repeated use and that the unidentified decomposition products might often act as oxygen evolving catalysts[55].

4 Oxygen Evolving Center Composed of a Polymer Membrane

In the oxygen evolving center of photosynthesis, more than two Mn ions form a protein complex, which couples the four-electron water oxidation with photo-excitation. Oxygen evolves during photosynthesis in a heterogeneous phase of the Mn protein. The protein part provides the Mn ions with appropriate environment and redox potentials by complexation. Another important role of the protein must be to stabilize the highly oxidized state of the catalyst which otherwise can oxidize organic moieties present around the complex instead of oxidizing water molecules. To realize an efficient and stable four-electron water oxidation center in an artificial system, one approach is to design a catalyst in a condensed system such as a polymer membrane[5,6]. Our continuing effort is to design a model system where the involvement of more than one or two metal

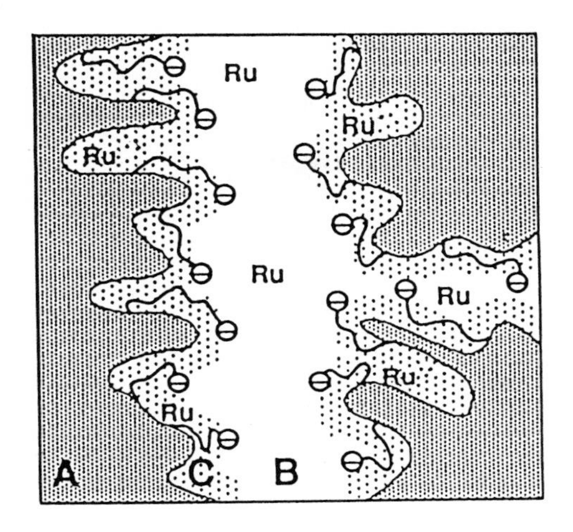

Fig. 7. A schematic view of Nafion membrane showing the microheterogeneous environment. A: hydrophobic fluorocarbon phase; B: hydrophilic sulfonate ionic clusters; C: interfacial region formed between A and B and Ru: adsorbed ruthenium complex water oxidation catalyst

centers lead to the formation of an active multielectron transfer site in a membrane system rather than in a homogeneous solution. In artificial photosynthesis, attempts are currently being made to design molecular units that mimic the function of the oxygen evolving complex[6].

In the fabrication of the membrane, Nafion polymer is a promising material containing fluorocarbon chain and sulfonate ion end groups[57]. Most other commonly known polymers cannot hold water oxidation metal complex catalysts. The membrane structure of Nafion polymer is well characterized by many techniques including X-ray data[57]. The diameter of a sulfonate ionic cluster formed within a Nafion membrane is reported to be 40 Å, which is surrounded by an interfacial region (Fig. 7). The unique multi-domain structure of a Nafion polymer membrane made it as one of the most attracting materials for electrode modification or ion-exchange membrane. The structure of Nafion membrane involves three domains, i.e., a hydrophobic fluorocarbon phase, interconnected ionic clusters having sulfonate head groups and an interfacial region[57, 58] (Fig. 7). The duality of the Nafion matrix comprising hydrophilic and hydrophobic domains in close proximity imposes a situation wherein both hydrophilic and hydrophobic interactions play an important role in determining the relative stability of the ions present in the polymer member in contact with solution. Nafion provides a low dielectric environment and has been shown to be much more selective for large organic cations and high charge metal complexes than for small inorganic cations like Na^+ or K^+.

4.1 Reversible Formation of Water Oxidation Catalyst in a Polymer Membrane

Meyer et al.[39, 41] reported the catalytic activity of oxo-bridged dimer ruthenium complex, $[bpy)_2(H_2O)RuORu(H_2O)(bpy)_2]^{4+}$, towards water oxidation to evolve dioxygen. Since this report, a substantial amount of work has been done by others[39–49] to understand the redox chemistry and its ability to oxidize water to dioxygen. The interesting aspects of this oxo-bridged dimer ruthenium complex are the presence of two metal centers with two aquo ligands and multiple redox states with suitable redox potential to catalyse water oxidation as well as pH dependent oxidation of the metal center accompanied with loss of protons to give hydroxo or oxo complex. In a homogeneous solution, the oxidation of *cis*-$[Ru(bpy)_2(H_2O)_2]^{2+}$ and other substituted bipyridine ruthenium complexes at appropriate oxidation potentials leads to the formation of corresponding *cis*-$[Ru^{III}(bpy)_2(H_2O)(OH)]^{2+}$ complex followed by dimerization to produce the corresponding dimer complex $[(bpy)_2(H_2O)RuORu(H_2O)(bpy)_2]^{4+}$, $H_2O{-}Ru^{III}{-}Ru^{III}{-}OH_2$[50,61].

As a general preparation procedure of the water oxidation catalyst of dinuclear ruthenium complex, $[(bpy)_2(H_2O)RuORu(H_2O)(bpy)_2]^{4+}$, the corresponding mononuclear ruthenium complex *cis*-$[Ru(bpy)_2(H_2O)_2]^{2+}$ has been used as a precursor complex[39,41,48,50]. However, other reports claimed that the mononuclear ruthenium complex itself can mediate water oxidation[53,59] under homogeneous and heterogeneous conditions. It was also reported[48,50,60] that the mononuclear ruthenium complexes dimerize under homogeneous as well as in heterogeneous conditions to give active water oxidation catalysts when oxidized under appropriate conditions. However, it was demonstrated later that such mononuclear ruthenium complexes did not oxidize water to evolve dioxygen[41,52]. It has recently been reported[62] that the oxidation of the monomeric ruthenium complex, *cis*-$[Ru(bpy)_2(H_2O)_2]^{2+}$, incorporated in a Nafion membrane leads its dimerization to form a water oxidation catalyst $[(bpy)_2(H_2O)RuORu(H_2O)(bpy)_2]^{4+}$.

The monomeric ruthenium complex *cis*-$[Ru(bpy)_2(H_2O)_2]^{2+}$ was adsorbed into a Nafion film coated onto an Indium Tin Oxide (ITO) electrode. The *in situ* spectrocyclic voltammetric investigations showed that the monomeric complex in the Nafion film dimerizes upon oxidation with clear spectral changes to yield corresponding oxo-bridged dimer complex $[(bpy)_2(H_2O)RuORu(H_2O)(bpy)_2]^{4+}$ which is a water oxidation catalyst[62]. The *in situ* absorption spectra were recorded for the monomeric ruthenium complex coated electrode during the scan between 0.4 and 1.4 V (vs. SCE) at slow scan rate of 0.5 mVs^{-1} (Fig. 8A and 8B). During the oxidative scan from 0.4 to 0.9 V (vs. SCE), the absorption at 480 nm decreased due to the oxidation of *cis*-$[Ru(bpy)_2(H_2O)_2]^{2+}$, the oxidation of the monomeric complex leads to its dimerization to form $Ru^{IV}{-}Ru^{III}$ complex (Fig. 8A), and in the reductive potential scan from 0.9 to 0.6 V (Vs. SCE), the absorbance at 645 nm due to dimer ruthenium complex ($Ru^{III}{-}Ru^{III}$) increased due to the reduction of its $Ru^{IV}{-}Ru^{III}$ complex (Fig. 8B). When a potential of -0.1 V (vs. SCE) was applied to the coated electrode, the $Ru^{III}{-}Ru^{III}$ complex was reduced to the highly unstable $Ru^{II}{-}Ru^{II}$ complex which underwent decomposition to produce the corresponding monomeric *cis*-$[Ru(bpy)_2(H_2O)_2]^{2+}$ within the coating (Fig. 9)[63]. The reoxidation of the *cis*-$[Ru(bpy)_2(H_2O)_2]^{2+}$ in the Nafion membrane has led to the reformation of the

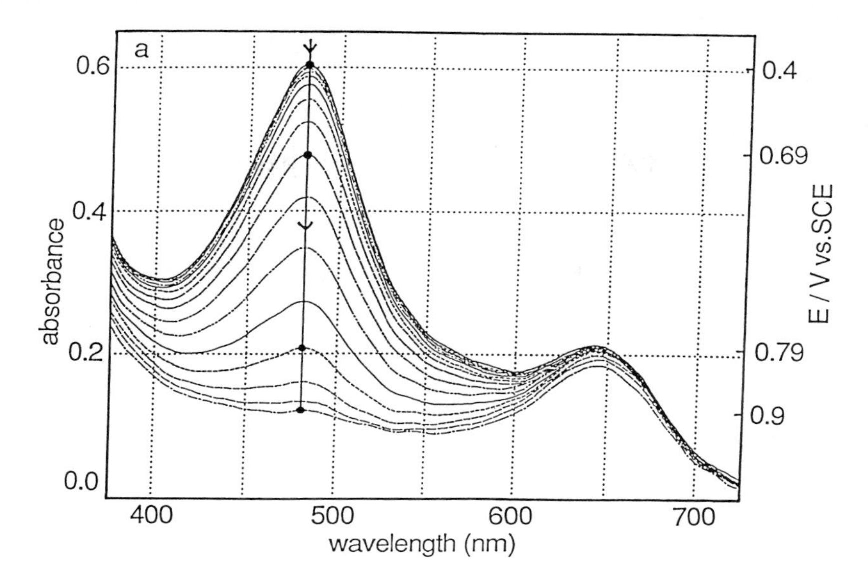

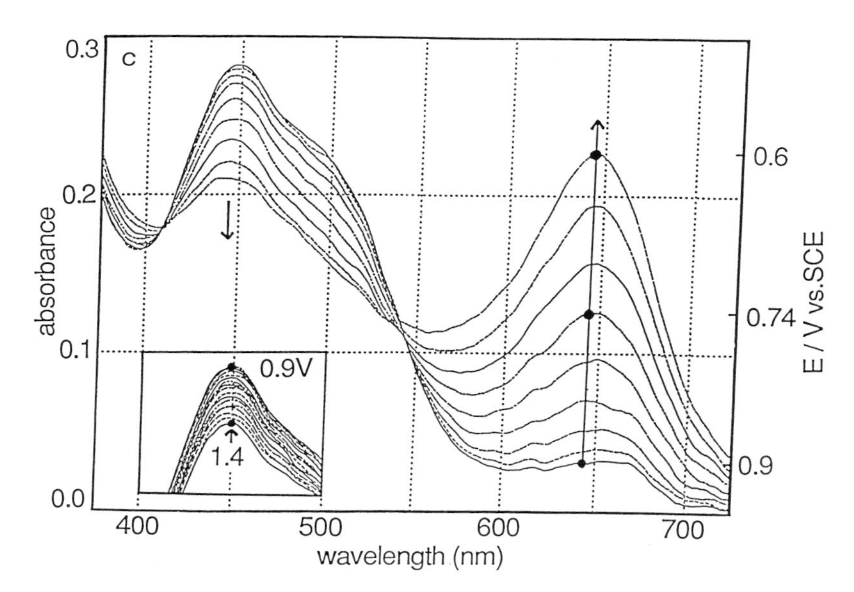

Fig. 8. Changes in the absorption spectra of ITO/Nf/*cis*-$[Ru(bpy)_2]^{2+}$ in 0.1 M HNO_3 at a scan rate of 0.5 mVs^{-1}: (A) positive scan from 0.4 to 0.9 V/SCE and (b) negative scan from 0.9 to 0.6 V/SCE (the inset shows changes in the absorption spectra in the negative scan from 1.4 to 0.9 V/SEC. The direction of absorbance change is indicated in the figure. (Reproduced with permission from Elsevier Sequoia S.A)

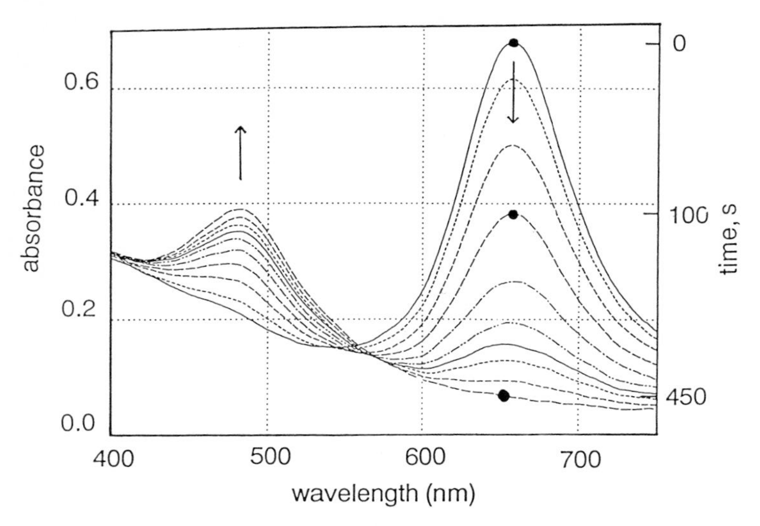

Fig. 9. Absorption spectral changes of ITO/Nf/Ru-dimer in 0.1 M HNO_3 at an applied potential of − 0.1 V/SCE. The direction of absorbance change is indicated in the figure

dimer ruthenium complex in the Nafion film, and oxygen gas evolution was confirmed by chromatography when the electrode was kept at 1.4 V (vs. SCE).

This work[62)] is of importance in relevant to the water oxidation reaction in the photosynthetic membrane. The oxidation of two water molecules to evolve dioxygen in photosystem II is understood to be realized by an enzymatic organization of a special Mn cluster in a protein architecture[31−33)]. This oxygen evolving center (OEC) of photosynthesis resides near the inner side of the thylakoid membrane. The OEC, the key constitutent of the plant photosystem II, is composed of at least two/four manganese ions and acts as a storage center for four redox equivalents to bring about oxidation of two loosely bound water molecules to release dioxygen. The above work [62,63)] showed that the oxidation of the two molecules of the precursor monomeric complex inactive for water oxidation confined in a Nafion membrane leads to the formation of a catalytically active dimeric ruthenium complex. Further oxidation of this active complex in the membrane leads to the accumulation of four positive charges to form a higher oxidation Ru^{V}–Ru^{V} complex which oxidizes two water molecules to liberate dioxygen. At appropriate reduction potentials the active catalyst returned back to the catalytically inactive monomeric precursor complex in the Nafion membrane. It has not been known if such structural reorganization of the water oxidation catalyst occurs accompanying redox process of the Mn ions in the photosynthetic oxygen evolving center, but the present work might suggest a possible involvement of similar reorganization of the active site in the oxygen evolving center induced by redox reactions.

The absorption spectra of the dimer complex, H_2O–Ru^{III}–O–Ru^{III}–OH_2, recorded between pH 1 and pH 4.5 are rather similar with an absorption band at

650 nm whereas in the pH region from 5 to 9, the absorption of the dimer complex showed two bands at 430 and 670 nm[63]. The different redox reactions of the dimer $H_2O-Ru^{III}-Ru^{III}-OH_2$ accompanied with deprotonation has already been established[39,41]. The absorption spectra recorded for the dimer ruthenium complex adsorbed in a Nafion film coated on an ITO electrode showed very similar spectral changes with pH and the absorption maxima are also similar to those observed in solution[41,43,64]. The *in situ* spectrocyclic voltammetry studies were performed with the dimer complex adsorbed in a

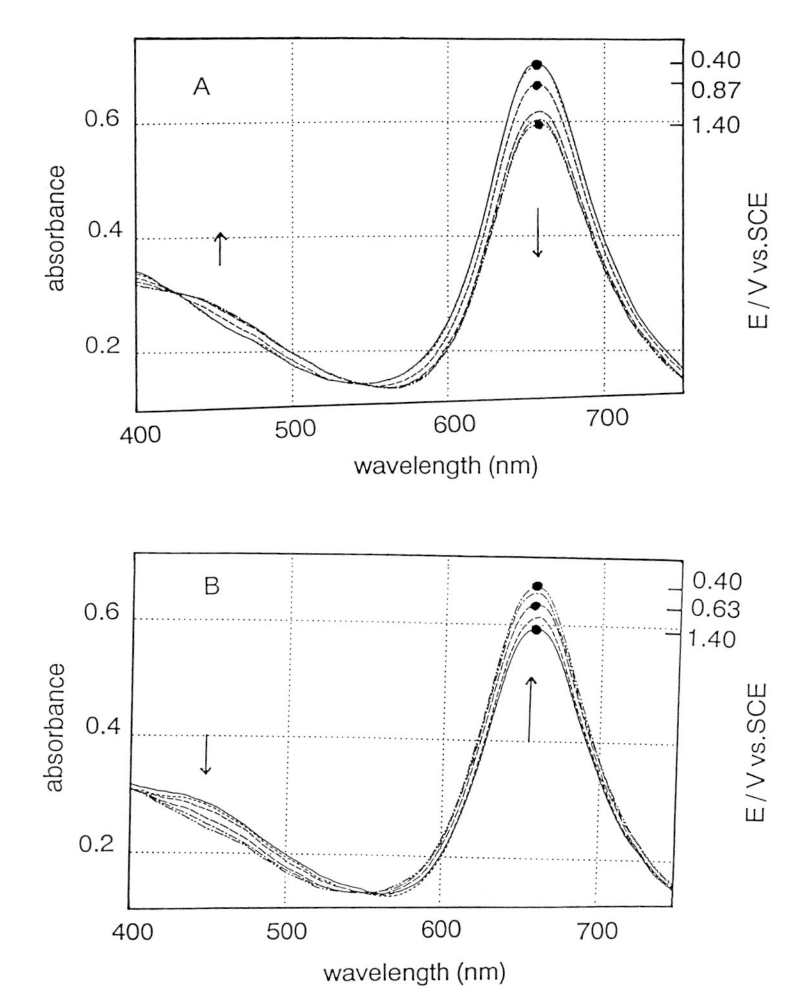

Fig. 10. Absorption spectral changes of ITO/Nf/Ru-dimer in 0.1 M HNO_3 at 0.5 mVs^{-1} scan rate. The directions of absorbance changes and the potential range are indicated in the figures. (A) Absorption spectral changes in the oxidative scan from 0.4 to 1.4 V/SEC and (B) absorption spectral changes in the reductive scan from 1.4 to 0.4 V/SCE

Nafion film coated on an ITO electrode to understand the structural transformations of the dimer in the Nafion coating during the catalytic water oxidation process[63]. The absorption spectral changes observed during the oxidation scan from 0.4 to 1.4 V (s. SCE) (Fig. 10A) showed a decrease in the absorbance at 655 nm with simultaneous increase in the absorbance at around 450 nm with clear isosbestic points at 430 and 545 nm. In the reductive scan from 1.4 to 0.4 V (vs. SCE) (Fig. 10B), the absorbance at 655 nm increased and a simultaneous decrease in the absorbance at 450 nm was observed with an isosbestic point at 555 nm. The absorbance at 655 nm was almost recovered back. Initially the oxidation of the dimer complex $H_2O-Ru^{III}-Ru^{III}-OH_2$ leads to the formation of $H_2O-Ru^{III}-Ru^{IV}-OH_2$ with an absorption maximum at around 450 nm and further oxidation at higher positive potentials must lead to $Ru^{V}-Ru^{V}$ formation in a successive oxidation process. The $Ru^{V}-Ru^{V}$ would be rapidly reduced by water molecules to produce $H_2O-Ru^{III}-Ru^{IV}-OH_2$ at pH 1. The same *in situ* spectrocyclic voltammetry experiments at pH 9.3 showed an absorption maximum at around 500 nm with the formation of $H_2O-Ru^{III}-Ru^{IV}-OH$ in the Nafion film. In relevant to the absorbances at 450 and 500 nm, Meyer et al.[41,64] reported that the complexes $H_2O-Ru^{III}-Ru^{IV}-OH_2$ and $H-O-Ru^{III}-Ru^{IV}-OH$, each absorbing at around 444 and 495 nm respectively, are formed at pH 0 and 2.

The reactions observed for the dimer complex adsorbed in a Nafion film coated on an ITO electrode at different pH by *in situ* absorption spectral measurements are summarized as shown in Fig. 11. At higher positive potentials and at potentiostatic conditions, a band at around 450 nm was observed indicating the formation of $H_2O-Ru^{III}-Ru^{IV}-OH_2$ at acidic conditions and formation of $H_2O-Ru^{III}-Ru^{IV}-OH$ at basic conditions in addition to the absorbance at 655 nm. This shows that during the catalytic water oxidation process, the diaquo dimer complex exists as an intermediate. In a Nafion polymer membrane, the metal complex is isolated and experiences a microheterogeneous environment imposed by hydrophobic fluorocarbon moiety and

(A). at pH 1

$$\underset{655\ \mathrm{nm}}{H_2O\text{-}Ru^{III}\text{-}Ru^{III}\text{-}OH_2} \underset{1.4\ \mathrm{to}\ 0.4\ \mathrm{V}}{\overset{0.4\ \mathrm{to}\ 1.4\ \mathrm{V}}{\rightleftharpoons}} \underset{450,\ 655\ \mathrm{nm}}{H_2O\text{-}Ru^{III}\text{-}Ru^{IV}\text{-}OH_2}$$

(B). at pH 9.3

$$\underset{430,\ 670\ \mathrm{nm}}{H_2O\text{-}Ru^{III}\text{-}Ru^{III}\text{-}OH} \underset{0.3\ \mathrm{to}\ 1.4\ \mathrm{V}}{\overset{1.4\ \mathrm{to}\ 0.3\ \mathrm{V}}{\rightleftharpoons}} \underset{450,\ 670\ \mathrm{nm}}{HO\text{-}Ru^{III}\text{-}Ru^{IV}\text{-}OH}$$

Fig. 11. Reactions of dinuclear ruthenium complex at Nafion membrane coated on ITO electrode at different pH conditions (ITO/Nf/Ru-dimer)

hydrophilic sulfonate ion. In a microheterogeneous environment, the isolated $Ru^{V}-Ru^{V}$ complex would act as a stable four-electron oxidant to evolve dioxygen[62,63] whereas such a high valent complex present in a homogeneous solution would easily lead to oxidative decomposition of another complex to monomeric complex[43] by diffusion and collision.

4.2 *Water Oxidation Center Model Using Trinuclear Ruthenium Complexes*

Polynuclear metal complexes are more suited for water oxidation catalyst because of their nature to act as multielectron transfer reagents in addition to the fact that charge delocalization can lead to stabilization of the catalyst rather than decomposition during the process. The trinuclear ruthenium complexes Ru-red and Ru-brown, $[(NH_3)_5Ru-O-Ru(NH_3)_4-O-Ru(NH_3)_5]^{6+}$ $(Ru^{III}-Ru^{IV}-Ru^{III})$ and $[(NH_3)_5Ru-O-Ru(NH_3)_4-O-Ru(NH_3)_5]^{7+}$ $(Ru^{IV}-Ru^{III}-Ru^{IV})$, respectively, have been shown to be efficient water oxidation catalysts for oxygen evolution with high turnover numbers[46,47]. When Ru-red was dissolved in an acidic aqueous solution, it underwent one-electron oxidation with the formation of Ru-brown. When Ru-brown was dissolved in a basic solution, the complex underwent reduction to produce Ru-red. The one-electron oxidation and reduction of the Ru-red and Ru-brown has already been well established (Eq. 11)[46,65-67].

$$Ru^{III}-Ru^{IV}-Ru^{III} \underset{+e}{\overset{-e}{\rightleftharpoons}} Ru^{IV}-Ru^{III}-Ru^{IV} \qquad (11)$$

Rud-red (in neutral or alkaline solution) — Ru-brown (in acidic solution)

Eventhough Ru-brown was reported to oxidize OH^- ion[68,69], actual evolution of O_2 was not observed. However, Ru-brown was suggested as an inorganic model for water oxidation process (Eq. 12).

$$Ru^{IV}-Ru^{III}-Ru^{IV} + OH^- \longrightarrow Ru^{III}-Ru^{IV}-Ru^{III} + 1/2H_2O + 1/4\,O_2 \qquad (12)$$

The cyclic voltammetric studies of Ru-red in aqueous solution[46,47] showed that the trinuclear ruthenium complex undergoes various redox processes to produce $Ru^{V}-Ru^{V}-Ru^{V}$. The stability of the trinuclear ruthenium complex in the various oxidation states can be interpreted on the basis of intramolecular charge delocalization among the ruthenium centers.

The water oxidation experiment to evolve dioxygen was reported using trinuclear ruthenium complexes[46,47] with higher turnover numbers. The higher oxidation state Ru^{V}–Ru^{V}–Ru^{V} produced upon the addition of excess Ce(IV) oxidizes water to oxygen (Fig. 6). However, in an another report, it has been suggested that the Ce(IV) ions oxidized the Ru-red (Ru^{III}–Ru^{IV}–Ru^{III}) to Ru-Brown (Ru^{IV}–Ru^{III}–Ru^{IV}) and with excess of Ce(IV), decomposed the Ru-brown irreversibly to products and the decomposed products oxidized water to O_2[54]. It has also been suggested that the active catalyst might be a hydrolysed Ru^{IV} polymeric species. In order to realize an effecient and four-electron water oxidation center in an artificial system mimicking membraneous water oxidation center (OEC), the trinuclear ruthenium complexes were used in a condensed system such as a polymer membrane rather than in a homogeneous solution[70–72].

The cyclic voltammogram recorded for an aqueous solution of Ru-red at pH 7.4 with a scan rate of 5 mVs^{-1} using a thin layer ITO cell is shown in the inset of Fig. 12A[73]. In addition to the oxidation waves, it shows a large anodic current at higher positive potentials due to catalysed water oxidation. In the reverse scan, the cyclic voltammogram does not show any distinguishable reduction waves showing the reduction of the higher oxidation state complex by water molecules[38]. The *in situ* absorption spectra recorded during the scan from − 0.1 to 0.32 V (vs. Ag wire) in pH 7.4 solution are shown in Fig. 12A. The absorption spectra recorded in the potential range from − 0.1 to 0.32 V (vs. Ag wire) show a decrease in absorbance at 535 nm with simultaneous increase in absorbance at 470 nm with an isosbestic point. The decrease in the absorbance at 535 nm is assigned to the oxidation of Ru-red (Ru^{III}–Ru^{IV}–Ru^{III}) and the increase in absorbance at 470 nm to the formation of Ru-brown (Ru^{IV}–Ru^{III}–Ru^{IV})[46]. When the applied potential was increased above 0.32 V (Ag wire) the absorbance at 470 and 535 nm starts decreasing (Fig. 12B) and when the potential reached 1.0 V (vs. Ag wire), they almost completely disappeared. In the reverse scan, there was no recovery of the absorbance at 470 and 535 nm showing the decomposition of trimer Ru-red probably to form monomeric ruthenium complexes in homogeneous solution. A very similar spectral changes were also noticed for Ru-brown at pH 1 in the *in situ* spectrocyclic voltammetry studies in a homogeneous solution[72].

The cyclic voltammogram recorded for Nafion adsorbed Ru-red at ITO coated electrode at pH 7.4 is shown in the inset of Fig. 13. Electrocatalytic oxygen evolution by water oxidation with this complex/polymer coated electrode has already been reported[71]. The Ru-brown complex incorporated into Nafion coated ITO electrode was found to be very stable upon scanning between 0.5 to 1.4 V (vs. SCE) at different scan rate. The *in situ* absorption spectra at the scan rate of 0.5 mVs^{-1} recorded during the scan from 0.5 to 1.4 V (vs. SCE) (Fig. 13) show a decrease in the absorbance at 540 nm with a simultaneous absorbance increase at 470 nm with an isosbestic point[71,72]. The absorbance decrease at 540 nm is assigned to the oxidation of Ru-red (Ru^{III}–Ru^{IV}–Ru^{III}) and the increase in the absorbance at 470 nm to the formation of Ru-brown (Ru^{IV}–Ru^{III}–Ru^{IV}). During reverse scan, the absorption spectra does not show any change demonstrating the stability of Ru-brown (Ru^{IV}–Ru^{III}–Ru^{IV}) in the

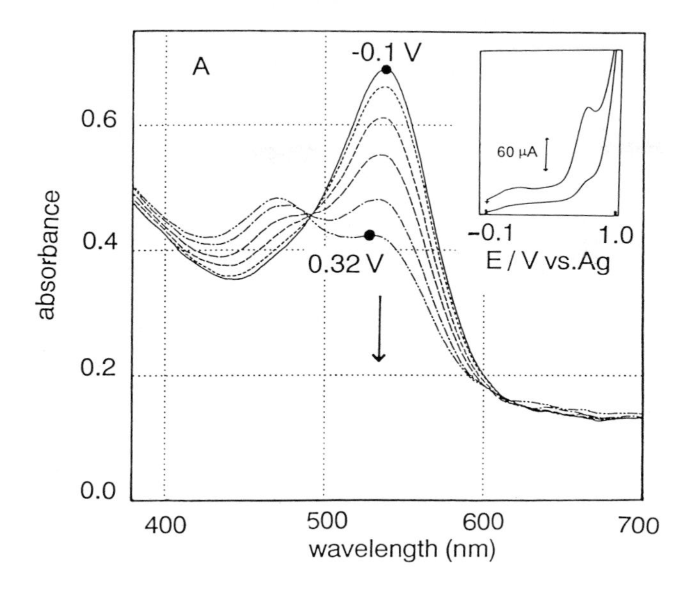

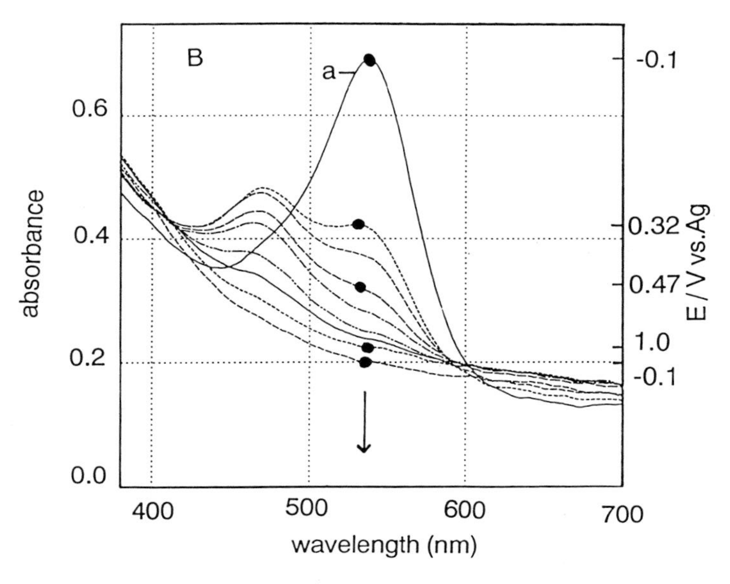

Fig. 12. *In situ* absorption spectra of Ru-red in an aqueous solution (pH 7.4 phosphate buffer) measured during a voltammetric scan at 5 mVs^{-1} using a thin layer ITO cell. The direction of the absorbance change and the potential range are indicated in the figure. (A) Absorption spectra recorded during the oxidative scan from -0.1 to 0.32 V/Ag wire. Inset: cyclic voltammogram recorded during the *in situ* absorption spectral measurements at 5 mVs^{-1}. (B) Absorption spectra recorded during the oxidative scan from 0.32 to 1.0 V/Ag wire and during the reductive scan from 1.0 to -0.1 V/Ag wire. Spectrum a: Absorption spectrum recorded in the beginning of the scan. (Reproduced with permission from The Royal Society of Chemistry)

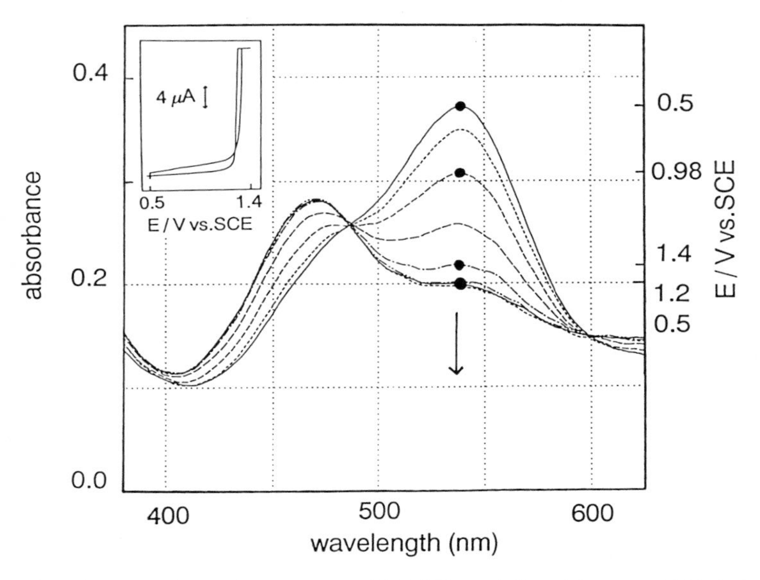

Fig. 13. *In situ* absorption spectra of ITO/Nf/Ru-red dipped in water (pH 7.4 phosphate buffer) at 0.5 mVs^{-1} scan rate. The directions of the absorption and the potential range are indicated in the figure. Inset: Cyclic voltammogram recorded during the *in situ* absorption spectral measurements. (Reproduced with permission from The Royal Society of Chemistry)

Nafion film during the catalytic water oxidation. The absorption spectra did not show any further changes in repeated cyclic scans. When an applied potential of 1.2 V (vs. SCE) was applied, Ru-brown underwent oxidation to produce $Ru^{V}-Ru^{V}-Ru^{V}$ complex and this $Ru^{V}-Ru^{V}-Ru^{V}$ complex is reduced by water molecules to produce Ru-brown ($Ru^{IV}-Ru^{III}-Ru^{IV}$) by four-electron process. Independent of pH condition, the cyclic reaction between $Ru^{IV}-Ru^{III}-Ru^{IV}$ and $Ru^{V}-Ru^{V}-Ru^{V}$ showed only the spectrum of Ru-brown[72]. The reactions involved at Nafion membrane is summarized in Fig. 14. This report clearly shows that in the cyclic catalysed water oxidation reaction the Ru-red and Ru-brown complexes combined in a Nafion membranes act as four-electron catalyst, independently of pH, and the complex is stabilized against decomposition in the membrane.

The Ru-red or Ru-brown complex adsorbed in the Nafion membrane is probably present in a microheterogeneous environment imposed by hydrophobic cluster made of fluorocarbon moiety and by hydrophilic cluster made of sulfonate ions, and the Ru-red or Ru-brown is electrostatically held by the sulfonate ions. In a polymer membrane, the metal complex molecules would be isolated and the microheterogeneous environment would alter the complex-solvent interaction. Such effects are well characterized for macromolecular metal complexes[5,6]. Since Ru-red and Ru-brown water oxidation catalysts are strong oxidants in their higher oxidation states, they would attack organic ligands of the

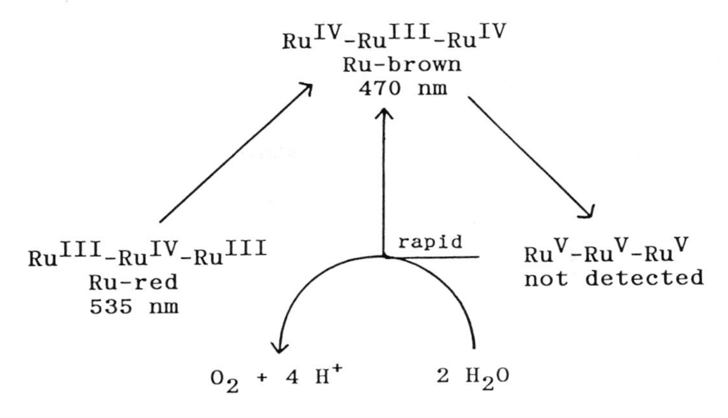

Fig. 14. Reactions of Ru-red and Ru-brown at Nafion membrane coated on ITO electrode at different pH conditions (ITO/Nf/Ru-red or Ru-brown)

other complex molecules in a homogeneous solution and the decomposition products would be solvated and stabilized by water molecules. This kind of degradative oxidation is probably prevented by the microheterogeneous environment imposed by the polymer membrane on the isolated metal complex entities. This work not only demonstrates realization of an efficient four-electron water oxidation system utilizing a polymer membrane, but also shows remarkable stabilization of the water oxidation catalyst against decomposition in a membrane.

4.3 Transformation and Catalysed Water Oxidation by a Schiff Base Mn Complex in a Polymer Membrane

In chloroplasts there is a loosely bound pool of manganese ions which act as a four-electron transfer catalysts for the oxidation of water to oxygen[1–3]. It is unlikely that a one/two manganese ion could remove all four electrons from two water molecules to liberate one oxygen molecule, although monomeric manganese complexes of Schiff base ligands were reported as model complexes for photosynthetic oxygen evolution[34, 35]. An oxygen electrode was usually used for the detection of the dissolved oxygen[33]. It has recently been reported that Schiff base complex of manganese becomes insoluble upon addition of Ce(IV), and the insoluble complex oxidizes water by the Ce(IV) oxidation with the formation of visible oxygen gas bubbles[38].

The monomeric manganese complex ($[Mn(salen)(H_2O)]^+$, salen H_2 = bis(salicyl aldehyde)ethylenediimine) of Schiff base ligand was incorporated by adsorption into a Nafion membrane coated on a Pt electrode. The cyclic voltammograms recorded for the manganese complex coated electrode showed much larger anodic current than for its homogeneous solution due to catalytic water oxidation by the manganese complex, and dioxygen gas bubbles were observed at higher positive potentials. The manganese Schiff base complex shows a less intense absorption band at 390 nm and an intense band at 355 nm in a homogeneous solution (Fig. 15a). The band at 390 nm was reported as being due to the monomeric structure, and that at 355 nm to the dimeric one[34]. When the complex was adsorbed into a Nafion membrane, these absorption bands were red shifted (Fig. 15b). The oxidation of the complex in the Nafion membrane at the applied potential of 1.3 V (vs. SCE) induced an increase of the band at 370 nm with a simultaneous decrease at 405 nm (Fig. 15c). This might be due to the transformation of the oxidized manganese Schiff base complex to a dimer or polymer[38]. The redox potential of the manganese Schiff base complex showed that the complex will act a one-electron oxidizing agent for water oxidation. Since water oxidation is a four-electron process, the one-electron process must be coupled in an efficient way. The dimer or polymer manganese Schiff base complex formed in the Nafion film after electrolysis at higher positive potentials would become active for water oxidation to evolve dioxygen. The role of dimer or polymer complex of the manganese complex would be to bring the oxygen atoms of two water molecules into close proximity to form an O–O bond which will lead to the generation of oxygen from two water molecules. The water oxidation catalyst dimer or polymer manganese complex in a Nafion membrane is of special interest because this might give an idea about the interaction of two/four manganese ions in a protein membrane which realizes the four-electron water oxidation in photosystem II[1–3,29,30,38].

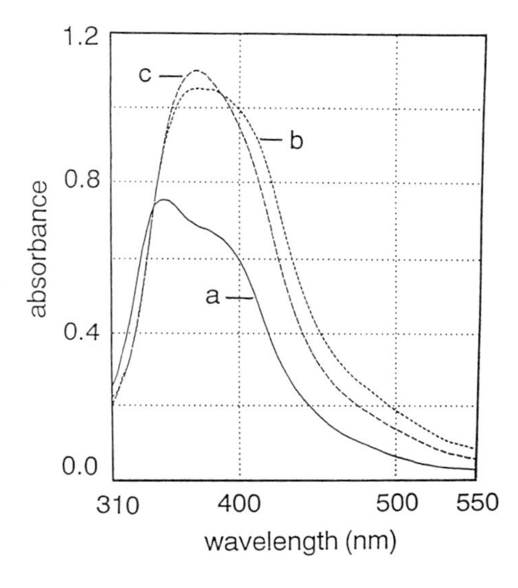

Fig. 15. Absorption spectra of complex $[Mn(Salen)\ (H_2O)]^+$: (a) 0.1 M HNO_3, (b) adsorbed onto Nafion film coated on ITO electrode and (c) after the complex was adsorbed on to ITO electrode held at an applied potential of 1.3 V/SCE for 10 min. (Reproduced with permission from Elsevier Science Publishers B.V.)

5 Photoinduced Water Oxidation by Metal Complex Incorporated in a Polymer Membrane

In the last decade attention has been focused on photoinduced electron transfer processes at semiconductor photoelectrode surfaces[4,73]. The relevant oxidizing and catalytic properties exhibited by n-type semiconductors such as n-TiO_2 under band gap illumination have been exploited to achieve various type of photooxidations and photoreductions. Water cleavage with a TiO_2 photoanode[74] occurs only with UV light and there are only a few reports appeared on water photolysis with visible light[4,73-77]. However, the unmodified narrow bandgap semiconductor electrodes put in aqueous electrolyte solution undergo photodecomposition. The potential of catalyst-coated (RuO_2) semiconductor electrodes for solar energy conversion and water splitting devices and more generally photocatalysis or artificial photosynthesis has been recognized by a rapidly growing number of workers[73-77].

In has been reported that the water oxidation catalyst Ru-red stabilized the n-CdS photoanode against photodecomposition, leading to efficient water cleavage to give hydrogen and oxygen simultaneously, when incorporated into a Nafion membrane coated on a narrow bandgap semiconductor photoanode[78]. A single crystal n-CdS electrode was coated with a Nafion membrane and used to adsorb Ru-red complex (referred as CdS/Nafion/Ru-red), and water splitting was carried out using this modified electrode, After irradiation, the gas chromatographic analysis showed the production of H_2 and O_2 with 2:1 ratio. In this water photolysis experiment, Ru-red acted as catalyst to evolve dioxygen at CdS

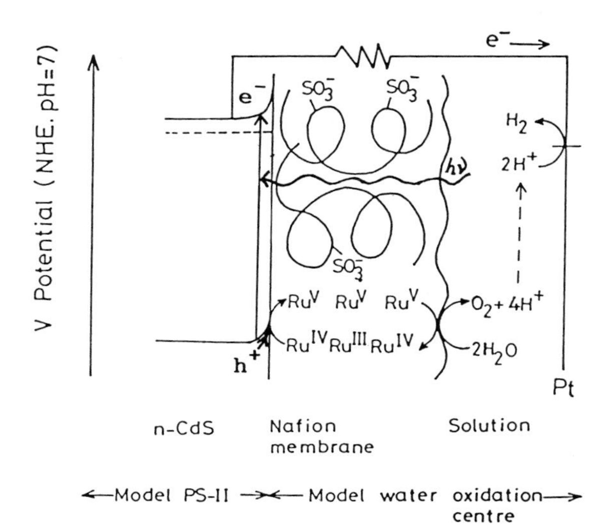

Fig. 16. A model water oxidation center composed of Nafion membrane coated n-CdS semiconductor electrode and metal complex catalyst

photoanode. As already explained in Section 4.2, the Ru-red in the first cycle converts into Ru-brown and in the further cycles it acts as four-electron transfer catalyst for water oxidation to evolve dioxygen (Fig. 16). The CdS photoanode modified with a Nafion polymer membrane incorporating a highly active water oxidation catalyst (trinuclear ruthenium complex) mimicks photosystem II oxygen evolving center.

6 Concluding Remarks

Models for photosynthetic oxygen evolving center have been reviewed especially in relevant to metal complexes incorporated in a polymer membrane. Polymer membranes provide active and stabile catalytic site; they often convert inactive metal complexes to active structure, and stabilize the active form as well. Water oxidation is the primary and the most important process to provide electrons to the whole photosynthetic system. This is a key reaction in order to develop an artificial photosynthetic system which is one of the most promising candidates for the 21st century's new energy resource. The use of polymeric materials is a crucial point for this approach.

Acknowledgement R.R. gratefully acknowledges the partial financial support from Department of Science and Technology, Government of India to work on Chemically Modified Electrodes.

7 References

1. Renger G, Photosynthetic Water Oxidation, Academic Press, London (1978)
2. Inoue T, Crofts CR, Govindjee, Murata N (eds) The Oxygen Evolving System of Photosynthesis, Academic Press, London (1983)
3. Biggins J (ed) Progress in Photosynthesis Research, Vol. 3, Martinus Nijhoff, Dordrecht (1987)
4. Graetzel M (ed) Energy Resources Through Photochemistry and Catalysis, Academic Press, London (1983)
5. Kaneko M, Woehrle D (1988) Adv Polym Sci 84: 141
6. Tsuchida E (ed) Macromolecular Complexes: Dynamic Interactions and Electronic Processes, VCH Publishers, New York (1991)
7. Govindjee (ed) Bioenergetics of Photosynthesis, Academic Press, London (1975)
8. Govindjee (ed) Photosynthesis-Energy Conversion by Plants and Bacteria, Vol. 1, Academic Press, London (1982)
9. Barber J (ed) Primary Processes of Photosynthesis, Elsevier/North Holland Biomedical Press, Amsterdam (1977)
10. Weighardt K (1989) Angw Chem Int Ed Engl 28: 1153
11. Pelizzetti E, Schiavello M (eds) Photochemical Conversion and Storage of Solar Energy, Kluwer Academic Publishers, Dordrecht (1991)
12. Deisenhofer J, Epp O, Miki K, Huber R, Michel H (1985) Nature 318: 618
13. Kalyanasundaram K, Graetzel M (1979) Angew Chem Int Ed Engl 18: 701
14. Lehn JM, Sauvage JP, Ziessel R (1979) Nouv J Chim 3: 423

15. Rillema DP, Dressick WJ, Meyer TJ (1980) J Chem Soc Chem Commun 247
16. Shafirovich V Ya, Khannanov NK, Strelets VV (1980) Nouv J Chim 4: 81
17. Kiwi J, Graetzel M (1978) Angrew Chem Int Ed Engl 17:.860
18. Harriman A, Mills A (1981) J Chem Soc Faraday Trans 1. 77: 2111
19. Kaneko M, Awaya N, Yamada A (1982) Chem Lett 619
20. Graetzel M (1980) Ber Bunsenges Phs Chem 84: 9181
21. Harriman A, Porter G, Walters P (1981) J Chem Soc Faraday Trans 2. 77: 2373
22. Creutz C, Suitn N (1975) Proc Natl Acad Sci USA. 72: 2858
23. Kaneko M, Takabayashi N, Yamada A (1982) Chem Lett 1647
24. Luneva NP, Shafirovich VYa, Shilov AE (1989) J Mol Catal 52: 49
25. Elizarova GL, Matvienko LG, Parmon VN (1987) J Mol Catal 43: 171
26. Calvin M (1874) Science 184: 375
27. Porter G (1978) Proc Royal Soc London. A362: 28
28. Cooper SR, Calvin M (1974) Science 185: 376
29. Ramaraj R, Kira A, Kaneko M (1986) Angrew Chem Int Ed Engl 25: 825
30. Ramaraj R, Kira A. Kaneko M (1987) Chem Lett 261
31. Lawrence LG, Sawyer DT (1978) Coord Chem Rev 27: 173
32. Sauer K (1980) Acc Chem Res 13: 249
33. Govindjee, Kambara T, Coleman W (1985) Photochem Photobiol 42: 187
34. Ashmawy FM, McAuliffe CA, Parish RV, Tames J (1985) J Chem Soc Dalton Trans 1391
35. McAuliffe CA, Parish RV, Abu-El-Wafa SM, Issa RM (1986) Inorg Chim Acta 115: 91
36. Ashmawy FM, McAuliffe CA, Parish RV, Tames J (1984) J Chem Soc Chem Commun 14
37. Matsushita T, Spencer L, Sawyer DT (1988) Inorg Chem 27: 1167
38. Gobi KV, Ramaraj R, Kaneko M (1983) J Mol Catal 81: L7
39. Gersten SW, Samuels GJ, Meyer TJ (1982) J Am Chem Soc 104: 4029
40. Meyer TJ (1984) J ELectrochem Soc 131: 221C
41. Gilbert JA, Eggeleston DS, Murphy Jr WR, Geselowitz DA, Gersten SW, Hodgson DJ, Meyer TJ (1985) J Am Chem Soc 107: 3855
42. Collin JP, Sauvage JP (1986) Inorg Chem 25: 135
43. Ramaraj R, Kira A, Kaneko M (1986) J Chem Soc Faraday Trans 1. 82: 3515
44. Honda K, Frank AJ (1984) J Chem Soc Chem Commun 1635
45. Lay PA, Sasse WHF (1985) Inorg Chem 24: 4707
46. Ramaraj R, Kira A, Kaneko M (1987) J Chem Faraday Trans 1. 83: 1539
47. Ramaraj R, Kira A, Kaneko M (1986) Angew Chem Intl Ed Engl 25: 1009
48. Rotzinger FP, Munavelli S, Comte P, Hurst JK, Graetzel M, Pern FJ, Frank AJ (1987) J Am Chem Soc 109: 6619
49. Nazeerudin MK, Rotzinger FP, Comte P, Graetzel M (1988) J Chem Soc Chem Commun 872
50. Comte P, Nazeerudin MK, Rotzinger FP, Frank AJ, Graetzel M (1989) J Mol Catal 52: 63
51. Kaneko M, Ramaraj R, Kira A (1988) Bull Chem Soc Jpn 61: 417
52. Goswami S, Chakravarthy AR, Chakravorty A (1982) J Chem Soc Chem Commun 1288
53. Nijs H, Crutz MI, Fripiat J, Van Damme H (1981) J Chem Soc Chem Commun 1026
54. Takuchi KJ, Samuels GJ, Gerstein SW, Gilbert JA, Meyer TJ (1983) Inorg Chem 22: 1409
55. Mills A, Russell T (1991) J Chem Soc Faraday Trans 87: 313
56. Hurst JK, Zhou J, Lei Y (1992) Inorg Chem 31: 1010
57. Eisenbery A, Yeager HL (1982) (eds) Perfluorinated Ionomer Membrane, Vol. 180, American Chemical Society, Washington DC
58. Yeager HL, Steck A (1981) J Electrochem Soc 128: 1880
59. Nijs H, Crutz MI, Fripiat JJ, Van Damme H (1982) Nouv J Chim 6: 551
60. Abdo S, Canesson P, Crutz MI, Fripiat JJ, Van Damme H (1981) J Phys Chem 85: 797
61. Dobson JC, Meyer TJ (1988) Inorg Chem 27: 3283
62. Ramaraj R, Kira A, Kaneko M (1993) J. Electronal Chem 348: 367
63. Ramaraj R, Kira A, Kaneko M (submitted to Polymers for Advanced Technologies)
64. Vining WJ, Meyer TJ (1986) Inorg Chem 25: 2023
65. Fletcher JM, Greenfield BF, Hardy CJ, Scargill D, Woohead JL (1961) J. Chem Soc A. 2000
66. Early JE, Fealey T (1973) Inorg Chem 12: 323
67. Earley JE, Fealey T (1971) Chem Commun 331
68. Early JE, Razari H (1973) Inorg Nucl Chem Lett 9: 331
69. Early JE (1973) Inorg Nucl Chem Lett 9: 487
70. Yao G.-J, Kira A, Kaneko M (1988) J Chem Soc Faraday Trans 1. 84: 4451

71. Ramaraj R, Kaneko M (1993) J Chem Soc Chem Commun 579
72. Ramaraj R, Kaneko M (1993) J Mol Catal 81: 319
73. Schiavello M (ed) Photoelectrochemistry, Photocatalysis and Photoreactors, Reidel, Dordrecht (1985)
74. Fujishima A, Honda K (1972) Nature 238: 37
75. Kayanasudnaram K, Graetzel M (1979) Angew Chem Int Ed Engl 18: 701
76. Frank AJ, Honda K (1982) J Phys Chem 86: 1933
77. Kaneko M, Okada K, Teratani S, Taya K (1987) Electrochim Acta 32: 1405
78. Kaneko M, Yao G.-J, Kira A (1989) J Chem Soc Chem Commun 1338

Editor: Prof. Abe
Received July 1994

Author Index Volumes 101-123

Author Index Vols. 1-100 see Vol. 100

Subject Index

Printing: Saladruck, Berlin
Binding: Buchbinderei Lüderitz & Bauer, Berlin